T0145316

Sustainable Agrochemistry

Sílvio Vaz Jr.
Editor

Sustainable Agrochemistry

A Compendium of Technologies

 Springer

Editor
Sílvio Vaz Jr.
National Research Center for Agroenergy
Brazilian Agricultural Research Corporation
Brasilia, Brazil

ISBN 978-3-030-17893-2   ISBN 978-3-030-17891-8   (eBook)
https://doi.org/10.1007/978-3-030-17891-8

This Springer imprint is published by the registered company Springer Nature Switzerland AG
The registered company address is: Gewerbestrasse 11, 6330 Cham, Switzerland

I would like to dedicate this book to my wife Ana and to my daughter Elena. They are my source of inspiration.

Preface

Agrochemistry was seen during many years as a source of environmental and health concerns due to, mainly, the pesticides used in agriculture around the world. Conventional pesticides produced negative impacts on the environment as the biota pollution and destruction near to the crop systems; furthermore, agricultural workers were contaminated by these agrochemicals, causing serious illness and death.

Nowadays, the demand from the modern society for sustainable production of food has promoted the development of a sustainable agrochemistry considering aspects such as reducing negative impacts on the environment- and health-friendly materials and molecules, bioactive compounds, etc.

Agriculture remains one of the most strategic sectors for the global economy and well-being. In this way, innovations and new paradigms are necessary for their sustainable exploitation, and agrochemistry can certainly continue to contribute to the generation of agricultural wealth for modern society.

This book intends to present a large variety of technologies for a sustainable agrochemistry, such as semiochemicals for pest management, nanomaterials, green chemistry principles for agriculture, among others, with their respective case study.

Chapter 1 defines sustainable agrochemistry, the main classes of agrochemicals as well as their usages and dynamic in the agriculture and the environment. Furthermore, some relevant aspects of sustainability in agriculture are presented and discussed. Chapter 2 presents historical aspects of crop protection and the use of pesticides to guarantee sustainable food supplies. Chapter 3 presents the principles of semiochemical use for monitoring and controlling pests and the way in which these natural molecules work is presented and discussed. Chapter 4 describes the major concepts related to nanoscience and nanotechnology, role of green nanotechnology as an essential part of a sustainable future of agriculture, and its applicability for the development of innovative solutions to challenging issues. Chapter 5 deals with the use of magnetic resonance techniques to improve agricultural systems, highlighting the obtaining of structural information about industrial biomass and soil organic matter for scientific and technological usages. Chapter 6 talks about chemical analyses and their important role in agriculture, as supporting technologies at all stages of agro-industrial chains as grains, forests, pulp and paper, waste and agricultural residues, among others sources of agricultural products; furthermore, a set of relevant analytical techniques are discussed in accordance with their application in the agriculture. Chapter 7 treats about the recent acknowledgement of the supramolecular nature of soil humic substances, that allowed to devise a fractionation procedure, called Humeomics, that enables detailed characterization of the structure of humic molecules in soil for their application in the improvement of soil functionality. Chapter 8 addresses some general aspects about the chemistry action of crop protection products against pest attack and the implications of these agrochemicals on the environment in order to produce food sustainably for a constantly growing world population. Chapter 9 deals with the more relevant information about the toxicity of agrochemicals for the biota and the human health. From this, the understanding of the impact from agrochemicals use. Chapter 10 presents and discusses the green chemistry principles, highlighting their application in agriculture. Moreover, from this, the understanding of methods to change the development and production of green agrochemicals. Chapter 11 deals with the more relevant tools for the ecological risk assessment of agrochemicals, what clarify the paramount importance of monitoring and control of agrochemicals in the environment using mathematical models. Finally, Chap. 12 deals with the more relevant strategies for the management of agrochemical residues in soil and water. Furthermore, the most advanced treatment technologies will be explored.

The presentation of case studies aims to expand the reader's knowledge and demonstrate examples of application of the technologies proposed by the authors.

Good lecture!

Acknowledgements The editor would like to thank all the authors for the efforts to prepare a high-quality text—this is not a simple task. Thanks also to Springer team for enabling the transformation of an idea into a book. Finally, thanks to colleagues from Embrapa Agroenergy for refining and improving some parts of the manuscript.

Brasilia, Brazil   Sílvio Vaz Jr.

Contents

Contributors

Maria C. Blassioli-Moraes Brazilian Agricultural Research Corporation, National Research Center for Genetic Resources and Biotechnology—Laboratory for Semiochemicals, Brasilia, Distrito Federal, Brazil

Cínthia Caetano Bonatto Brazilian Agricultural Research Corporation, Laboratory of Nanobiotechnology (LNANO), Embrapa Genetic Resources and Biotechnology, Brasilia, Distrito Federal, Brazil;
NanoDiversity, Applied Research, TecSinapse, São Paulo, SP, Brazil

Miguel Borges Brazilian Agricultural Research Corporation, National Research Center for Genetic Resources and Biotechnology—Laboratory for Semiochemicals, Brasilia, Distrito Federal, Brazil

Eloisa Dutra Caldas Department of Pharmacy, Faculty of Health Sciences, University of Brasilia, Brasilia, Distrito Federal, Brazil

Luiz Alberto Colnago Brazilian Agricultural Research Corporation, Embrapa Instrumentation, São Carlos, SP, Brazil

Vincenza Cozzolino Department of Agricultural Sciences, Interdepartmental Research Centre on Nuclear Magnetic Resonance for the Environment, Agro-Food and New Materials (CERMANU), University of Napoli Federico II, Portici, Italy

Paulo Marcos Donate Departamento de Química da Faculdade de Filosofia, Ciências e Letras, Universidade de São Paulo, Ribeirão Preto, SP, Brazil

Marios Drosos Faculty of Biology and Environment, Institute of Resource, Ecosystem and Environment of Agriculture (IREEA), Nanjing Agricultural University, Nanjing, China

Daniel Frederico Dinagro Agropecuária Ltda., Ribeirão Preto, SP, Brazil

Luciano Gebler Brazilian Agricultural Research Corporation, National Research Center for Grape and Wine (Embrapa Grape and Wine), Bento Gonçalves, Rio Grande do Sul, Brazil

Caroline Harris Exponent International Limited, Harrogate, North Yorkshire, UK

Gijs Kleter Wageningen Food Safety Research, Part of Wageningen University & Research, Wageningen, The Netherlands

Raúl A. Laumann Brazilian Agricultural Research Corporation, National Research Center for Genetic Resources and Biotechnology—Laboratory for Semiochemicals, Brasilia, Distrito Federal, Brazil

Mirian F. F. Michereff Brazilian Agricultural Research Corporation, National Research Center for Genetic Resources and Biotechnology—Laboratory for Semiochemicals, Brasilia, Distrito Federal, Brazil

Rafael Mingoti Brazilian Agricultural Research Corporation, Embrapa Territorial, Campinas, SP, Brazil

Yoshiaki Nakagawa Graduate School of Agriculture, Kyoto University, Kyoto, Japan

Etelvino Henrique Novotny Brazilian Agricultural Research Corporation, Embrapa Soils, Rio de Janeiro, Brazil

Alessandro Piccolo Department of Agricultural Sciences, Interdepartmental Research Centre on Nuclear Magnetic Resonance for the Environment, Agro-Food and New Materials (CERMANU), University of Napoli Federico II, Portici, Italy

Davide Savy Plant Biology Laboratory, University of Liège, Gembloux Agro-Bio Tech, Gembloux, Belgium

Luciano Paulino Silva Brazilian Agricultural Research Corporation, Laboratory of Nanobiotechnology (LNANO), Embrapa Genetic Resources and Biotechnology, Brasilia, Distrito Federal, Brazil

Riccardo Spaccini Department of Agricultural Sciences, Interdepartmental Research Centre on Nuclear Magnetic Resonance for the Environment, Agro-Food and New Materials (CERMANU), University of Napoli Federico II, Portici, Italy

Claudio A. Spadotto Brazilian Agricultural Research Corporation, Embrapa, Parque Estação Biológica, Brasilia, Distrito Federal, Brazil

John Unsworth Consultant, Chelmsford, UK

Sílvio Vaz Jr. Brazilian Agricultural Research Corporation, National Research Center for Agroenergy (Embrapa Agroenergy), Embrapa Agroenergia, Parque Estação Biológica, Brasilia, Distrito Federal, Brazil

Chapter 1
Introduction to Sustainable Agrochemistry

Sílvio Vaz Jr.

Abstract This chapter defines sustainable agrochemistry, the main classes of agrochemicals as well as their usages and dynamics in the agriculture and the environment. Furthermore, some relevant aspects of sustainability in agriculture are presented and discussed.

Keywords Agriculture · Environment · Impacts

1.1 Introduction

The practice of agriculture is one of the oldest activities developed by humans. In the Neolithic period, the constitution of the first techniques and materials used for the cultivation of plants and the confinement of animals was the main cause for what was denominated as the sedentarization of human. It allowed fixing residence in a given locality, although collection and hunting have long coexisted side by side with agriculture.

The development of agriculture, therefore, was directly associated with the formation of the first civilizations, which helps us understand the importance of techniques and the environment in the process of building societies and their geographical spaces. In that sense, as these societies further advanced their techniques and technologies, the more the evolution of agriculture was benefited.

Originally, the practice of farming was developed in the vicinity of large rivers, notably the Tigris and Euphrates, as well as the Nile, the Ganges and others. Not coincidentally, it was in these localities that the first great known civilizations emerged, because the practice of agriculture allowed the development of trade to a surplus in production.

S. Vaz Jr. (✉)
Brazilian Agricultural Research Corporation, National Research Center for Agroenergy (Embrapa Agroenergy), Embrapa Agroenergia, Parque Estação Biológica, s/n, Av. W3 Norte (final), Brasília, DF 70770-901, Brazil
e-mail: silvio.vaz@embrapa.br

© Springer Nature Switzerland AG 2019
S. Vaz Jr. (ed.), *Sustainable Agrochemistry*,
https://doi.org/10.1007/978-3-030-17891-8_1

One of the most important moments in the process of agricultural evolution throughout history was, without doubt, what became known as the Agricultural Revolution (British Broadcasting Corporation 2017). We can say that, over time, several Agricultural Revolutions have succeeded, but the main ones occurred after the Industrial Revolution.

The process of industrialization of societies allowed the transformation of the geographical space in the rural environment, which occurred thanks to the insertion of greater technological apparatuses in the agricultural production, allowing a greater mechanization of the field. This transformation materialized from the supply of inputs from the industry to agriculture, such as machinery, fertilizers and technical objects and practices in general.

The influence of the Agricultural Revolution in the world was also directly associated with the European maritime–colonial expansion, in which the European people disseminated their different cultures through the world by means of introducing new crops and novel agricultural practices. It is worth remembering that the interaction between settlers and colonizers also contributed to the agricultural evolution, inasmuch as previously little known techniques were applied and disseminated, such as the terracing practised in both ancient China and pre-Columbian civilizations.

As an effect of this revolution, Chemistry starts to have a fundamental role in the agriculture expansion and technological development by means of fertilizers, pest controllers, scientific knowledge, among other aspects.

In the twentieth century, more precisely after the Second World War, the evolution of agriculture reached one of its most important hallmarks, in what became known as the Green Revolution. It was based on a set of measures and promotion of techniques based on the introduction of genetic improvements in plants and the evolution of agricultural production apparatus to expand, above all, food production (Pingali 2012).

The introduction of techniques from the Green Revolution has led to a large-scale increase in grain and cereal production, significantly reducing the need for food in various regions of Asia, Africa and Latin America, even though hunger has not been eradicated, since its existence is not only due to lack of food. The impact on the world was so wide that the American agronomist Norman Borlaug, considered the "father" of the Green Revolution, was awarded the Nobel Peace Prize in 1990 (Nobel Prize 2018).

Although the Green Revolution is heavily criticized for its environmental impacts and the process of land concentration that accompanied its evolution due to policies that were used to promote rapid intensification of agricultural systems and increase food supplies (Pingali 2012), its importance for the development of agriculture in the world is undeniable. Furthermore, in the following decades, the improvements resulting from technology in the field, such as biotechnology, have increased in the following decades, which has been further increasing the productivity.

Figure 1.1 depicts a cultivated area with corn (*Zea mays* L.) in the Midwest Brazilian region. This is a region typically affected by the Green Revolution. The Brazilian Savannah was an inhospitable and low productive land during a couple

Fig. 1.1 Corn cropping in the Midwest region of Brazil, one of the largest producing regions of cereal in the world. Courtesy of Climatempo, São Paulo

Table 1.1 Global production of the five major crops, according to Food and Agriculture Organization of the United Nations (2015)

Crop	Production, in thousand tonnes
Sugarcane	1,877,110
Maize (corn)	1,016,740
Rice	745,710
Wheat	713,183
Potatoes	368,096

of centuries, but, nowadays, it is responsible for Brazil to be the major agricultural player in South America, and one biggest worldwide.

Table 1.1 describes the main crops cultivated around the world and their production. These values would not hardly be achieved without the use of agrochemicals, brought to the scene by the Green Revolution.

Nowadays, agriculture is constantly required to become more sustainable, with reduction of its negative impacts on environment allied to an increasing in their positive impacts on society and economy. These are challenges and, at same time, opportunities for new production systems.

1.2 Chemistry and Agriculture

Chemistry is a science closely related to agriculture since its origin. The term *agricultural chemistry* rises with the publication of *Elements of Agricultural Chemistry* by the chemist Humphry Davy (Fig. 1.2) in 1813 in England (Encyclopaedia Britannica 2018). The book was traduced to French, Italian and German. It was also published with great prestige in U.S. because it boosted the American agricultural community, which was based on efficient techniques that had taken into account scientific principles and observations.

Nowadays, we can define *agrochemistry* as the application of chemistry in agriculture. Its action, object of study and technical means are not only concerned with the production of agrochemicals but also with the analysis and prevention of the harmful effects of chemical substances on both crops and humans (farmers and consumers), and their impacts on the environment.

The contribution of chemistry goes back to the nineteenth century, with the synthesis of inorganic fertilizers and, by the middle of the last century, of a great number of compounds synthesized to control insects, diseases and weeds (Pinto-Zaervallos and Zarbin 2013).

This contribution can be seen clearly and decisively in the cycle of nitrogen, an essential element to some molecules that integrate the organic matter. Plants, with some exceptions, do not have the capacity to absorb this element from the

Fig. 1.2 Sir Humphry Davy, detail of an oil painting. Courtesy of the National Portrait Gallery, London

atmosphere (with 78% nitrogen), the opposite of what occurs with another essential element, carbon, which is absorbed as CO_2 by means the photosynthesis. So, the only natural way to close the nitrogen cycle is through the decomposition of organic material from dead animal or plant, or the excretion of living things, and this form of replenishment is naturally limited.

The capture and use of atmospheric nitrogen in the soil was only possible economically via the works of the German chemists Fritz Haber and Carl Bosch. They developed the Haber–Bosch reaction or process at the beginning of the twentieth century (Ritter 2008), which allows the ammonia synthesis from the small reactive atmospheric nitrogen and other abundant element, hydrogen, in industrial scale. Curiously, the incentive that led to this essential innovation was not initially the production of fertilizers, but the production of nitrates for military purposes (explosives) to be used in World War I.

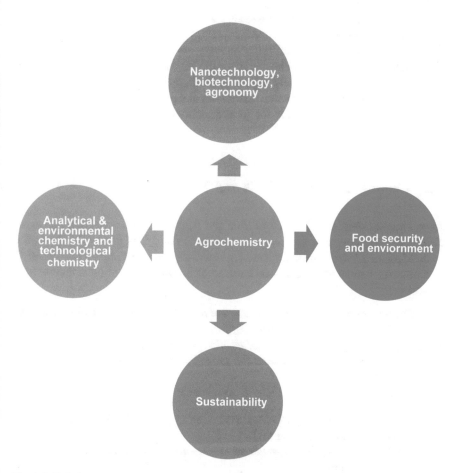

Fig. 1.3 Relation among agrochemistry and related areas of study and technological development

Ammonia, a molecule with about 82% by weight of nitrogen, can be absorbed by plants through the soil. However, for reasons of ease of application, it is preferred to use as nitrogenous fertilizers solid substances derived from ammonia, such as urea and ammonium nitrate. The world production of ammonia today reaches about 140 million tons per year (US Geological Survey 2017), and almost all global production is destined to the synthesis of industrial fertilizers. The percentage of the world population whose food depends on the use of synthetic nitrogen fertilizers is estimated at 53% (Liu et al. 2016).

Taking the nitrogen fertilizers as an example, we can also conclude that so-called organic farming, which advocates the exclusion of synthetic fertilizers, can function as a niche market in societies of abundance, but it is certainly not an alternative to feed the humanity. Then, we can see clearly the contribution of the agrochemistry to the well-being of modern society. Figure 1.3 depicts a relation among agrochemistry and related scientific and technological fields as an interdisciplinary theme.

Analytical and environmental chemistry allied to technological chemistry input techniques, technologies and knowledge to analyse, produce and monitoring agrochemicals. Nanotechnology and biotechnology are new technological approaches to be incorporated to the agrochemicals for the best agronomic usages. Food security and environment are closely related to the agrochemical uses and near to the consumer, implying in laws, market restrictions and public opinion. Finally, sustainability is a demand of the society for greater quality of life and greater transparency in the productive chains.

1.3 Classes of Agrochemicals and Their Uses

According to International Union of Pure and Applied Chemistry (2006), an agrochemical is an "*agricultural chemical used in crop and food production including pesticide, feed additive, chemical fertilizer, veterinary drug and related compounds*".

Currently, we can observe several agrochemical classes according to their uses in agriculture:

- Fertilizers—*any kind of substance applied to soil or plant tissues to provide one or more nutrients essential to plant growth*;
- Plant growth regulators—*(also called plant hormones)—several chemical substances that profoundly influence the growth and differentiation of plant cells, tissues and organs*;
- Phytosanitary products, pesticides or correctives:

  – Herbicides—*agents, usually chemicals, used for killing or inhibiting the growth of unwanted plants (i.e., weeds)*;
  – Insecticides—*pesticides formulated to kill, harm, repel or mitigate one or more species of insects*;
  – Fungicides—*pesticides that kill or prevent the growth of fungi and their spores*;

- Acaricides—*pesticides that kill members of the arachnid subclass* Acari, *which includes ticks and mites*;
- Bactericides—*a substance that kills bacteria.* Bactericides *are disinfectants, antiseptics, or antibiotics*;
- Rodenticides—*pesticides that kill rodents. Rodents include not only rats and mice, but also squirrels, woodchucks, chipmunks, porcupines, etc.*;
- Nematicides—*a type of chemical pesticide used to kill plant-parasitic nematodes*;
- Repellents—*chemicals that can help reduce the risk of being bitten by insects and therefore reduce the risk of getting a disease carried by mosquitos or ticks*;
- Fumigants—*any volatile, poisonous substance used to kill insects, nematodes, and other animals or plants that damage stored foods or seeds*;
- Disinfectants—*antimicrobial agents that are applied to the surface of non-living objects to destroy microorganisms that are living on such objects*;
- Antibiotics—*powerful drugs that fight bacterial infections; as a highlighted risk, their continuous use may result in resistance to them by some microorganisms*;
- Defoliants—*a chemical dust or spray applied to plants to cause their leaves to drop off prematurely*;
- Algaecides—*or algicide—a biocide used for killing and preventing the growth of algae.*

Agrochemicals move a huge global market, and it is expected that these markets achieve 250.5 billion USD by 2020 (Statista 2018). However, they are one of the main classes of chemical pollutants, with serious negative impacts on public health and environment. The search for alternatives to conventional agrochemicals presents

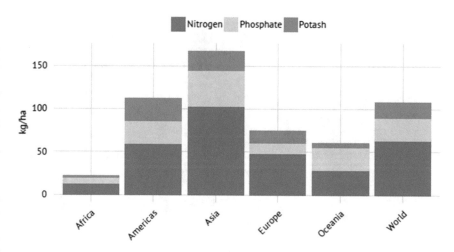

Fig. 1.4 Fertilizer consumption in nutrients per ha of arable land, according to Food and Agriculture Organization of the United Nations (2015)

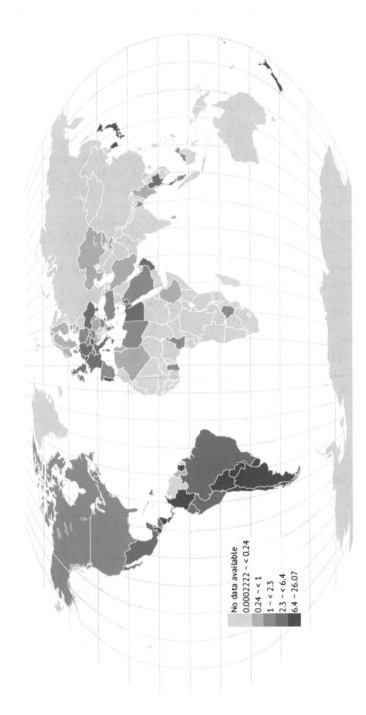

Fig. 1.5 Pesticides per ha of arable land (kg ha^{-1}), according to Food and Agriculture Organization of the United Nations (2015)

itself as an excellent opportunity for the development of sustainable agricultural technologies and for opening new businesses.

As an example of the demand for agrochemicals by the agricultural sector, in Fig. 1.4 it is presented the world consumption of fertilizers. Asia and Americas have higher fertilizer consumption because the most agricultural production is located in these continents.

In Fig. 1.5, the world pesticide is shown. Americas, Europe and part of Africa and Asia have higher pesticide consumption aiming plagues' control. Interestingly, small countries like Japan and Portugal show consumption comparable to larger countries such as Argentina.

1.4 Impacts of Agriculture on the Environment and Health

Governments, farmers and consumers show increasing concerns with the negative impacts on the environment and health caused by the large amount of inputs applied to produce different crops worldwide. Agrochemicals have a direct correlation with damages from agriculture, with pesticides being the main representative class with toxicological implications.

The main negatives from the agriculture impacts on the environment are:

- Water and air pollution due to the use of pesticides;
- Extinction of water bodies, due to high water demand;
- Erosion and soil degradation due to inadequate management during cultivation;
- Change in biota due to factors already listed;
- Changes in the quality of environmental resources, also due to factors already listed;
- Ecological risks for insects, plants and animals associated with the change of environment;
- Climate change due to the deforestation and biomass combustion.

Regarding those impacts on health, it can be highlighted:

- Poisoning due to pesticide use and food contaminated consumption;
- Occupational risks to farmers due to the exposition to pesticides;
- Human infections, or emerging infectious diseases, that do not respond to treatment due to the use of antimicrobials in agriculture (Grace 2018).

From these negative impacts, it is becoming paramount the development of a more environmentally and healthy friendly agriculture.

1.5   Sustainability and Agrochemistry

Agricultural chemistry is, undoubtedly, one of the fields of research and business whose impact is felt throughout the world, since we all need to eat to survive. Added to this is the fact that, increasingly, technology is intertwining with modern agriculture, both with regard to new production strategies and the reduction of negative environmental impacts (Herman 2015).

Sustainability can be seen and understood by means of its three components:

- Environmental impacts;
- Economic impacts;
- Societal impacts.

Impacts can be positive or negative according their direct or indirect effects upon the environment, economy and society. Considering that agrochemistry is the application of chemistry and its concepts and technologies to promote a better agriculture, economic and societal impacts are expected to be positive, specially the economic impacts. On the other hand, and due to a historic of environmental incidents at global level (e.g., water pollution), environmental impacts are expected to be negative; nevertheless, it could be positive if modern technologies and good agricultural practices are used. A more detailed evaluation of sustainability in agriculture can be seen in Quintero-Angel and González-Acevedo (2018).

Sustainable chemistry, a recent branch of chemistry, was defined as "*(…) a scientific concept that seeks to improve the efficiency with which natural resources are used to meet human needs for chemical products and services. Sustainable chemistry encompasses the design, manufacture and use of efficient, effective, safe and more environmentally benign chemical products and processes.*" (Organization of Economic Co-operation and Development 2018). From these statements, a relationship with agrochemistry can be constructed by means of design, manufacture and use of efficient, effective, safe and more environmentally benign agrochemicals, that is, by the establishment of a strong innovation drive in agriculture for the next decades as we can read in this book.

References

British Broadcasting Corporation (2017) Agricultural revolution in England 1500–1850. http://www.bbc.co.uk/history/british/empire_seapower/agricultural_revolution_01.shtml. Accessed Aug 2018

Encyclopaedia Britannica (2018) Sir Humphry Davy, Baronet. https://www.britannica.com/biography/Sir-Humphry-Davy-Baronet#ref172328. Accessed Aug 2018

Food and Agriculture Organization of the United Nations (2015) FAO statistical pocketbook. http://www.fao.org/3/a-i4691e.pdf. Accessed Aug 2018

Grace D (2018) Infectious diseases and agriculture. Reference module in food science. https://doi.org/10.1016/B978-0-08-100596-5.21570-9

Herman C (2015) Agricultural chemistry: new strategies and environmental perspectives to feed a growing global population. American Chemical Society, Washington

International Union of Pure and Applied Chemistry (2006) Glossary of terms related to pesticides. In: Stephenson GR, Ferris IG, Holland PT, Norberg M (eds) Pure and applied chemistry, vol 78, pp 2075–2154

Liu J, Ma K, Ciais P, Polasky S (2016) Reducing human nitrogen use for food production. Scientific reports. https://doi.org/10.1038/srep30104

Nobel Prize (2018) Norman Borlaug—biographical. https://www.nobelprize.org/nobel_prizes/peace/laureates/1970/borlaug-bio.html. Accessed Aug 2018

Organization of Economic Co-operation and Development (2018) Sustainable chemistry. http://www.oecd.org/env/ehs/risk-management/sustainablechemistry.htm. Accessed Aug 2018

Pingali D (2012) Green revolution: impacts, limits, and the path ahead. Proc Natl Acad Sci U S A 109:12302–12308

Pinto-Zaervallos DM, Zarbin PHG (2013) Química na agricultura: perspectivas para o desenvolvimento de tecnologias sustentáveis [Chemistry in agriculture: perspectives for the development of sustainable technologies]. Quim Nova 36:1509–1513

Quintero-Angel M, González-Acevedo A (2018) Tendencies and challenges for the assessment of agricultural sustainability. Agr Ecosyst Environ 254:273–281

Ritter SK (2008) The Harber-Bosh reaction: an early chemical impact on sustainability. Chemical & Engineering News. https://pubs.acs.org/cen/coverstory/86/8633cover3box2.html. Accessed Aug 2018

Statista (2018) Worldwide agrochemical market value in 2014 and 2020 (in billion U.S. dollars). https://www.statista.com/statistics/311957/global-agrochemical-market-revenue-projection/. Accessed in Aug 2018

U.S. Geological Survey (2017) USGS Minerals information: nitrogen. https://minerals.usgs.gov/minerals/pubs/commodity/nitrogen/. Accessed Aug 2018

Chapter 2
The Battle for a Sustainable Food Supply

John Unsworth, Yoshiaki Nakagawa, Caroline Harris and Gijs Kleter

Abstract Since the time that *Homo sapiens* took up farming, a battle has been waged against pests and diseases which can cause significant losses in crop yield and threaten a sustainable food supply. Initially, early control techniques included religious practices or folk magic, hand removal of weeds and insects, and "chemical" techniques such as smokes, easily available minerals, oils and plant extracts known to have pesticidal activity. But it was not until the early twentieth century that real progress was made when a large number of compounds became available for testing as pesticides due to the upsurge in organic chemistry. The period after the 1940s saw the introduction of important families of chemicals, such as the phenoxy acid herbicides, the organochlorine insecticides and the dithiocarbamate fungicides. The introduction of new pesticides led to significant yield increases, but concern arose over their possible negative effects on human health and the environment. In time, resistance started to occur, making these pesticides less effective. This led agrochemical companies putting in place research looking for new modes of action and giving less toxic and more environmentally friendly products. These research programmes gave rise to new pesticide families, such as the sulfonylurea herbicides, the strobilurin fungicides and the neonicotinoid insecticide classes.

This work formed part of a project (Project No. 2012-019-1-600) supported by the International Union of Pure and Applied Chemistry (IUPAC).

J. Unsworth (✉)
Consultant, Vellacotts,
Chelmsford CM1 7EA, UK
e-mail: unsworjo@aol.com

Y. Nakagawa
Graduate School of Agriculture, Kyoto University, Kyoto 606-8502, Japan

C. Harris
Exponent International Limited, The Lenz, Hornbeam Business Park, Harrogate, North Yorkshire HG2 8RE, UK

G. Kleter
Wageningen Food Safety Research, Part of Wageningen University & Research, Akkermaalsbos 2, 6708 WB Wageningen, The Netherlands

© Springer Nature Switzerland AG 2019
S. Vaz Jr. (ed.), *Sustainable Agrochemistry*,
https://doi.org/10.1007/978-3-030-17891-8_2

Keywords Pest management · Crop yield · Natural pesticides · Inorganic pesticides · Synthetic organic pesticides · Pest resistance

2.1 Introduction

Homo sapiens, or Latin for wise man, was a term first applied in 1758 by Linnaeus for modern humans. They are thought to have first appeared roughly 200,000 years ago in what is now Ethiopia (McDougall et al. 2005), although there is some evidence that they emerged substantially earlier than this (Hublin et al. 2017). Until the practice of agriculture some 190,000 years later, the only way to obtain a sufficient food supply was by foraging, i.e., hunting wild animals and gathering edible plants and seeds. Hunter-gatherer societies were generally nomadic, moving several times each year to new areas when the returns of forays from their current camp dropped below those to be expected from another camp (Kelly 1992). The transition from hunter-gatherer to agriculture can be considered as having occurred in four steps (Olsson 2001):

(i)   Wild plant–food procurement—hunter-gatherers in this stage occasionally burn the vegetation, gather and protect useful plants and fruit.

(ii) Wild plant–food production with some tillage—maintenance of plant populations in the wild is carried out, and both planting and sowing of wild plants are undertaken, as well as weeding. Seeds from selected plants with desirable characteristics are propagated in new habitats. When harvesting is completed, some of the seed is stored for future use.

(iii) Cultivation with systematic tillage—land is cleared and the food-producing activities that were introduced in the previous stage (sowing, weeding, propagation of plants, etc.) are intensified.

(iv) Plant domestication—as the selective cultivation of plants proceeds, new genotypes eventually appear that more efficiently serve human needs.

Several theories have been put forward to explain why the change from hunter-gatherers to farmers took place (Price and Bar-Yosef 2011; Weisdorf 2005), including changes in climate (Richerson et al. 2001; Dow et al. 2005; Bettinger et al. 2009), population density and technological sophistication (Baker 2008), over-hunting of large animals leading to their extinction (Smith 1975) and social evolution (Bender 1978). It has also been postulated that as societies began to use ever more varied equipment for foraging with containers, stone mortars, baskets, fishnets and traps, drying racks, etc., plus starting to store food in pits and embarking on agriculture, the necessary equipment became increasingly burdensome. Moving regularly from one area to another became more difficult and mobility was thus reduced, with people tending to stay in one area. This increasing and practically enforced sedentarism of hunter-gatherers increased their involvement in agriculture (de Saulieu and Testart 2015). Whatever the reason, or combination of reasons, agriculture, as in the cultivation of plants for food, was adopted between 5000 and 12,000 years ago. Independent points of origin included the Near East, north and south China, sub-Saharan Africa,

Central Mexico, South-Central Andes and the eastern USA, with subsequent diffusion to most parts of the globe (Richerson et al. 2001). Once adopted, agriculturally based communities increased significantly in size (Gignouxa et al. 2011) with the corresponding requirement to increase the food supply.

The practice of agriculture first began in the Fertile Crescent of Mesopotamia (part of present-day Iraq, Turkey, Syria and Jordan) where edible seeds were initially gathered, as in stage (i) above, by a population of hunter-gatherers (Kislev et al. 2004). Interestingly, the change from hunting and gathering to farming as a subsistence method does not seem to be a logical choice since, with a limited population and plentiful resources, hunting and gathering are less demanding work than farming. Agriculture requires cooperation, whilst living in settlements has social impacts such as diseases, social inequality and division of labour (Hirst 2017). The wide-scale transition of many human cultures from a lifestyle of hunting and gathering to one of agriculture and settlement, allowing the ability to support an increasingly large population, has been called the Neolithic Revolution or Neolithic Demographic Transition (Bellwood and Oxenham 2008). The adoption of agriculture in the Stone Age is one of the major curiosities in human cultural history since studies carried out among contemporary primitive people suggest that the first farmers worked harder to attain subsistence than their foraging predecessors (Weisdorf 2009). Although the population explosion that followed the Neolithic Revolution was initially explained by improved health for agriculturalists, empirical studies of societies shifting subsistence from foraging to primary food production have found evidence for deteriorating health from an increase in infectious and dental disease and a rise in nutritional deficiencies (Latham 2013). Hence, early farmers appear to have been smaller and less healthy than hunter-gatherers (Cohen and Crane-Kramer 2007; Armelagos et al. 1991). This counterintuitive increase in nutritional diseases resulted from factors such as seasonal hunger, reliance on single crops deficient in essential nutrients, crop blights and social inequalities. The impact of agriculture, accompanied by increasing population density and a rise in infectious disease, was observed to decrease stature in populations across the entire globe, regardless of the temporal period during which agriculture was adopted, including Europe, Africa, the Middle East, Asia, South America and North America (Mummert et al. 2011).

As the population became more settled, and farming became the way of life, it has been suggested that eight plants, the so-called eight founder crops formed the basis of the origins of agriculture, with all eight coming from the Fertile Crescent region. The eight crops include three cereals (einkorn wheat, emmer wheat and barley), four legumes (lentil, pea, chickpea and bitter vetch) and one oil and fibre crop—flax or linseed (Hirst 2018; Weiss and Zohary 2011). Somewhat later, rice and millet were domesticated in China (Deng et al. 2015), whilst about 7500 years ago rice and sorghum were being farmed in the Sahel region of Africa. Local crops were domesticated independently in West Africa and possibly in New Guinea and Ethiopia, whilst several crop plants were domesticated in the Americas including corn, squashes, potato and sunflowers (Petruzzello 2019). It is clear that farmed crops could suffer from pests and diseases causing a large loss in yield, with the ever-present possibility of famine for the population. There was thus a great incentive to find ways of overcoming the problems caused by pests and diseases.

There are a large number of pests and diseases that can affect crops, thereby reducing crop yield with the concomitant possibility of hunger and in extreme cases famine for the population. It has been estimated that there are over 70,000 different pest species that can damage agricultural crops, including 9000 species of insects and mites, 50,000 species of plant pathogens and 8000 species of weeds. Although, in general, less than 10% of these organisms are considered to be major pests, this still leaves a considerable number that can have a significant effect on crop productivity (Pimental 2009). In addition, farmers also have to contend with damage caused by nematodes, slugs, snails, birds and mammals. Damage to crops can occur in different ways, and pests and diseases can be classified according to the types of damage that they cause (Boote et al. 1983). These classifications are:

- Stand reducers—e.g. damping-off fungi;
- Photosynthetic rate reducers—e.g. fungi, bacteria, viruses;
- Leaf senescence—e.g. various pathogens;
- Light stealers—e.g. weeds, some leaf pathogens;
- Assimilate sappers—e.g. nematodes, pathogens, sucking arthropods;
- Tissue consumers—e.g. chewing animals, necrotrophic pathogens;
- Turgor reducers—e.g. nematodes, root-feeding insects, root rot pathogens.

Initially, pest control was probably a manual occupation, e.g. weeding or removal of caterpillars and insects by hand, followed by attempts to protect crops by means of religion, folk magic and the use of what may be termed early chemical methods. For example, pyrethrum is recorded as being used in China 2000 years ago (McLaughlin Gormley King Company 2010) and is still used today (Market Research.com 2015). Up until the mid-nineteenth century, various plant, animal or mineral derivatives (Smith and Secoy 1975, 1976a) were used for pest control; these were later augmented by compounds from the dyestuff industry, e.g. Paris Green, and by-products from the burgeoning chemical industry based on coal tar with phenol-based chemicals (Gray 1916). The growth in synthetic pesticides accelerated in the 1940s with the discovery of the pesticidal effects of numerous organic chemicals, which led to the development of, for example, chlorinated hydrocarbons such as DDT as insecticides and phenoxy acid derivatives like 2,4-D as herbicides. As the use of these "new" pesticides increased, concerns for their effect on human health and the environment were raised and attention was given to the development of low-rate, lower-risk organic compounds as pesticides. It can be argued that there have been five significant phases in the evolution of pesticides. With some inevitable overlap of the individual phases, they can be summarised as:

- Before 1000—early pest management;
- 1000–1850—plant, animal or mineral derivatives;
- 1850–1940—inorganic products, industrial by-products;
- 1940–1970—synthetic organic compounds;
- 1970–present—lower-risk synthetic organic compounds.

Given the significant number of pests and diseases that can impact on crop yields, and the variety of ways by which this can happen, it is not surprising that since the dawn of agriculture, farmers have been engaged in a constant battle against them. This battle continues today, and it has been estimated that potential crop losses worldwide would be between 50 and 82%, depending on the crop, in the absence of pest control (Oerke 2006), and the economic loss has been estimated at US $250 billion (Agropages 2015).

As an example of the deleterious effects of weeds on crop yield, it has been estimated that at a density of only two weeds per foot of row corn yield was reduced 10% by giant foxtail, 11% by common lambsquarters, 18% by velvetleaf and 22% by common cocklebur (Moechnig et al. 2013). Given the fact that approximately 1 in 9 of the global population is suffering from undernourishment (FAO 2018) and that the global population is estimated to rise from 7.3 billion today to 9 billion in 2050, it is essential that food shortages caused by reductions in crop yields due to pests and diseases are kept to an absolute minimum. Chemistry will continue to have a significant role in the control of fungal pathogens, insect pests and weeds which, together with modified farming practices, new crop varieties, increased irrigation and optimum fertiliser use, will play a key role in achieving increased and sustainable productivity.

2.2   Early Efforts at Pest Management

Ancient farmers would have tried every method available to them at the time to maintain or increase crop yields. For example, in the Mediterranean region repeated fertilisation of the soil, careful tillage, intelligent selection of crops suited to climate and soil, importation of foreign seed and plants, painstaking seed selection, with crop protection techniques as part of the mix were all used (Semple 1928). However, despite this, the yield per hectare remained relatively low. In Mesopotamia, irrigated cereal crops have been reported as giving a yield of almost 2 tonnes per hectare, although this declined to only about one-third as much by 1700 BC, probably due to increasing salinity of the irrigated areas (Evans 1980). A maximum return known in Roman Italy about 50 BC gave a yield of eight to tenfold of planted seed. This was the yield of winter wheat from rich volcanic soils in eastern Sicily, and it amounted to 20 or 24 bushels to the acre, or approximately 1.4–1.7 tonnes per hectare. A yield of 15-fold or 36 bushels of wheat to the acre (approximately 2.5 tonnes per hectare) was quite extraordinary and restricted to a few favoured spots (Semple 1928), although wheat yields of up to 3.6 tonnes per hectare were probably obtained in Galilee around the beginning of the first century AD (Evans 1980). These figures can be compared with about 5 tonnes/hectare achieved in Greece and Italy and 7 tonnes per hectare in Egypt today (World Bank 2018). Nevertheless, whilst reasonable yields could be obtained under some favourable conditions, the possibility of disease significantly reducing crop yield was always present. Cereal stem rust was particularly feared and could lead to 100% crop loss. In the Bible, there are several references to epidemics of rusts and smuts inflicted on the Israelites as a punishment for their sins (Leonard and Szabo 2005).

2.2.1 Religion and Folk Magic

In many agricultural societies, there was a widespread belief that pests were evil and were sent as a punishment from the gods; as a result, many early pest management strategies were based on religious practices. Thus, the Sumerian goddess Ninkilim was invoked to protect growing barley from damage by rodents, and a Babylonian tablet contains an incantation asking the sun and moon gods to prevent ergot from developing on ears of grain (Secoy and Smith 1977). In Egypt, reliefs in tombs dating as early as the Old Kingdom, about 2649–2150 BC, show that illiterate shepherds and farmers could recite spells to protect herds and crops (Zucconi 2007). In ancient Rome, the deity Robigus and the goddess Flora were believed to be responsible for rust and mildew, but when they were appeased, these diseases would not harm the grain and trees. In honour of Robigus, the solemn feast of the Robigalia was thus established, during which a rust coloured dog was sacrificed, while for Flora, there were games called Floralia (Smith and Secoy 1975). The Greek god Apollo was believed to give protection to crops against summer heat, blight, mildew and other pests such as field mice and grasshoppers (Secoy and Smith 1977). In India the Atharvaveda, a sacred Hindu text, described the chanting of mantras to protect crops from insects such as grasshoppers; one such mantra from around 400 BC evokes the Hindu deity Rama to protect grain. The mantra had to be written with the red lac-dye on a leaf and tied in the field (Goyal 2003). Similarly, in Sri Lanka, mantras have been used in agriculture to obtain higher yields and also to protect crops from damage by pests and wild animals (Upawansa and Wagachchi 2018). In Japan, farmers would entreat deities through ceremonies such as *mushi-oi* or *mushi-okuri*, meaning driving away noxious insects. For this, the local population walked around rice paddies in procession shouting, with bells and torches to drive away rice pests and pray for that year's farming. In addition, farmers would place beef or phallic-shaped ritual equipment in the water inlets to fields to ward off insect pests (Ohta 2013). Besides the invocation of deities to keep pests from crops, there were a host of folk remedies that could also be practiced. Thus, in Roman times, it was thought that trees could be protected against caterpillars by touching the tops with the gall of a green lizard and that a crayfish hung up in the middle of a garden would afford protection against caterpillars. Mildew would be controlled by placing branches of laurel in the ground, near to growing wheat, thus causing the mildew to pass from the fields to the foliage of the laurels. Diseases of millet could be prevented by carrying a toad around the field at night before burying the creature in the middle of the field in a pot (Smith and Secoy 1975). According to Columella (4–70 AD), the Etruscans set the flayed head of an ass at the edge of a field to protect grain from rust (Secoy and Smith 1977). It was also hoped that amulets, sometimes in the form of a protective deity, and others shaped like the pest itself, would ward off a danger, e.g. locust amulets which have been discovered in ancient Egyptian tombs (Kenawy and Abdel-Hamid 2015). Many other religious and folk remedies have been described, and more information can be found in the above references.

2.2.2   Manual and Physical Techniques

Whilst religious and folk remedies were of dubious value other management practices involving, for example, removal of weeds by hand, hoeing or ploughing and the picking of insects from crops would have been successful, if time-consuming. Chasing or throwing stones at birds and small mammals would have been effective, and, although probably less effective, scarecrows were also used to deter these pests. The first scarecrows are reported in Ancient Egypt, dating back over 3000 years, where they were used to protect wheat fields along the river Nile from quail. The Greeks, Romans, Japanese and many other cultures also used scarecrows to protect their crops.

The practice of burning cereal stubble after harvest would also have contributed to the destruction of weed seeds. In addition, physical methods such as the use of tar on tree trunks to trap crawling insects and the adding of fine sand to stored grain would prevent much of the damage caused by insects. In stored grain, the addition of fine sand inhibits movement of the insects and reduces the oxygen content, thereby making it hard for the insects to get enough oxygen to survive; in addition, the sand is abrasive and would scratch the hard covering that surrounds the bodies of beetles, leading to desiccation and death (De Groot 2004). This method is still used today, particularly in Africa, and was used in ancient Egypt. It can be assumed that it was sufficiently effective to exterminate a pest such as the lesser grain borer, recently arrived in the region (Pharaonic Egypt 2000). Similarly in 750 BC, Homer noted the use of wood ash spread on land as a form of pest control where it acts as a physical poison, usually causing abrasion of epicuticular waxes and thus exposing pests to death through desiccation. It also interferes with the chemical signals emanating from the host plants, thus obstructing the initial host location from pests. When foliage is treated with wood ash, it also becomes unpalatable for foliage feeders like cutworms, caterpillars and grasshoppers (Verma 1998).

Biological methods were also used; for example, mention is made in a document reputed to have been written in 304 AD of the use in ancient China of citrus ants against insects found on citrus trees (Huang and Yang 1987). Similarly, around 1000 AD in Arabia, ants were moved from nearby mountains to oases to control phytophagous ants which attack date palms.

2.2.3   Use of Natural Products as Pesticides

Management practices using what may be termed chemical methods were also tried for the control of pests and diseases, the first recorded instance, in 2500 BC, being the control of mites and insects by the ancient Sumerians using sulphur. As there was no chemical industry, any products used had to be either of plant or animal derivation or, if of a mineral nature, easily obtainable. Thus, control of pests was tried either using everyday products, such as cow dung, milk and olive oil, or products derived from plants which were known to have pharmacological properties and were used in human medicine. In 2000 BC, the Hindu book, the *Rig Veda*, makes reference to the use of poisonous plants for pest control, and it is known that plants were used as sources of insecticidal compounds by the Egyptians during the time of the Pharaohs. In 970 AD, the Arab scholar Abu Mansur described over 450 plant products with toxicological and/or pharmacological properties (Thacker 2002). At the time, of course, it was not known why or how these products worked, but in some circumstances may have given a measure of control. Smokes were used in ancient Roman and Indian agriculture (Smith and Secoy 1975; Goyal 2003) against various pests, including mildews. The principle was to burn some material such as ghee, straw, chaff, hedge clippings, crabs, fish, dung, animal horn and hair to windward so that the smoke, preferably malodorous, would spread throughout the orchard, crop or vineyard. It is unlikely that this proved to be entirely efficacious although it may have had a temporary effect in repelling insects, small mammals and snakes. A probably more effective method was the heating of sulphur with bitumen to produce a smoke which, blowing over vines and fruit trees, would repel various insects (Secoy and Smith 1983). The use of cow dung for dressing seeds has been suggested since the time of Kautilya (c.300 BC). The bacteria present in the dung quickly colonise the area around the seed and compete with pathogenic fungi and bacteria, thus preventing them from attacking the seed (Upawansa and Wagachchi 2018).

The use of milk and milk products is mentioned in a text by the Indian Varahamihira, about 600 AD, for use against various pests (Goyal 2003); interestingly, it has been shown that milk is efficacious against powdery mildew (Bettiol 1999). Elemental sulphur, which can be found in nature, is one of the oldest fungicides and is still commonly used, when applied in powdered form, for grapes, strawberries and many vegetables such as peas and beans. It has a good efficacy against a wide range of powdery mildew diseases as well as black spot and has an effect on some rusts and smuts (Williams and Cooper 2004). Various treatments against fungal diseases to improve germination and yields made use of alkaline conditions such as the mixing of seed with ashes which is mentioned by Someshwara Deva c.1126 AD (Goyal 2003), and Pliny (23-79 AD). Pliny also suggested the steeping of seed in a mixture of amurca (olive oil lees) and native soda and the soaking of beans in stale urine (Smith and Secoy 1975). These treatments would give an alkaline environment, and alkaline seed treatments have been shown to be effective in reducing the incidence of the fungal disease of wheat now known as bunt. For similar reasons vines, figs and other plants when sprinkled with a mixture of ashes in water could have been given

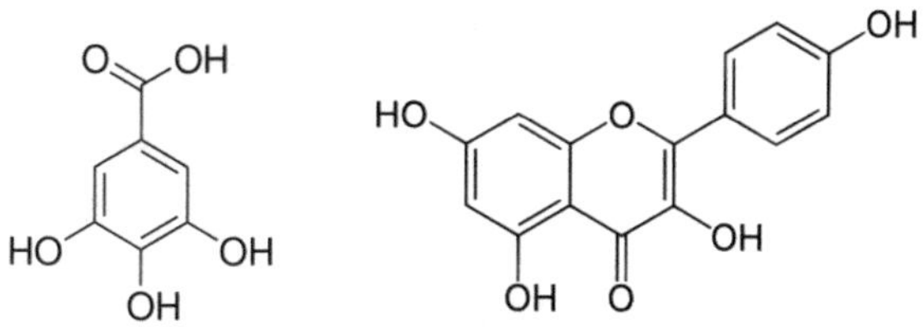

Fig. 2.1 Molecular structure of gallic acid (left) and kaempferol (right), natural pesticides

a small measure of protection against certain diseases (Smith and Secoy 1975). This might have been more effective when mixed with sandarach (red sulphide of arsenic) which Pliny mentioned during the first century AD as a means of preventing grape rot.

For weeds, the watery residue amurca, obtained when the oil is drained from pressed olives, was used. Amurca contains various phenolic compounds including oleuropein, gallic acid, 3-hydroxyphenol, sinapic acid, kaempherol, isopropyl-5-methyl phenol and luteolin (Janakat and Hammad 2013), some of which have herbicidal properties. For example, gallic acid (3,4,5-trihydroxybenzoic acid) is phytotoxic to a variety plants (Rudrappa et al. 2007), whilst kaempferol (3,5,7-trihydroxy-2-(4-hydroxyphenyl)-4H-chromen-4-one) has also been shown to be a phytotoxic compound (Bais et al. 2003) (Fig. 2.1).

However, the weed killing properties of amurca may have been due more to the salt contained in it which was sometimes added during the pressing of olives (Smith and Secoy 1975). Salt and sea water were known to have phytotoxic properties (Janek 2008), but interestingly, this does not seem to be referred to in the context of any uses as a weed killer (Smith and Secoy 1975).

For insects, Egyptians, followed by Greeks and Romans, used sulphur to purify their houses and established it as fumigant when it was burnt to give sulphur dioxide. The Chinese controlled lice and insects with mercury and arsenic 2500 years ago, and it was recorded in 900 AD that arsenic sulphides were used to control insects.

As described above, smokes were used against insect pests, but a variety of plant extracts were also used to kill or repel insects and their larvae. Many plants, such as absinthe, asafoetida, and other aromatic plants like bay, cassia, cedar, citron, cumin, elder, fig, garlic, heliotrope, hellebore, ivy, oak, origanum, pomegranate, rhododaphne and squill were reputed to have properties of being able to kill or repel insects and their larvae (Smith and Secoy 1975). This also included extracts of wild cucumber and bitter lupin which were used extensively against a wide variety of pests including locusts, cantharides and fleas (Smith and Secoy 1975). Wild cucumber, also known as squirting cucumber (*Ecballium elaterium*), has been used for thousands of years as a medicinal plant with extracts having many pharmacological properties (Attard and Cuschieri 2009). There are accounts of the constituents of the extracts having insecticidal properties, probably due to the presence of tetracyclic triterpenoids such as cucurbitacin E glycoside, against, for example, the red flour beetle, *Tribolium castaneum* (Pascall-Villalobos and Miras 1999) and the aphid pest *Aphis craccivora* (Torkey et al. 2009) Fig. 2.2.

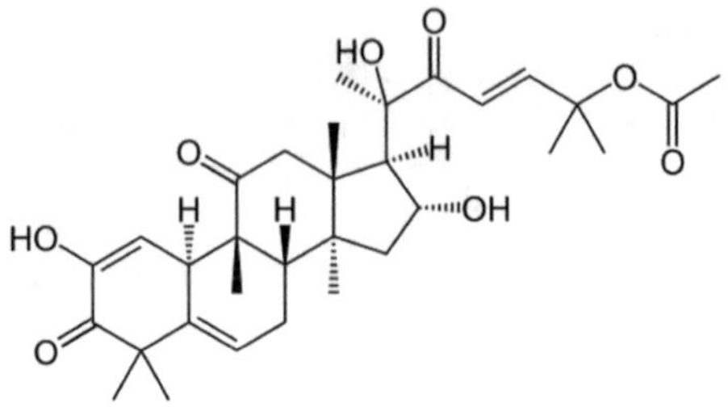

Fig. 2.2 Molecular structure of Cucurbitacin E, a natural insecticide

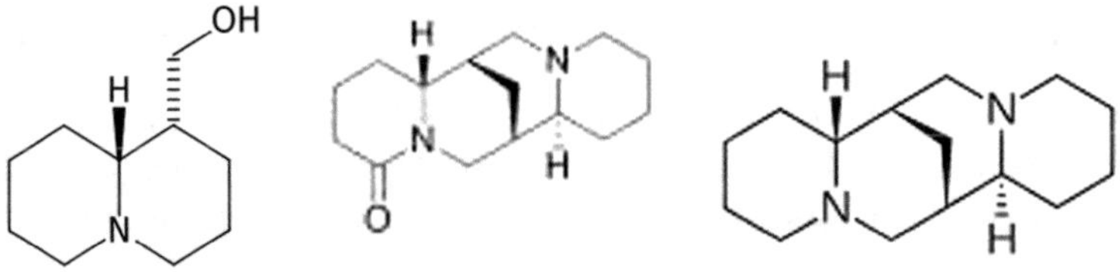

Fig. 2.3 Molecular structures of, from left to right, lupinine, lupanine and sparteine, natural insecticides

Lupins contain significant amounts of quinolizidine alkaloids (Wink et al. 1995; Boschin et al. 2008), which have a broad range of pharmacological properties (Bunsupa et al. 2012). However, in plants, these alkaloids, for example, lupinine, lupanine and sparteine, Fig. 2.3, play a significant role in the chemical defence against various predators. They have been shown to be feeding deterrents for a number of oligo- and polyphagous insects, including aphids, moth and butterfly larvae, beetles, grasshoppers, flies, bees and ants. In addition, they are also toxic to several insects and their larvae (Wink 1994; Bermúdez-Torres et al. 2009).

Indeed, modern science has only recently come to recognise that many plant extracts used by ancient apothecaries contain compounds with useful activity against insects (Koul et al. 2008).

Interestingly, there are two pesticides which were used in ancient times and are still in use today, and these are products from pyrethrum and neem. The pyrethrum daisy, better known as a chrysanthemum, has long been known to have insecticidal properties. It was noted during the Chou Dynasty of the first century AD in China that the flowers naturally repelled a number of pests (McLaughlin Gormley King Company 2010). Pyrethrum is a natural plant oil that is found mainly in tiny oil-containing glands on the surface of the seed case in the tightly packed flower head and is the plant's own insecticide that it has evolved to keep insects away. Pyrethrum is made up of six complex chemical esters known as pyrethrins which work in combination to repel and kill insects. Pyrethrins can be considered to be a single pesticide active ingredient having six components with insecticidal activity: pyrethrin I, pyrethrin II, jasmolin I, jasmolin II, cinerin I and cinerin II. Pyrethrin I, jasmolin I

pyrethrin I : R1 CH$_3$, R2 CH=CH$_2$ pyrethrin II : R1 CH$_3$O$_2$C , R2 CH=CH$_2$

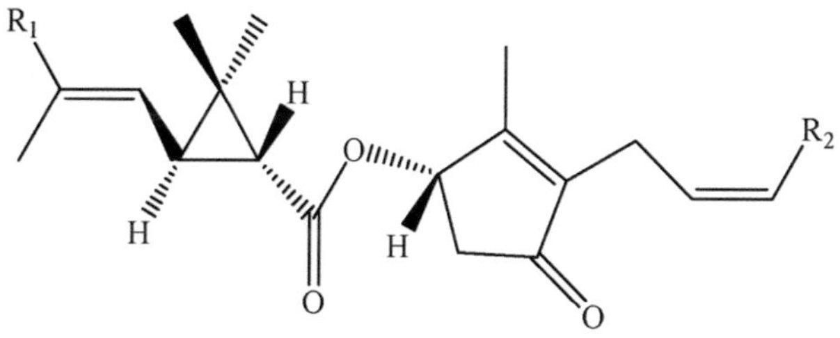

jasmolin I : R1 CH$_3$, R2 CH$_2$CH$_3$ jasmolin II : R1 CH$_3$O$_2$C , R2 CH$_2$CH$_3$

cinerin I : R1 CH$_3$, R2 CH$_3$ cinerin II : R1 CH$_3$O$_2$C , R2 CH$_3$

Fig. 2.4 Six pyrethrins, insecticidal constituents of pyrethrum (from Pan et al. 2017)

and cinerin I are esters of chrysanthemic acid, whereas pyrethrin II, jasmolin II and cinerin II are esters of pyrethric acid Fig. 2.4.

Pyrethrins kill insects by disrupting their nervous systems as they are toxic to the "sodium channel", the cellular structure that allows sodium ions to enter a cell as part of the process of transmitting a nerve impulse. This leads to repetitive discharges by the nerve cell which causes paralysis and death (Crosby 1995). The synthetic pyrethroid insecticides, first developed in the 1960s, have chemical structures adapted from those of the pyrethrins to increase their stability to sunlight.

Ancient Sanskrit medical writings created more than 1000 years ago speak about the usefulness of the neem tree. This tree which is grown widely in India and Africa is fast growing and normally reaches a height of 7–15 m; it has fragrant white flowers which give rise to the fruit which comprises a fibrous pulp around a hard inner shell containing the neem seeds (Kureel et al. 2009). Various parts of the neem tree have been used as traditional ayurvedic medicine, a system of Hindu medicine native to the Indian subcontinent, from time immemorial (Biswas et al. 2002). The practice of mixing neem materials with stored products became rooted as part of traditional wisdom and culture. Mixing of neem leaves (2–5%) with wheat, rice or other grains is even now practised in many villages in India and Pakistan (Ogbuewu et al. 2011). The Upavanavinod, an ancient Sanskrit treatise dealing with forestry and agriculture, cites neem as a cure for ailing soils, plants and livestock. Neem is a potent pesticide, effective against locusts, brown planthoppers, nematodes, mosquito larvae, Colorado beetles and boll weevils. These properties and in particular the multitude of medicinal uses, known to Indians for millennia, have given the tree its Sanskrit name, *sarva roga nivarini*, "the curer of all ailments" or, in the Muslim tradition, *shajar-e-mubarak*, "the blessed tree" (Shiva and Hollabhar 2008). All parts of the tree such as leaf, flower, fruit, seed, kernel, bark, wood and twig are biologically active, the maximum activity being associated with the seed kernel. A broad spectrum of activity against insects, phyto-nematodes, plant pathogens, etc., is

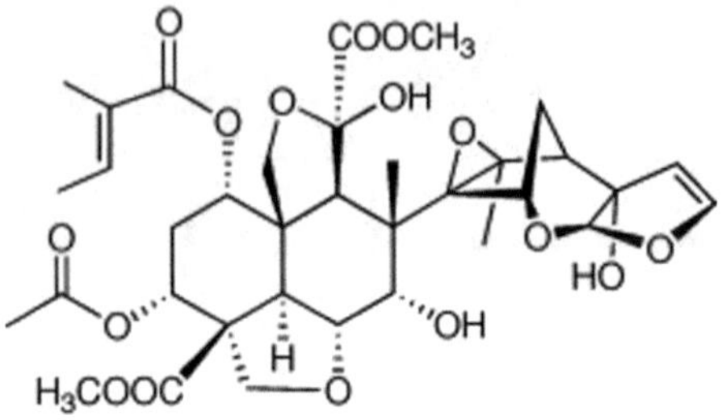

Fig. 2.5 Molecular structure
of azadirachtin, a naturally
occurring insecticide from
the neem seed (from Morgan
2007)

exhibited with multifarious modes of actions. Worldwide, more than 500 pest species
are controlled. Neem has a multipronged effect against insects acting as a repellent,
antifeedant, oviposition deterrent, moulting or growth disruptor and sterilant with
ovicidal activity. These properties help to control effectively a variety of farm and
household insect pests and pathogens infesting agricultural, plantation and cash crops
(IARI 2010). The seed is an important source of biopesticidal compounds. Neem
seed kernel consists of azadirachtin and other limonoids such as nimbin, salanin and
meliantriol. However, azadirachtin is the most important limonoid and is known to
possess antifeedant, attractant, repellent, growth disrupting and larvicidal properties
against a large number of pests (Kaushik et al. 2007). The structure of azadirachtin
(Fig. 2.5) is immensely complex, with 16 adjacent chiral centres and its elucidation
took 17 years, whilst the complete chemical synthesis took another 22 years (Veitch
et al. 2007).

Whilst natural products, containing active compounds, were undoubtedly used for
pest control, there was no understanding of why these substances did or did not work.
In the absence of a chemical industry, the only recourse for controlling pests was
either to pray to the gods or to use readily available materials, often using substances
that had been shown to have some beneficial effect on humans.

2.3 Pest Control to the Early Twentieth Century

Farmers continued to suffer from pests and diseases that caused a significant loss
of yield, and, with a lack of effective means of control, there was always the possibility of famine, particularly where significant numbers of the population relied
on one staple crop. Examples are the Kyoho famine in Japan in 1732, caused by
cold weather and planthoppers which destroyed the rice crop, with an estimated
one million deaths due to starvation (Ohta 2013). In Ireland, the famine, caused by
late blight (*Phytophthora infestans*) on potatoes was implicated in a million deaths
from malnutrition and disease, with almost another million emigrating abroad to
escape the famine, for an absolute population decline of 21%, between 1845 and
1851 (Turner 2005). In Europe, there was little progress in pest management after
the collapse of the Roman Empire; generally, advice from classical authorities was

followed and knowledge progressed slowly. Consequently, pest control was based on techniques described by earlier authors. Early essays on agriculture, written by the classical scholars Cato, Varro and Columella, were still being used in Europe right up until the sixteenth century or even later (Thacker 2002). As in earlier times, methods were mainly based on manual effort and readily available inorganic and natural materials. Divine intervention was still called on for pest control even up to the nineteenth century, with limited scientific observation on the behaviour and control of pests. However, some progress was made, and in the seventeenth century, more scientific measures were being adopted to combat insect pests. Francis Bacon (1561–1626), Lord Chancellor of England, is widely credited. Although classical authorities continued to be acknowledged as sources, their accounts no longer took precedence over direct observation and experimentation. Most of the magical practices fell into abeyance and were virtually forgotten (National Academy of Sciences 2000). In the mid-1700s, the great Swedish botanist and taxonomist Carolus Linnaeus catalogued and named many pests and his writings were key to the future study into pests (as well as plants and animals generally). In the early 1800s, books and papers devoted entirely to pest control were published and, at the same time, the agricultural revolution began in Europe and heralded a more widespread application of systematic pest control. Apart from weeds and fungal problems, farmers had to battle with many other pests that could cause *"the destruction of the whole or portions of their crops, through the medium of various living creatures, who derive from their stalks, roots, or leaves, a predatory existence"* as described in a book published in 1847 (Richardson 1847). This was divided into three sections, quadrupeds, voracious birds and insects, and not only gave a description of the pest but also gave methods for their control. A similar volume, published in 1852, concentrated on pests found in the USA (Richardson 1852). At this time, natural products used by native populations in the Americas and Asia were being discovered, and at the beginning of the nineteenth century, the chemical industry began to produce chemicals in substantial amounts that could be tested for their ability to kill pests. However, taking the historical trends in the yield of wheat in England as an example, it was only towards the end of the nineteenth century that productivity began to increase rapidly. This being due to changing agricultural processes, increased mechanisation, increased use of fertilisers, such as Chilean nitrate and superphosphate, and crop protection products. Before these changes took place, average wheat yields in England and in other European countries during the Middle Ages were quite low at 0.5–0.75 tonne per hectare (Evans 1980). Before about 1600 yields for wheat have been estimated as being relatively constant, estimates of wheat production in England calculated for the period 1590–1860 gave yields of 0.7 tonnes ha^{-1} in 1590, 0.9 tonnes ha^{-1} in 1700, 1.5 tonnes ha^{-1} in 1810 and 1.9 tonnes ha^{-1} in 1860 (Clark 1991)—assuming that 1 bushel $acre^{-1}$ is 0.0673 tonnes ha^{-1}. There would, of course, have been variations due to different growing conditions in different areas. The nineteenth century also saw the development of large-scale chemical industries, particularly in Great Britain and Germany, with an explosion in both the quantity of production and the variety of chemicals that were manufactured. This coincided with trials with a large number of different products as pesticides such as metals, and their salts, and chemicals derived from the coal tar and dyestuff industries.

2.3.1 Divine Intervention

As in more ancient times and, in the absence of a real understanding of the dynamics and causes of pest infestations, ignorance and superstition thrived and it was still often thought that pests were sent as a punishment from the heavens. As a result, some pest management strategies were still based on religious practices. For example, when infestations of pests became an epidemic, the church in Europe sometimes invoked divine intervention to deal with the problem. In the twelfth century, the Bishop of Leon excommunicated mice and caterpillars, whilst in 1479, cockchafers were indicted before the ecclesiastical court at Lausanne and condemned to banishment. In 1485, the High Vicar of Valence commanded caterpillars to appear before him, gave them a defence counsel and finally condemned them to leave the area, and in 1488, the High Vicar of Autun commanded the weevils in neighbouring parishes to stop their attacks on crops and grain and excommunicated them (Robertshaw 1998). The last recorded indictment was in 1830, in Denmark, where insects were excommunicated for crop damage (Secoy and Smith 1977). With the Renaissance in Europe, however, pests began to be viewed less as a punishment from the heavens and more as members of the natural world that could be studied, understood and controlled.

2.3.2 Manual and Physical Techniques

Techniques probably used since the beginning of agriculture continued to be used during the following centuries. Hand weeding, or weeding by ploughing with oxen if the crops of wheat and barley were sown in widely spaced rows, was used extensively. For example, in eastern Norfolk, around 1300 hand weeding was intensive and expensive with at least 60% of the crops being weeded (Zadoks 2013). As late as 1850, 65% of the population lived on farms in the USA and removing weeds was one of the main farm chores (Freed 1980). Insects and caterpillars were removed by hand, and, for example, hand picking in medieval vineyards was routinely done by women and children (Zadoks 2013). This was very labour intensive, but the discovery of mechanical instruments made the task easier, with the development of the earliest mechanical insect trap in the early 1700s. In 1670, in Japan, it was found that leafhoppers on rice could be controlled by pouring whale oil into paddy fields and then shaking the plants so that the insect pest fell into it, whereby they suffocated (Ohta 2013).

Trapping has been used to remove small mammals with typically glazed pots sunk into the ground as the simplest method. In Medieval Britain, young boys, known as bird scarers, patrolled fields and chased birds away by waving their arms and throwing stones. Bird scarers continued to patrol fields until the early 1800s. If there were insufficient children as bird scarers, for example, after the "Great Plague", scarecrows in human form were used. Scarecrows have been used since ancient times and in 1592 were described as: "That which frightens or is intended to frighten without doing physical harm", literally that which scares away crows, hence the name—scarecrow. Many of these techniques are still used today and are compatible with the organic farming; thus, scarecrows are used in some situations, with "high-tec" versions now available with moving limbs and acoustic possibilities. Specialist traps using, for example, pheromones, are also available for removing insects.

2.3.3 Natural Products

In ancient times, it was known that several plant species could act against insects, the best known of these were substances derived from pyrethrum daisies and the neem tree, as previously described. Later, from the fifteenth century onwards, exploration by Europeans of the Americas and Asia led to the discovery of more plant-derived pesticides which had been used by native populations for many years prior to the coming of the Europeans (Richardson 1847). The efficacy of these plant-derived pesticides depended on the active constituents in their leaves, stems, roots and tubers and had been discovered by trial and error, as the native populations found ways to protect their crops. The actual compounds involved were, of course, unknown until techniques became available to elucidate their complex formulae, in many cases not until the twentieth century. One example is sabadilla, which is derived predominantly from the sabadilla lily (*Schoenocaulon officinale*), and has been used in Europe from the sixteenth century. It was noticed by the first explorers of the Americas that the local population was using the powdered seeds to protect crops against insect attack. When sabadilla seeds are aged, heated or treated with alkali, several insecticidal alkaloids are formed or activated. The active compounds are ceveratrum alkaloids, mainly cevadine and veratridine (Fig. 2.6), which affect nerve cell membrane action, causing loss of nerve function, paralysis and death (Richardson 1847; Texas A&M 2019).

Tobacco was introduced into Europe in 1559, and by 1690, extracts of tobacco leaves were being used as a wash on pear trees to control pear lace bug and later in 1773 as a fumigant by heating tobacco and blowing the smoke on infested plants (McIndoo 1943). The active principle in tobacco extracts was later shown to be the alkaloid nicotine (Fig. 2.7), an extremely fast acting nerve toxin which was first isolated in 1828, its structure being elucidated later in 1893.

In the 1600s, rotenone containing plants were recorded by explorers in Peru as being used to poison fish and its first documented use against leaf-eating caterpillars was in 1848. Products containing rotenone are typically preparations from plant species of the genus *Derris* or *Lonchocarpus* (Leguminosae) with the majority from Cubé resin, a root extract of *Lonchocarpus utilis* and *Lonchocarpus urucu* (Duke

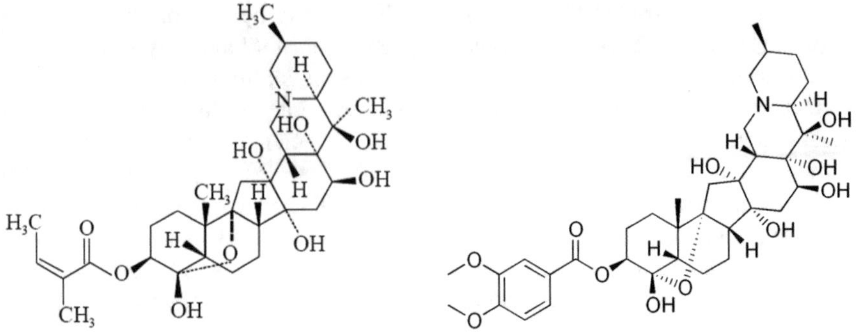

Fig. 2.6 Molecular structure of cevadine (left) and veratridine (right), naturally occurring insecticides derived from the seeds of sabadilla (*Schoenocaulon officinale*)

Fig. 2.7 Molecular structure of nicotine, a naturally occurring insecticide, derived from tobacco extracts

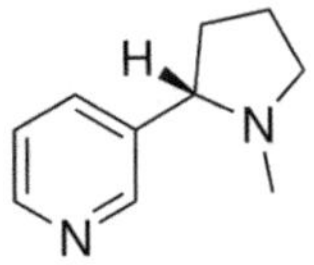

Fig. 2.8 Molecular structure of rotenone, a naturally occurring contact and stomach insecticide

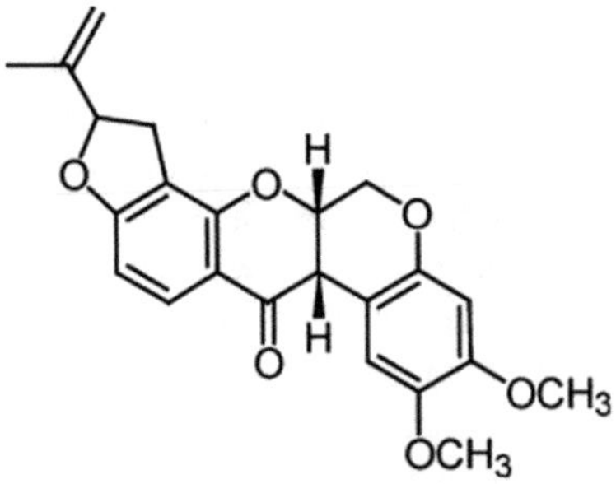

et al. 2010). Rotenone itself was first isolated in 1895, and its structure determined in 1932 (Laforge et al. 1933), Fig. 2.8. Rotenone acts as a contact and stomach insecticide.

The wood or bark of *Quassia amara* L. (Surinam Bitter wood) or *Picrasama excelsa* (Jamaica Bitter wood) family *Simaroubaceae* was first introduced into Europe in the middle of the eighteenth century for medicinal use. In 1825, it was found that extracts of plants of the genus *Quassia* were useful in killing flies, and in 1885, it was recorded that these plant extracts were efficient against hop aphids. Subsequently, these extracts were shown to be effective against various other insects (McIndoo and Sievers 1917). The extracts contain quassinoids, a group of compounds which are the bitter principles of the Simaroubaceae family, of which quassin is one of the active principles. Quassin was isolated in 1937, and its structure was determined in 1961 (Guo et al. 2009), Fig. 2.9.

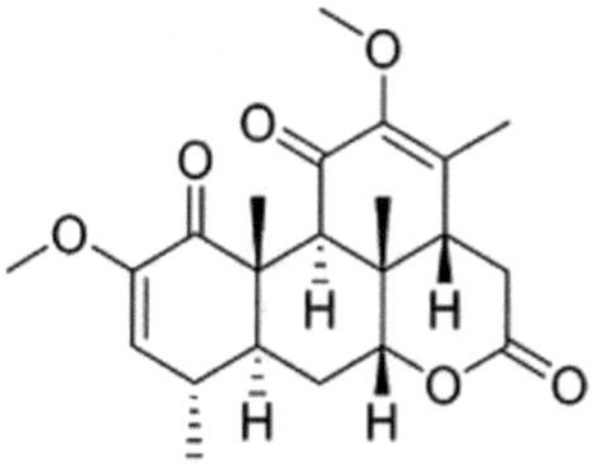

Fig. 2.9 Molecular structure of quassia, a naturally occurring insecticide derived from the wood or bark of Surinam or Jamaican Bitter Wood (*Quassia amara* L. or *Picrasama excels*)

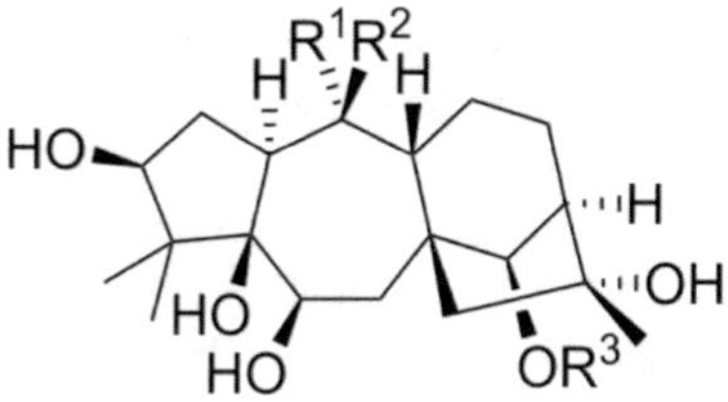

Grayanotoxin	R¹	R²	R³
Grayanotoxin I	OH	CH₃	Ac
Grayanotoxin II		CH₂	H
Grayanotoxin III	OH	CH₃	H
Grayanotoxin IV		CH₂	Ac

Ac = acetyl

Fig. 2.10 Molecular structure of grayanotoxin, a naturally occurring insecticide derived from Japanese andromeda (*Pieris japonica*)

In the relatively closed society of Japan during the Edo era from 1600 to 1868, there was a similar situation regarding the use of natural products as pesticides. In 1697, a ten-volume "Compendium of Agriculture" was published which recorded methods for eradicating insect pests using extracts or smokes based on various plants. This was followed in 1731 by another guide on eradicating insects entitled "Rich Treasury of Records". All of the plants listed in these two books are types of medicinal plants and are known to have insecticidal or insect repelling components; thus, for example, tobacco contains nicotine, chinaberry (*Melia azedarach*) contains azadirachtin, and Japanese andromeda (*Pieris japonica*) contains grayanotoxin, Fig. 2.10.

Soaps have also been used as insecticides in China as long ago as 1101, and there is documentation that they were used in Europe in 1787. Prior to 1900 fish or whale oil soaps were the most commonly used insecticidal soaps (Baldwin 2005).

2.3.4 Early Chemical Control of Pests

The chemical industry, particularly in Europe but also in the USA, came of age during the nineteenth century. In 1800, many chemicals were produced in gram quantities and only to special order. However, by 1900, similar chemicals were being produced in large amounts with excellent quality and at a vastly cheaper price (Reilly 1951). Chemicals typically tested as herbicides were sodium chloride, copper sulphate, ferrous sulphate, sulphuric acid, sodium nitrate, ammonium sulphate, potassium sulphide, phenol and formaldehyde (Pammel 1911). A book published in 1913, with E. Bourcart as author, gives a good indication of the large number of chemicals used as pesticides in the late nineteenth and early twentieth centuries, with information on 152 different materials (Bourcart 1913). As little was known about modes of action, this increase in available chemicals, and their cheaper price, allowed research on a trial and error basis to find those chemicals that would be useful as pesticides. In addition, careful observation of the effects of certain products on pests also led to the discovery of viable pesticides. For example, Paris Green, or Emerald Green, was first produced as a brilliant green dyestuff which was used for dyeing cotton fabrics and as a pigment in paint. It was found to be an excellent insecticide after, it is said, a farmer threw the remaining paint, after painting his shutters, on potato plants infested with Colorado potato beetle. Copper sulphate was found to be active against downy mildew when it was noted that the copper sulphate, used to protect stakes in vineyards, when washed from the stakes on to the vines controlled the disease. An alternative story is that it was noticed that vines closest to the roads did not show mildew, while all other vines were affected. It was found that those vines had been sprayed with a mixture of copper sulphate and lime to deter passersby from eating the grapes, since this treatment was both visible and bitter-tasting.

Many of the techniques using chemicals for controlling pests were still in use many centuries after they were first used in agriculture. Around the fifteenth century toxic substances, based on arsenic, lead and mercury, were used as pesticides, and many of these were still being used in the early years of the twentieth century. In the case of inorganic compounds, a range of different substances were used in efforts to control pests and weeds (Smith and Secoy 1976a).

Sulphur has been used since ancient times to control pests with its first recorded use being in 2500 BC. Thus, the use of burning bitumen and sulphur, as previously described, was mentioned in the twelfth and fourteenth centuries in different translations of the collected works of Greek and Latin writers (Secoy and Smith 1983). Burning lime and sulphur as a fumigant was cited as a means of controlling caterpillars on trees in the sixteenth century, whilst in the seventeenth and eighteenth centuries, sulphur was used as a fumigant to control adult and larval insects on plants and in greenhouses (Smith and Secoy 1976a). In the early nineteenth century sulphur mixed with other substances, including quicklime, was recommended for the control of powdery mildew on grape vines, roses and fruit trees, including peach trees (Tweedy 1981). When sulphur was mixed with quicklime and boiling water, a clear solution, "lime sulphur", could be obtained, the active ingredient of which was identified as the pentasulphide of calcium. Lime sulphur was effective against diseases and insect pests, including in the 1850s powdery grape mildew, and continued in use into the 1940s (Secoy and Smith 1983).

Sodium chloride, which was mentioned by ancient writers as killing plants non-selectively as a way of punishing vanquished enemies, is recorded in 1594 as killing weeds when sown with cereals. In 1733, the practice in Scotland of sowing ten or twelve bushels of salt to an acre of wheat in winter in order to kill the weeds, but not the winter wheat, was described. Selective control of the weed coltsfoot (*Alopecurus pratensis*) in a foxtail ley with salt as a soil herbicide has also been described (Mukula and Ruuttunen 1969). The use of salt for the killing of weeds in paths and gravel was described during the latter part of the seventeenth century. During 1905–1915, herring brine, i.e. salt solution, was used in the coastal towns of southern Finland to keep footpaths clear of weeds (Mukula and Ruuttunen 1969). In 1826, it was reported that thistles as well as *Plantago, Leontodon, Rumex* and other rosette-form weeds could be killed by applying salt, to the area of a half-crown, on the plant centre and grinding it in with the heel. This killed not only the visible plant but also the deepest roots (Smith and Secoy 1976b). Similar spot treatments whereby salt was used to control mugwort (*Artemisia vulgaris*), cow parsley (*Anthriscus silvestris*) and dandelion (*Taraxacum officinale*) were described in 1904 by von Essen (Mukula and Ruuttunen 1969). His method was to cut the weeds at the base of the stem and to apply the salt to the cut surface of the root. This was particularly recommended for the control of dandelions. However, it often needed to be used in large quantities rendering the soil unfit for growing subsequent crops (Pammel 1911). In about 1650 brine, an aqueous solution of salt, was also used as a fungicide by steeping wheat seed in it against wheat smut. According to Jethro Tull, a prominent figure in agricultural methods in the eighteenth century, this process originated when a shipload of wheat was beached near Bristol at high tide and, after being salvaged, was used as seed (Buttress and Dennis 1947). The crop from seed treated in this way was clean, whilst fields sown with untreated seed suffered from smut. However, this treatment was not always successful and the apparent fungicidal effect might have arisen because the salvaged seed came from a locality where there was no smut (Woolman and Humphrey 1924).

Mercury compounds have also been used since early times, and mercuric chloride, known as corrosive sublimate, has been used as a fungicide, insecticide and herbicide. It was recorded in 1807 as being added to seed steeps to control smut in cereals (Klittich 2008), although later studies showed that it was not successful in treating stinking smut in wheat (*Tilletia foetans* or *T. iriiici*) when compared to copper sulphate or hot water treatments (Kellerman and Swingle 1890). In 1822, an aqueous solution was described as killing insects in the wood and on the walls of greenhouses (Pammel 1911). Similarly, in the 1860s, mercuric chloride solutions were used to control soil-borne pests and earthworms. A more personal use, although not related to food production, was the control of bedbugs with mercuric chloride in alcohol (Potter 2008) and of mosquitoes using sublimated mercuric chloride (Guiteras 1909). Mercuric chloride was also tested as a herbicide, for example, in 1896 by Bolley in the USA, but the results proved to be unsatisfactory (Mukula and Ruuttunen 1969).

Mercuric chloride was used historically as an antibacterial agent in medicine, and this property has been utilised in the treatment of potato scab (Reilly 1951) caused by bacteria of the genus *Streptomyces*, notably *S. scabies* (Johnson and Lambert 2010).

Similarly, arsenic and its compounds have a long history of use as pesticides (Bourcart 1913), and in 400 AD Ko Hung reported the use of white arsenic (arsenic trioxide) as a root application to protect rice against insect pests when transplanting (Jentsch n.d.). In the 1500s, during the Ming dynasty, Pen Ts'ao Kan-Mu compiled his great work on the natural world and noted the toxicity associated with arsenic compounds and their use as pesticides in rice fields (Royal Society of Chemistry 2019). In the sixteenth century, the Chinese used arsenic disulphide as an insecticide. Arsenic trioxide (arsenous oxide) was used as a rodenticide and in 1669 is the earliest known record of arsenic as an insecticide in the Western World when it was mixed with honey and used in a bait to control ants (Papworth 1958). Arsenic trioxide was also used as a soil sterilant in the early 1900s at rates of 400–800 lbs acre^{-1} (448–996 kg ha^{-1}), the highest rate being required because of binding of the arsenic to the soil, whilst sodium arsenite was widely used as a herbicide and was more effective than arsenic trioxide as it is translocated to some extent in plants (National Academy of Sciences 1968).

As previously mentioned, Paris Green, or Emerald Green, was first produced as a brilliant green dyestuff which was in great demand by the Lancashire cotton industry. This pigment, copper acetoarsenite [(CH$_3$COO)$_2$Cu.3Cu(AsO$_2$)$_2$], was found to have insecticidal properties and in the 1860s was used successfully against the Colorado potato beetle (*Leptinotarsa decemlineata* (Say)). Colorado potato beetles were causing devastating losses to the potato crop in the midwestern states of the USA and during the next 15 years extended its range north to Canada and east to the Atlantic coast. The use of Paris Green against this pest was the first successful example of the insecticidal control of a crop pest (Raman and Radcliffe 1992). In the 1870s, Paris Green was used against the codling moth (*Cydia pomonella*) on apples in the USA (Schooley et al. 2008). However, because of faulty manufacturing methods and because arsenious acid was cheaper than the other constituents large amounts of the free acid could be present in some commercial products, which caused severe scorching when used (Haywood 1902). Another widely used insecticide was London Purple, an arsenic-containing by-product from aniline dye preparation. This was produced by boiling a purple residue from the dye industry, containing free arsenious acid, with slaked lime and was also used against the Colorado potato beetle in the late 1800s (Haywood 1902). Another widely used arsenic-containing insecticide was lead arsenate, introduced in 1892 against the gypsy moth (*Lymantria dispar* (Linnaeus)), replacing Paris Green because of its superior performance. Later, growers began using it to combat the codling moth (*Cydia pomonella* (Linnaeus)) in apple, plum and peach orchards. This is a destructive insect pest which, without treatment, would regularly damage 20–95% of the apples in every orchard (Schooley et al. 2008). Lead arsenate when applied in a foliar spray also adhered well to the surfaces of plants, so its effect as a pesticide was longer lasting. It was initially prepared by farmers at home who reacted soluble lead salts with sodium arsenate, a practice that continued in some countries through the 1930s and likely 1940s. Lead arsenate pastes and powders were also sold commercially. Their formulations became more refined over time, and eventually, two principal forms were marketed: basic lead arsenate [$Pb_5OH(AsO_4)_3$] for use in certain areas in California, USA, and acid lead arsenate [$PbHAsO_4$] in all other locations (Peryea 1998).

By the late 1800s, insecticides containing arsenic were widely used against biting insects in the USA, with Paris Green and London Purple being the most commonly used, leading to a significant increase in crop yield (Marlatt 1894). In the USA, it was estimated that in the case of the apple crop spraying protected 50–75% of the fruit, which otherwise would have been "wormy", i.e. containing the larvae of the codling moth (Marlatt 1903).

Copper compounds have been extensively used in agriculture, and their first recorded use was in 1761, when it was discovered that seed grains soaked in a weak solution of copper sulphate inhibited seed-borne fungi (Copper Development Association 2019). By 1807, the disinfection of cereal seeds with copper sulphate solution was being practiced, and this was improved in 1873 through the addition of a limewater bath following the copper sulphate treatment (Johnson 1935). This became the standard farming practice for controlling stinking smut or bunt of wheat, which by then was endemic wherever wheat was grown. This procedure overcame the potentially injurious effect of soluble salts of copper on germination and subsequent growth due to the formation of insoluble cupric hydroxide between the copper sulphate and lime (Evans 1896). In 1838, in France, Boucherie found that copper sulphate could be used to protect wood and this was widely adopted in French vineyards in the preservation of grape stakes. The first specific recommendation of a combination of copper sulphate and lime as a fungicide, prompted by an epidemic of late blight (*Phytophthora infestans*) on potatoes, was that of Professor Morren, at the University of Liege in 1845. However, this was not taken up, and it was not until 1885 that Professor Millardet of the Academy of Sciences at Bordeaux first made public his famous discovery that a mixture of copper sulphate and lime gave effective control for one of the most destructive fungal diseases in France, the downy mildew of grapes (*Peronospora viticola*). This mixture became known as Bordeaux mixture and is still used today (Johnson 1935). Morren's hypothesis that copper sulphate and lime could be used against late potato blight (*Phytophthora infestans*) was finally shown to be correct at the Vermont Agricultural Experimental Station in 1889 when it demonstrated that this disease could be effectively controlled with Bordeaux mixture (Copper Development Association 2019). Unfortunately, this was too late to prevent the Irish Potato Famine caused by the loss of 40% of the potato crop in 1845 and almost 100% in 1846 after infection by potato late blight. In Ireland, there was a widespread dependency on the potato and it eventually became the sole subsistence food for one-third of the country. As previously discussed, the loss of crops due to the late potato blight disease resulted in the death of nearly 1 million people. A similar preparation to Bordeaux mixture introduced as a fungicide in the 1880s was Burgundy mixture prepared from copper sulphate and sodium carbonate.

Although copper sulphate is best known for its fungicidal properties, it has also been used as a herbicide. In 1896, Bonnet in France showed that a 6% solution could be used to selectively control wild radish (*Raphantu raphanistrum*) and charlock

(*Sinapis arvensis*) in oats (Mukula and Ruuttunen 1969). At this time, researchers not only in France, but also in Germany and the USA, studied various metallic salts to kill weeds in cereals (Zimdahl 1969). These included copper sulphate, copper nitrate, ferrous sulphate, for example, a 15% solution of ferrous sulphate was found to be effective (Mukula and Ruuttunen 1969), and sodium arsenite. Incidentally, ferrous sulphate also has weak fungicidal activity in that it will inhibit the germination of bunt spores when used in large amounts (Guiteras 1909). Sulphuric acid itself has also been tested as a herbicide (Brown and Streets 1928; Ball and French 1935) and fungicide (Kraemer 1906).

2.3.5   Early Organic Chemicals as Pesticides

In the early 1800s, the coal gas industry also became important, with the production of significant quantities of coal tar which was a by-product from the process of heating coal to produce gas and coke. The tar and ammoniacal liquor then condense from the gas stream. Whilst coal tar was initially deemed to be a nuisance by-product, it contains an extremely complex and variable mixture of more than 10,000 compounds of which about 500 have been identified, and it contains a complex mixture of polycyclic aromatic hydrocarbons (PAHs) with naphthalene and anthracene as the major components.

Distillation of coal tar yields creosote, a volatile, heavy, oily liquid, which is a highly complex mixture containing several hundred, probably a thousand, chemicals, but only a limited number of them are present in amounts greater than 1%. There are six major classes of compounds in creosote: aromatic hydrocarbons, including PAHs and alkylated PAHs (which can constitute up to 90% of creosote); tar acids/phenolics; tar bases/nitrogen-containing heterocycles; aromatic amines; sulphur-containing heterocycles; and oxygen-containing heterocycles, including dibenzofurans (Melber et al. 2004). Whilst creosote was little used in food production, it was proposed as an insecticide (Bourcart 1913), including a role as a winter wash for fruit trees (Wardle and Buckle 1923); however, its main use was as a wood preservative treatment. In 1838, a UK patent was granted which employed, amongst other things, coal tar thinned with "dead oil" or creosote oil as it is known today to preserve timber (Holland 2002). This use protected timber against wood destroying insects and wood rotting fungi. Treatment was by pressure injection, the Bethell process, and was used initially for protecting ships' timber and then used extensively for treating railway sleepers (railway ties) in the great railway boom of the 1840s in the UK. PAHs in creosote include phenanthrene, acenaphthene, fluorene, anthracene, phthalene and pyridine, Fig. 2.11, of which anthracene and naphthalene have pesticidal properties. In 1819–1820, naphthalene was described as a white solid with a pungent odour derived from the distillation of coal tar, and in 1821, the chemist John Kidd described many its properties and the means of its production. Naphthalene is best known for its use against moths, and its use as an insect repellent and insecticide was noted in 1850. Naphthalene has also been proposed for the control of various insects includ-

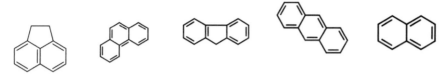

Fig. 2.11 Molecular structure of some polycyclic aromatic hydrocarbons found in creosote derived from coal tar. From left to right: acenaphthene, phenanthrene, fluorine, anthracene and naphthalene

ing, in 1882, the grape phylloxera which had come close to destroying the French wine industry after its accidental introduction in 1863. However, it was reported that naphthalene had no action on this pest and that large-scale trials also gave negative results when naphthalene was placed in proximity to the contaminated roots. On the other hand, it was shown that Balbiani's ointment, a mixture of naphthalene, tar and lime, when applied in the winter, destroyed the eggs of the phylloxera laid on the aerial parts of the plant (Bourcart 1913).

In 1856, William Perkin serendipitously synthesised the dyestuff mauve, or mauveine, while attempting to produce quinine from products derived from coal tar. By 1857, a factory had been set up to manufacture this dyestuff, which is the prototype of aniline or coal tar dyes. This event marked the first step in the industrialisation of organic chemistry and the onset of commercialisation of scientific invention (Travis 1990). The study of coal tar constituents, particularly with the burgeoning interest of coal tar dyes, gave rise to significant advances in organic chemistry throughout the nineteenth century (Fay 1919) including in 1865 the publication by Kekulé of his benzene theory, which placed the whole chemistry of the aromatic compounds on a scientific basis. With the interest in dyes produced from products found in coal tar, particularly in Germany which became the leading manufacturer of these dyes, came the development of industrial research laboratories and scientific research began to play a decisive role in the development of the dyestuff industry. Bayer AG, founded in 1863, and still a major chemical company today, is a good example of how companies developed their own research laboratories in order to find and develop new products (Meyer-Thurow 1982).

As competition increased in the dyestuff industry, companies began to look at other areas in which they could expand, for example, pharmaceuticals and pesticides, thus paving the way for the production of pesticides based on synthetic organic chemistry (Jarman and Ballschmiter 2012). One of the constituents of coal tar is ortho-cresol, and this was a precursor to 4,6-dinitro-o-cresol (DNOC), Fig. 2.12, which was originally marketed as a dyestuff under the name of Victoria Yellow or Saffron substitute (Nietzki 1892). In 1892, this became the first synthetic organic insecticide when used against nun moth [*Lymantria monacha*] larvae which threatened to destroy large tracts of German pine forest. This signalled the start of the involvement of Bayer in the production of synthetic pesticides (Ferguson 1992). In 1893, DNOC became available in the USA for the control of gypsy moth and general pest management against rats, insects and as a wood preservative (Jentsch n.d.). Later, DNOC was

Fig. 2.12 *o*-cresol (left) a
precursor to
4,6-dinitro-*o*-cresol (right), a
multiuse pesticide

found to have herbicidal, as well as fungicidal and insecticidal properties, and with various structural modifications gave rise to the dinitrophenol family of pesticides.

2.4 Pest Control to the 1940s

During the first part of the twentieth century, there was a lack of new pesticides and farmers relied on tried and tested remedies; indeed, in 1915, it was suggested that to control fungal diseases and insect pests in orchards, only four standard spray materials were required, i.e. lime sulphur, lead arsenate, Bordeaux mixture and a nicotine preparation Black Leaf 40 (Reynolds 1915). Even in 1945, at the end of World War II, farmers were still mainly relying on these remedies to control insect pests and diseases, although experimental trials were being carried out with dithiocarbamates as fungicides and dichlorodiphenyltrichloroethane (DDT) as an insecticide (Barss and Andre 1945). However, the available pesticides were not always effective, and studies were ongoing to find new ways of improving the efficacy of those available. For example, mercuric chloride (corrosive sublimate) was used for many years to control fungal problems, but in the 1920s, studies were still ongoing to improve its efficacy for the control of rhizoctonia, scab and blackleg by seed treatment of potatoes (White 1928). However, this period also saw increasing research into diseases of plants with, for example, the formation of the American Phytopathological Society in 1908 (APS 2019), with its first Annual Meeting held in 1909. The development of synthetic organic insecticides was stimulated by research during World War I (Perkins 1982). In addition, early in the twentieth century, discoveries made by fledgling pharmaceutical companies studying the medicinal aspects of compounds made from metals and dyestuff intermediates stimulated plant pathologists to look for compounds that could control plant diseases (Klittich 2008). In the late 1920s and 1930s, many chemical companies became research orientated and, at the same time, governments were encouraging agriculture research. With increasing interest in organic chemistry, many thousands of new compounds were synthesised and many of these chemicals were tested, rather haphazardly, for their pesticidal properties. For example, in looking for substitutes of derris and pyrethrum, Imperial Chemical Industries tested several thousand organic and other chemicals coming from various parts of the company (Achilladelis et al. 1987). Commercialisation of pesticides became important during this period, and to limit the sale of substandard or fraudulent products, the Federal Insecticide Act was passed in 1910 in the USA. There was also a growing awareness that some of the insecticides in use, particularly those containing arsenic and lead, were extremely toxic and residues left on the crops were a public health concern. Following cases in 1900 of poisoning from arsenic in beer in the UK a tolerance, i.e. the maximum allowable amount of a pesticide, was set at 0.01 grains per gallon in liquids and per pound in solids, equating to 0.14 ppm (Emsley 1985). This became the generally accepted tolerance for arsenic in food. In 1927, this was adopted in the USA after exported apples containing higher residues of arsenic caused some cases of poisoning, leading to a threatened embargo of American fruit in the UK. Arising from this the concept of tolerance (a level of poison that most people could consume daily without becoming ill) was developed and was gradually incorporated into legal standards (Morone and Woodhouse 1986).

Progress in weed science was less marked in this period compared to the work on fungicides and insecticides with greater reliance on manual and cultural techniques. Even in the 1930s, farmers were still using many of the same simple chemicals that had been used since the late 1800s to try and control weeds. These methods often proved to be relatively expensive and not very effective. However, this was all to change with the discovery of 2,4-D, first synthesised in 1941 (Pokorny 1941) and its subsequent commercial acceptance as an effective herbicide. Until this point, research was limited in funding as well as in interest by the scientific community (Kelton and Price 2009).

2.4.1   Fungicides

Prior to the development and adoption of effective fungicides, uncontrolled plant disease epidemics significantly reduced crop yields in years favourable for infection. Amongst the many examples is the loss of 38% of wheat production in the USA due to stem rust in 1916, a reduction of 10–70% in Florida cucumber production due to downy mildew in the 1920s and a 20% loss of potato production in Maine in the 1930s due to pink rot (Gianessi and Reigner 2005).

2.4.1.1   Sulphur Compounds

Lime sulphur, also known as sulphuret of lime, Grison's compound or hydrosulphate of lime, was recognised in the 1850s as being very effective against powdery mildew. Lime sulphur was widely used in the early part of the twentieth century, particularly after the development of equipment which made the large-scale spraying of orchards and fields feasible (Secoy and Smith 1983). The use of sulphur as a fungicide was important, and in California, for example, there was the possibility that complete loss of the grape crop due to powdery mildew could occur without sulphur sprays. Apart from grapes, sulphur preparations were used in the USA on a wide variety of fruits, such as treatment for brown rot of peaches, powdery mildew of apples, gooseberries, hops, ornamentals, peaches and strawberries. After its registration in the 1920s, spraying of lime sulphur for scab (*Venturia inaequalis*) in apple orchards in the USA became a universal practice, and it was impossible to grow apples for the fresh market without fungicide sprays (Gianessi and Reigner 2006). However, it was recognised that the spray caused more injury to the fruit and foliage than was desired, and studies were carried out to find ways of reducing phytotoxicity without reducing its effectiveness as a fungicide. Modifications to the lime sulphur solution by adding chemicals having an acidic reaction so that the polysulphide sulphur present was precipitated as elemental sulphur did not seem to have any significant beneficial effect when compared to a weaker lime sulphur solution or the finely divided elemental sulphur sprays that were available. For example, lime sulphur, containing elemental sulphur as the active ingredient, which was the first successful fungicide for the control of brown rot (*Sclerotinia fructicola*) in peaches was almost completely superseded by sprays containing finely divided sulphur as a suspension in water (Roberts 1936a). These "wettable sulphurs", designed to give better coverage of the treated crop, were obtained in various ways, for example, as a by-product in the production of "illuminating gas"; by mixing or grinding very finely ground sulphur with colloids such as casein or saponin as spreaders (Fisher 1921) or glue; by fusing the sulphur with a colloidal clay such as bentonite and grinding the product to an extreme fineness; by precipitating sulphur from lime sulphur solution; and by the interaction of sulphur dioxide and hydrogen sulphide (Roberts 1936a). Lime sulphur and sulphur remained the standard fungicides for the control of such diseases as apple scab (*Venturia inaequalis*) and the powdery mildews (*Erysiphaceae*) affecting many species of plants. Sulphur was used for the control of brown rot (*Sclerotinia fructicola*) and scab (*Cladosporium carpophilum*) in peaches. As the sulphur preparations were developed, the dusts and sprays covered and adhered better compared to earlier preparations, as they were more finely ground and the use of organic "conditioners" prevented the particles from sticking together (Roberts 1936b). However, by the 1940s, lime sulphur began to be replaced by synthetic organic fungicides which caused less damage to the crop's foliage.

2.4.1.2 Mercury Compounds

Mercury-containing chemicals were also still being widely used as fungicides, particularly mercuric chloride (corrosive sublimate) which was introduced in 1910 as the first remedy against snow mold (*Fusarium nivale*) in cereals (Egli and Sturm 1981). Mercuric chloride also had a good efficiency against corn smut (*Ustilago maydis*) and rhizoctonia in potatoes (MacMillan and Christensen 1927). In 1914, methylmercury became commercially important as a crop fungicide, and there was a shift towards the less phytotoxic organomercury seed dressings with the launch of chlorophenol mercury in Germany by Bayer under the trade name Uspulun (Bayer SeedGrowth 2014). The organomercury compounds gave a broader spectrum of control of seed-borne disease in cereals, compared to copper seed treatments which controlled only bunt and could be phytotoxic (Klittich 2008). Various compounds including phenylmercury acetate, methoxyethylmercury and tolylmercury acetate formulations became more widely used in the 1920s and 1930s. In 1929, ethyl mercuric chloride was introduced, under the trade name Ceresan, which gave more effective protection against a broad spectrum of fungal diseases compared to previous mercury compounds. Incidentally, the widespread use of Ceresan as a dry seed treatment was one of the main reasons why contract seed treatment became an economically viable business, using the recently introduced commercial seed treaters (Bayer Seed Growth 2014). However, liquid treatments, using alkylmercury active ingredients, were reintroduced about the time of World War II. This was because of the reduced hazards and inconvenience to operators dressing the grain in specially designed machines, although alkylmercury compounds are more toxic than arylmercurials (Smart 1968). By the early 1940s, seed treatments of a wide range of vegetable crops, with various mercury compounds, were recommended to protect the seeds and young, developing seedlings against excessive damping off (Haskell and Doolittle 1942).

2.4.1.3 Copper Compounds

Bordeaux Mixture continued to be a widely used copper based fungicide and remained the standard treatment. The mixture could be "home-made" by dissolving granulated copper sulphate in water in the spray tank and adding hydrated lime, although commercially prepared Bordeaux mixture was physically much better and contained a much higher percentage of copper. Crop losses could be significantly reduced when copper-based fungicides were used. For example, in the USA, Bordeaux mixture controlled shot hole (*Stigmina carpophila*) in almonds increasing yields 200%; four to five applications of Bordeaux mixture reduced cranberry fruit rots by 50%; sugar beet yields were increased by 20% with control of cercospora; and in tomatoes, phoma rot was reduced from 35 to 6%. As previously mentioned, in the case of the Irish potato famine, potato crops were significantly affected by late blight (*Phytophthora infestans*) and a summary of twenty years of experimental data in Vermont (1890–1910) showed an average yield increase of 64% in potatoes as a result of controlling it with Bordeaux mixture (Gianessi and Reigner 2006). Copper-based fungicides were also recommended for peach leaf curl which could completely reduce the yield of peaches and kill young trees (Pierce 1900). Some examples show that the use of Bordeaux mixture was not without its problems; thus, although it controlled downy mildew on cucumbers, it also damaged the crop and was not widely used. Significant yield losses of 55% in sections of New York and 30% in South Carolina and Florida due to downy mildew on cucumbers were reported for 1924 (Weber 1931; Gianessi and Reigner 2005). It also damaged celery plants, due to excess lime left on them, when used to control early blight.

Bordeaux mixture, like the older copper-based sprays such as Burgundy mixture and ammoniacal copper carbonate, tended to fall out of favour due to their phytotoxicity (Roberts 1936a), and in the 1930s, there was a significant amount of experimental work to develop copper-based fungicides which were less phytotoxic. The work was based on the idea that "insoluble" copper compounds would cause less injury to plants compared to "soluble" copper compounds but would still act on the spores of the fungi (Roberts 1936a; McCallan 1930). A wide range of copper compounds were subject to experimentation, and by the mid-1930s, Bordeaux mixture was being replaced by basic copper sulphate, cuprous oxide, copper oxychloride and copper phosphate (Richardson 1997).

2.4.1.4 Synthetic Organosulphur Compounds

Following the "inorganic" sulphur fungicides were the early group of the so-called organosulphur compounds which were discovered somewhat serendipitously when, as reported in the Michigan Technic Magazine, "*a rubber chemist suggested to a plant pathologist that derivatives of dithiocarbamic acid, parent substance of a well known group of rubber accelerators, be tested as insecticides. His suggestion was based on the possibility that sulphur combined in this form might be more effective than free sulphur*". In due course, derivatives were tested as both insecticides and fungicides and it was found that they were excellent fungicides, with one of the first compounds tested, sodium dimethyldithiocarbamate, shown to be a powerful fungicide, although somewhat injurious to the plants (DuPont Digest 1948). In 1934, Tisdale and Williams patented the use of derivatives of dithiocarbamic acids as disinfectants, including tetramethylthiuram disulphide subsequently known as thiram, although at that stage they had not been tested against fungi living on plants. A systematic research programme looking at other derivatives led eventually to the dithiocarbamate family of fungicides, including the more active metal salts (Goldsworthy et al. 1943). Despite the patent of 1934, thiram was first introduced in 1942 followed in 1943 by zineb and nabam, zinc and sodium ethylenebis (dithiocarbamate), respectively. Later additions to the family, after World War II, were maneb, manganese ethylenebis (dithiocarbamate) in 1955 and mancozeb, a manganese ethylenebis (dithiocarbamate) (polymeric) complex with zinc salt, in 1961 (Russell 2006). The development of the dithiocarbamates in the 1930s set the scene for fungicide research, but it was not until the 1950s and 1960s that fungicide research really took off.

2.4.2 Insecticides

Losses due to insect pests could be quite significant with, for example, in the early 1900s losses due to the grape berry moth of 25–50% in New York, whilst in Wisconsin, it was common for 35–40% of cabbage heads to be unfit due to caterpillar damage. Losses due to borers in sugarcane in Louisiana were estimated at 8–30% (Gianessi 2009). During this period, particularly in the USA, there was an increasing trend towards monoculture agriculture on farms which resulted in potentially greater crop losses from an insect pest complex that was less diverse but more abundant than previously (Matson et al. 1997). This provided ideal feeding conditions for certain insect pests, and, beginning in the 1930s, there was a need for more effective insecticides. However, up until the 1940s, insecticides still tended to be based on natural products, petroleum-based products, sulphur and arsenic.

2.4.2.1 Natural Products

Over this period, work centred around trying to improve the efficacy of natural products that were currently in use. Nicotine used either as a smoke, or as extracts from tobacco leaves, had been used as an insecticide since the late seventeenth century. In the period 1900–1934, various nicotine salts or compounds were prepared some of which entered worldwide markets as insecticides (McIndoo 1943), notably nicotine sulphate, one of the most stable derivatives of nicotine, which was introduced into the United States in 1912 as Black Leaf 40 (McIndoo 1943; Fusetto and O'Hair 2015). This heralded the beginning of the modern nicotine business as the product, containing 40% nicotine, could be more easily shipped compared to the large volumes of the weaker extracts used previously. In the 1930s, the so-called fixed nicotine combinations, such as nicotine tannate and nicotine bentonite, were introduced. Up until this time, nicotine derivatives were mainly used to give quick action as contact insecticides. However, the fixed nicotine derivatives were active over a much longer period and could be used against leaf chewing insects (Beinhart 1951).

Pyrethrum has a long history of use as an insecticide, and one aspect of its development was to develop "pyrethrum extenders" which could be added to pyrethrum formulations to increase their efficacy. Today, such compounds, with little or no intrinsic action on their own, are known as synergists.

Between February 1938 and June 1941, 75 patents were granted in the USA listing 1400 synthetic or natural organic compounds which were proposed as synergists (Tozzi 1998). However, it was not until 1947 that piperonyl butoxide, now widely used as an insecticide synergist, was synthesised and its efficacy demonstrated.

2.4.2.2 Sulphur Compounds

As well as having fungicidal properties, sulphur and lime sulphur also have insecticidal and miticidal properties. Flowers of sulphur were widely used against mites such as the red spider, the six-spotted spider mite and the rust mite of citrus. The sulphur was applied as a dry powder dust or dissolved in sodium hydroxide solution (Marlatt 1903). Sulphur was particularly important in controlling false spider mites (*Brevipalpus phoenicis, B. obovatus* and *B. californicus*) which vector several virus and virus-like diseases of citrus and other crops. The most notable of these diseases is leprosis which nearly destroyed the Florida citrus industry in the early 1920s (Sauls 1999), but which began to disappear in the late 1920s which coincided with the increased use of sulphur (Knorr et al. 1968). Crop losses due to soil pests, such as grubs, wireworms, cutworms, root maggots and root aphids, were also significant with an estimated loss of upwards of 12 million dollars in the USA in 1912.

Anecdotal evidence suggested that sulphur had some activity against these economically important pests; however, an extensive research programme showed this not to be the case (Bulger 1928). In the USA, fruit orchards suffered from various insect pests such as the two-spotted spider mite (*Tetranychus urticae*), the peach twig borer (*Arnasia linetella*) and San Jose scale (*Quadraspidiotus perniciosus*). San Jose scale was a major pest of peach and pome fruit after being accidentally introduced into California from China in 1870. In 1890, it was found in New Jersey and by 1920 had virtually wiped out the peach industry in that state (Shearer 2001). This pest, together with the two-spotted spider mite (*Tetranychus urticae*) and the peach twig borer (*Arnasia linetella*), could be controlled using lime sulphur. However, it is interesting to note that lime sulphur gave rise to the first instance of insecticide resistance with San Jose scale in 1908 (Cloyd 2013).

2.4.2.3   Arsenic Compounds

The most widely used insecticides in the early part of the twentieth century were the arsenic-containing compounds first introduced in the nineteenth century, such as Paris Green, Scheele's Green and London Purple. Lead arsenate, first introduced in 1892, was also popular and replaced Paris Green for certain applications, since it was less phytotoxic and adhered better to the treated foliage, giving a longer insecticidal effect. As a result, lead arsenate was rapidly adopted as an insecticide throughout the world, particularly where codling moth was a pest of apples (Peryea 1998). However, during World War I, lead salts became very expensive and calcium arsenate replaced lead arsenate for several uses, because of its lower cost, particularly for the control of boll weevil (*Anthonomus grandis*) in cotton and for forest insect pests. Lead arsenate remained the preferred treatment against codling moth on apples which, if not treated, would lead to damage of between 20 and 95% to apples in every orchard. A variety of other metal arsenates were also evaluated as insecticides, in particular aluminium arsenate which was developed and used in France in the 1930s (Peryea 1998). As mentioned previously, there was growing concern over residues of lead and arsenic on treated fruit, which could not be easily removed by washing, and attempts were made to find alternative insecticides.

2.4.2.4   Synthetic Organic Compounds

Whilst 4,6-dinitro-*o*-cresol (DNOC) was first used as an insecticide in 1892 and had been used as a winter wash for the control of dormant insects in orchards, the search for organic insecticides only became a major goal in the 1930s. Screens were developed for several easily reared insect species in order to check the insecticidal activity of the many compounds being synthesised (Casida and Quistad 1998). For example, many sulphur-containing compounds were prepared and tested, including dithiocarbamic acid derivatives, isothiocyanates, thiazines, including phenothiazines, (Smith 1937) and thiols. A comprehensive list was published by the United States

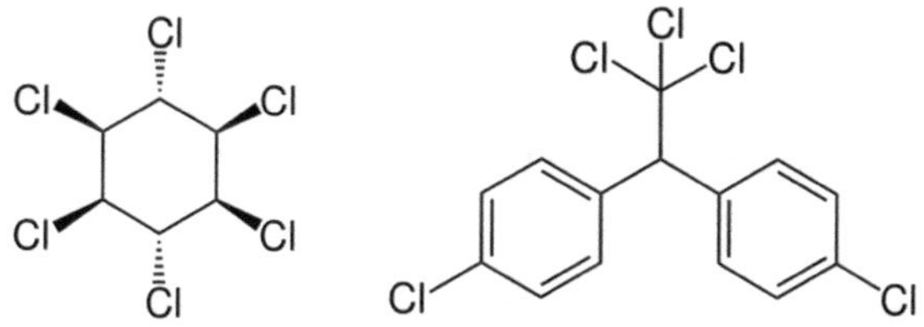

Fig. 2.13 Molecular structure of lindane (left) and DDT (right), both potent insecticides

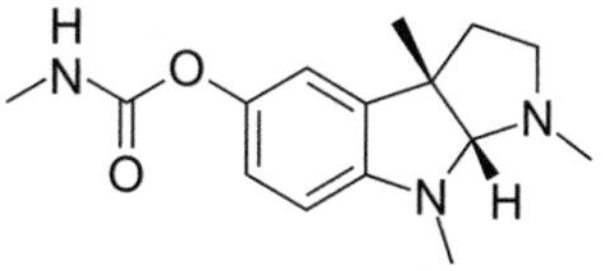

Fig. 2.14 Molecular structure of physostigmine, a naturally occurring methylcarbamate found in the Calabar bean

Bureau of Entomology and Plant Quarantine, Division of Insecticide Investigations in 1941 (Vivian and Acree 1941).

Although γ-hexachlorocyclohexane (lindane) and dichlorodiphenyl-trichloroethane (DDT), Fig. 2.13 had been known for many years, a significant breakthrough in the control of pests came with the discovery of their insecticidal properties. Lindane is one of the several stereoisomers of the so-called benzene hexachloride, or hexachlorocyclohexane, first prepared by Michael Faraday in 1825, but it was not until 1943 when its insecticidal properties were discovered by Imperial Chemical Industries and linked to the γ isomer (Lipnick and Muir 2001) which was much more active than the other isomers. DDT was first prepared by Othmar Zeidler in 1873, but it was not until 1939 that its insecticidal properties were discovered by Paul Müller, working in the laboratories of J. R. Geigy (Lipnick and Muir 2001). In 1948, Müller was awarded the Nobel Prize in Physiology or Medicine "for his discovery of the high efficiency of DDT as a contact poison against several arthropods".

The fact that these compounds were potent insecticides led to the synthesis, through the 1940s and early 1950s, of other chlorinated compounds in the search of new, cheap and persistent contact insecticides (Casida and Quistad 1998). Another two families of insecticides which are inhibitors of acetylcholinesterase (AChE), the methylcarbamate and organophosphate families also originated in the late 1930s and 1940s. The methylcarbamate insecticides came from the original tests on the botanical alkaloid physostigmine, a naturally occurring methylcarbamate, Fig. 2.14, found in the Calabar bean, following work by Gysin at J. R. Geigy in the 1940s, whilst Gerhard Schrader, working at I. G. Farbenindustrie, discovered insecticidal organophosphates in 1937 (Casida and Quistad 1998). The organophosphates were kept secret during World War II because of their potential as chemical warfare agents, and neither of these two insecticide families reached their full potential until the 1950s.

2.4.3   Herbicides

Despite different cultivation practices and hand weeding significant losses can be attributed to the presence of weeds reducing crop yields, with, for example, a 10% reduction in corn, 12–15% in spring cereals and 5–9% in winter cereals (Cates 1917). Another example is that of a study conducted on winter wheat at Rothamsted Research Centre in the UK where there was a decline in yield of 40% between 1900 and 1925. This loss of yield was almost certainly due to increasing competition from weeds as hand weeding was stopped during World War I and afterwards due to labour shortages (Moss et al. 2004). In a study in Illinois published in 1932, it was estimated that 10% of the cropland had reductions in yield of 50% or more in a "normal" year (Case and Mosher 1932). Weeds reduce crop yields by competing for nutrients and space and in addition, they can harbour insect and disease organisms that attack crop plants. For example, black stem rust can use the common barberry, quackgrass or wild oats as a host before attacking cereal crops (Monaco et al. 2002). Some weeds proved difficult to eradicate, one of which was field bindweed (*Convolvulus arvensis*) which in the western states of the USA could reduce the value of farms by more than 50% (Latshaw and Zahnley 1927). The search for effective and economically viable herbicides continued during the early part of the twentieth century, and many different preparations were tested (Mukula and Ruuttunen 1969; Gasquez 2015). For example, a bulletin published by the United States Department of Agriculture listed 37 different preparations tested for the eradication of the common barberry (*Berberis vulgaris* L.) (Thompson and Robbins 1926). However, although some chemicals were said to be useful for total weed control for tennis courts, driveways and railways, it was suggested that in general that the use of chemicals was a complex, expensive and uncertain process (Cates 1917).

2.4.3.1   Inorganic Compounds

Crop losses due to weeds in growing crops were still significant, and a large number of inorganic compounds were investigated as potential herbicides. In the early part of the twentieth century, chemical weed control still mostly relied on those preparations that were in use in the late 1800s. In order to control persistent perennial weeds, typical treatments consisted of high rates of sodium chlorate or sodium arsenite. However, as these total weed treatments were relatively expensive and had an adverse effect on subsequent crops, their use was limited to preventing the spread of small patches of weeds (Holm and Johnson 2009). The use of sodium chlorate was particularly prevalent in New Zealand for the control of ragwort (*Senecio jacobaea*) which was a particular problem for dairy farmers. However, this led to several unfortunate incidents as related in "The Significance of Mr. Richard Buckley's Exploding Trousers". When sodium chlorate is mixed with organic matter such as wool or cotton fibres, an exothermic reaction can take place leading to clothing exploding and catching fire; this was a particular problem when applicators without protective clothing sprayed the chemical for weed control (Watson 2004). Inorganic compounds, for example, sulphuric acid (Brown and Streets 1928; Gasquez 2015), were also being investigated for the selective control of annual broadleaf weeds in field crops. In a Canadian study, several inorganic compounds, sulphuric acid, copper nitrate and sulphate, iron sulphate, sodium chloride, sodium dichromate, sodium chlorate, ammonium thiocyanate and bisulphate, were investigated for the control of broadleaf weeds. Of the inorganic compounds tested over four years in the Canadian prairies, only copper nitrate, copper sulphate and sulphuric acid gave selective control of weeds; however, after considerable research, it was concluded that none of the compounds tested gave a viable combination of efficacy, crop safety, user safety and low cost (Holm and Johnson 2009). Thus, although the inorganic compounds had some use in total weed control, their use in selective control of weeds was far from being ideal.

2.4.3.2   Organic Compounds

Whilst DNOC (dinitro-*o*-cresol) was first used as an insecticide in 1892, it was not until 1932 that it was patented as a selective herbicide in cereals by Truffaut and Pastac in France (Streibig 2003). Whilst investigating the fungicidal activity of organic dyestuffs, it was noticed that the susceptibility of plants varied according to the species and their age. This led to the realisation that by choosing a suitable dyestuff and an appropriate dose rate, a selective herbicide could be achieved. This discovery was followed up by Ripper and co-workers who established that DNOC was the most effective dyestuff to control weeds in cereals (Holmes 1958). Under the trade name of SINOX, it was used throughout Europe as a selective spray in cereals and flax. It was introduced into the USA in 1938, mainly for the control of wild radish (*Raphanus sativus*) and wild mustard (*Brassica arvensis*), and very significant increases in yield were demonstrated between treated and untreated plots for both cereals and flax. For example, in cereals increases from 26 lb per acre, in untreated plots, to 1870 and 2570 lb per acre, in treated plots, were recorded in some trials (Westgate and Raynor 1940). Its selectivity was ascribed to the fact that in field crops such as cereals, flax, alfalfa, peas and onions, the spray will almost entirely run off the leaves and stems, whilst for broad-leaved and hairy-leaved weeds, there is significant contact between the plants and the spray. In addition, for cereals and onions, the growing point is protected by leaves, which is not the case for common broad-leaved weeds (Holmes 1958). With various structural modifications, derivatives of DNOC were prepared giving, amongst others, the better known dinoseb ((RS)-2-sec-butyl-4,6-dinitrophenol), introduced in France in 1932, and dinoterb (2-tert-butyl-4,6-dinitrophenol) herbicides, Fig. 2.15.

In the early 1940s, the discovery by four groups of researchers in the USA and the UK of the first "hormone" herbicides, 2,4-D ((2,4-dichlorophenoxy)acetic acid), 2,4,5-T ((2,4,5-trichlorophenoxy)acetic acid) and MCPA ((4-chloro-2-methylphenoxy)acetic acid), initiated an agricultural revolution and modern weed science (Troyer 2001). Their discovery can be traced back to early work between

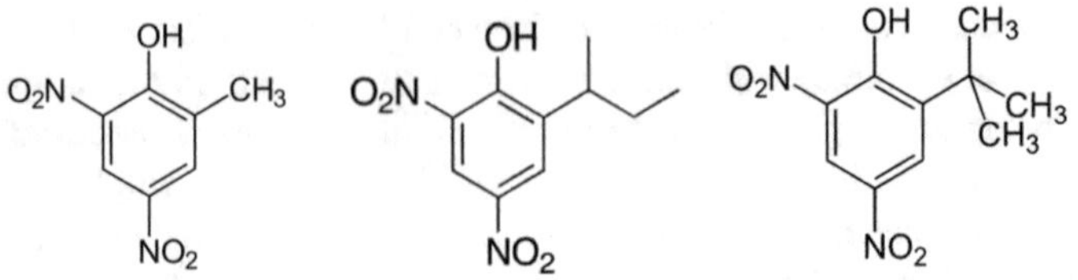

Fig. 2.15 Molecular structure of, from left to right, DNOC, dinoseb and dinoterb, herbicides selective against broad-leaved weeds

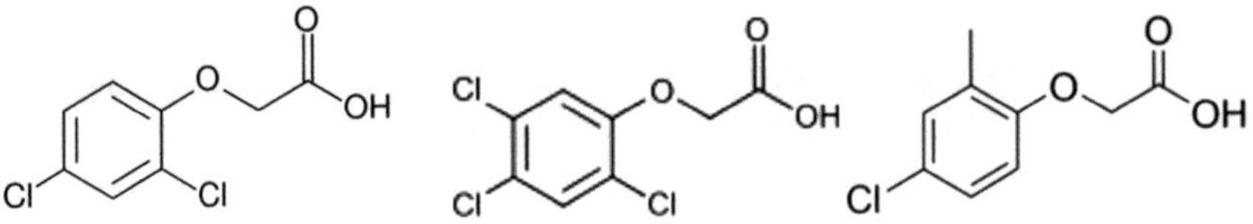

Fig. 2.16 Molecular structure of, from left to right, 2,4-D, 2,4,5-T and MCPA, the first "hormone" herbicides

1880 and the mid-1930s on substances promoting plant growth or auxins, such as indoleacetic acid. Further research demonstrated that synthetic, and not naturally occurring, chemicals could also act as auxins. This eventually led to the discovery of the phenoxyacetic acids (Troyer 2001; Peterson 1967) which are selectively active against broad-leaved weeds. Unlike DNOC and its derivatives, the selectivity of the growth regulating compounds does not only depend on differences in external plant form but also in a fundamental difference in the biochemistry of mono- and dicotyledonous plants. This family of herbicides (Fig. 2.16), first introduced in the mid-1940s, found widespread use in agriculture by the 1950s.

2.5 The Golden Age of Pesticides (1940s–1960s)

During the 1940s and 1950s, there was increasing interest in organic chemistry and thousands of compounds were routinely screened for their pesticidal properties. The period after World War II saw new chemistry, new compounds based on prewar chemistry and the first modern organic, synthetic pesticides. These new pesticides marked, at least in terms of their selective activity, a quantum leap in the progress of pest control (Müller 2002). Hence, for example, the introduction of the dithiocarbamate family of fungicides, the methylcarbamate and organophosphate families, as well as lindane and DDT for insecticides, and the dinitro and phenoxy acid families of herbicides. With the discovery of these compounds which had excellent properties as pesticides, there was increasing interest in the chemical control of pests and diseases leading to the so-called Golden Age of Pesticides in the 1940s and 1960s. The chemicals being introduced as pesticides were far superior to those that went before and which were generally rapidly adopted by farmers. World War II proved to be a catalyst in the testing and production of new pesticides; for example, it was noted in 1946 (Essig 1946) that *"the many war uses for inorganic compounds and the necessity for organic insecticides—especially nicotine, rotenone and pyrethrum for the control of human disease-bearing insects have created serious shortages for the control of insects injuring agricultural crops and domestic animals. These scarcities have resulted in a great production of new and promising insecticides"*. Amongst these new insecticides was DDT which at the time was referred to as *"a substance the properties of which are so remarkable that it has been accorded the doubtful honour of becoming 'news'. Very little of the vast amount of experimental work done has yet been published, and the door has thus been opened wide to the wildest specula-tion and exaggeration"* (Shaw 1946). These new insecticides included not only DDT but also, amongst others, members of the chlorohydrocarbon and organophosphate families resulting from research during World War II.

2.5.1 Fungicides

In the 1940s, many chemicals were tested for their efficacy as fungicides; for exam-ple, between 1940 and 1946, some 35 different formulations of sprays and dusts made from 17 different chemicals were tested in Florida against downy mildew in cabbage (Eddins 1947). Interestingly, the major new organic fungicides in this period came from compounds related to uses in the rubber industry, a good example of how different areas of chemistry overlap. The main crops treated were high-value crops such as orchard crops or those where fungal disease caused obvious damage and reduced the market value, for example, potato late blight, cereal seed treatment for smuts and bunts and leaf spot of vegetables (Russell 2005). Late blight (*Phytoph-thora infestans*) is a significant fungal disease of potatoes and was responsible for the Irish Potato Famine of 1845 and 1846. Spores of late blight are spread rapidly by the wind, and the disease can totally kill a crop within a few weeks. Significant losses still occur, particularly in developing countries, and complete control of late blight would especially impact food security and income in these countries (Haverkort et al. 2009). The beginning of the era of synthetic organic fungicides can be considered to be in 1934 when Tisdale and Williams working for E. I. du Pont de Nemours and Co. patented the use of dithiocarbamates, dithiocarbamic acid derivatives, relating to *"...disinfectants, and more particularly to methods and means of controlling and preventing growth of fungi and microbes"* (Tisdale and Williams 1934). After com-prehensive testing on various crops, these compounds were found to be effective as non-systemic, protectant fungicides, as were the other major organic fungicides introduced during this period. It was not until the 1960s that systemic fungicides began to be commercialised in an important way.

2.5.1.1 Dithiocarbamates

The use of dithiocarbamates as accelerators in the vulcanisation of rubber was described in an extensive review published in 1923 (Bedford and Winkelman 1923); however, it was not until the late 1930s that derivatives of dithiocarbamic acid, thiuram disulphides and zinc diethyl dithiocarbamate were tested against fungi growing on plants. In the early 1940s, trials were carried out with some metal dialkyl dithiocarbamates, including sodium, iron, lead, zinc, copper, silver and mercury dimethyl, diethyl and dibutyl dithiocarbamates and selenium diethyl dithiocarbamate, against various fungal diseases. In general, the dimethyl derivatives appeared to be most promising having the greatest fungicidal action, whilst the iron, lead and zinc dimethyl derivatives were the least phytotoxic (Goldsworthy et al. 1943). The first commercial dithiocarbamate was tetramethylthiuram disulphide, or thiram, which was an effective seed dressing and could also be used to protect fruit and vegetable crops from a variety of fungal diseases (Sharma et al. 2003). However, thiram was not a particularly strong product when applied as a foliar spray although it was an effective seed dressing (Brandes 1953; Gullino et al. 2010). Following on from thiram came ferbam (ferric dimethyldithiocarbamate), used on apples, tobacco and ornamentals, and ziram (zinc dimethyldithiocarbamate), used for certain diseases on tomatoes and other vegetables (Gullino et al. 2010. As an example of the benefits of the dithiocarbamates, in the USA, the crop yield for apple trees treated with ferbam gave a 41% increase, whilst losses due to black rot in apples of 25–50% were reduced to just 1% (Gianessi and Reigner 2006). The initial dithiocarbamates were prepared from a monoamine and carbon disulphide but for later compounds a diamine was used instead of the monoamine (Gullino et al. 2010). The compounds produced, ethylene bis-dithiocarbamates, gave the first true ethylene bis-dithiocarbamate, disodium ethylene bis-dithiocarbamate (nabam) and zinc bis-dithiocarbamate (zineb) which were patented in 1943 (Hester 1943) (Fig. 2.17). Nabam and zineb were found to have significant fungicidal properties and were particularly useful against late blight in tomatoes and potatoes. Trials in 1943–44 in Florida with nabam plus zinc sulphate, later shown to form zineb, under severe late blight conditions gave better control compared to standard copper fungicides and also gave a significant increase in yield, whilst trials in Delaware showed that this combination was effective against early blight (Ruehle 1944). This new liquid product was commercialised in 1944 as Dithane-14, whilst later, zineb itself was commercialised as Dithane-Z78. These new fungicides consistently gave good control of early and late blight with increased yields normally obtained compared to copper and other dithiocarbamate fungicides. By 1953, nabam and zineb were used on approximately 75% of the area used for potatoes in the USA. Whilst potatoes were important in the initial development of the dithiocarbamates, tests showed that these fungicides were effective against a wide range of disease organisms, on a wide variety of crops (Brandes 1953). Research into more effective dithiocarbamates continued with manganese ethylene bis-dithiocarbamate (maneb) being patented in 1950, and in 1962, the zinc ion complex of maneb (mancozeb), Fig. 2.18, was registered which was to become the most important and commercially significant of all the ethylene bis-dithiocarbamates.

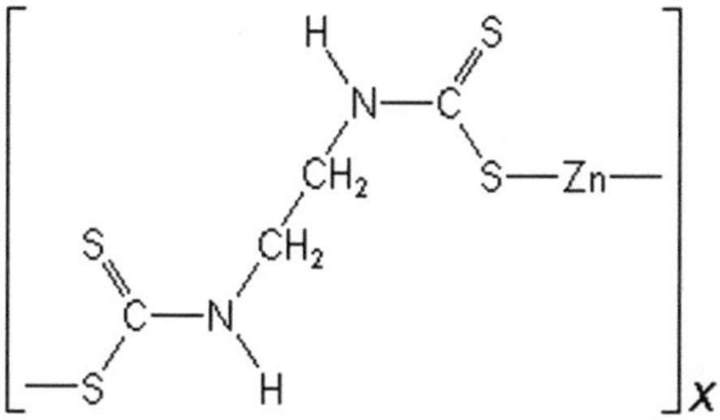

Fig. 2.17 Molecular structure of zineb, one of the first bis-dithiocarbamate fungicides

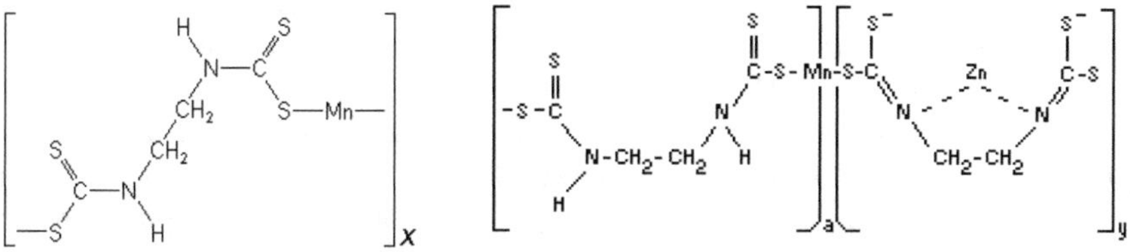

Fig. 2.18 Molecular structure of maneb (left) and mancozeb (right), more effective bis-dithiocarbamate fungicides compared to zineb

By the mid-1960s, the ethylene bis-dithiocarbamate fungicides were considered to be the most important and versatile group of organic fungicides yet discovered (Gullino et al. 2010), although powdery mildew control is missing from their spectrum of activity. In general, dithiocarbamates were less phytotoxic and as effective or, depending on the situation, more effective than copper-, mercury- and sulphur-based fungicides. However, as with these earlier fungicides, they do not significantly penetrate into the plants and so cannot control established infections. In addition, treatments must be repeated at frequent intervals to replace losses from the plant surface due to weathering and to protect new growth (Brent 2003).

2.5.1.2   Phthalimides

Interest in sulphur-containing compounds, as potential fungicides, continued and phthalimides which, like the dithiocarbamates, were useful as accelerators in the vulcanisation of rubber were investigated as fungicides. A patent published in 1951 gave details of the preparation of compounds containing the $-NSCCl_3$ grouping (Kittleson 1951); this was followed up in the open literature with a report listing sixteen compounds prepared by reacting perchloromethyl mercaptan with several alkali metal salts of amides and imides. Each of the compounds was found to have fungicidal properties with one, N-(trichloromethylthio)-tetrahydrophthalimide, commonly known as captan, showing promising results as an agricultural fungicide (Kittleson 1952). Field trials in 1952–1954 confirmed that treatments with captan could control various

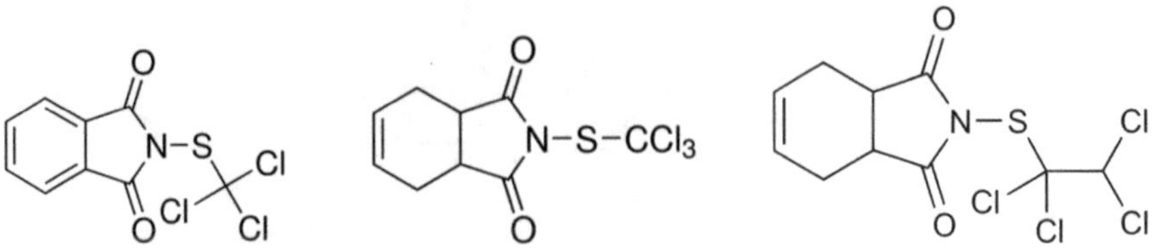

Fig. 2.19 Molecular structure of, from left to right, folpet, captan and captafol, fungicides derived from the rubber industry

fungal diseases in fruit crops (Diener et al. 1955; Andes and Epps 1956). Later, in 1962, a closely related compound, N-(1,1,2,2-tetrachloroethylthio)cyclohex-4-ene-1,2-dicarboximide (captafol), was also introduced as a fungicide, Fig. 2.19.

Captan came to be a widely used fungicide to protect against a variety of fungal problems on a very wide range of fruit and vegetable crops (Cornell University 1986). For example, in 1993 in the USA, it was estimated that 50% of all apple acreage, 60% of almond acreage and 100% of Florida strawberry production were treated with captan (EXTOXNET 1996a).

2.5.1.3 Quinones

Just as the dithiocarbamates and phthalimide fungicides arose from research with compounds originating in the rubber industry, another family, the quinones, came via a similar route. As related by Horsfall, in discussions with the United States Rubber Company (Horsfall 1975), *"the current dogma was that copper in Bordeaux Mixture killed by an oxidizing action. As copper acts as an oxidant in rubber as well why not try tetrachloroquinone, an organic pro-oxidant?"*. Trials confirmed that tetrachloroquinone, known as chloranil or commercially as Spergon, worked as a seed protectant against, for example, powdery and downy mildew on vegetables, including cabbage, cauliflower and broccoli. As chloranil is hydrolysed in the environment, it is not effective as a foliar fungicide; however, another quinone, 2,3-dichloro-1,4-naphthoquinone, known as dichlone or commercially as Phygon, was found to be more stable and was particularly effective against brown rot of stone fruit and scab on apples and pears, as well as blossom blights (Cornell University 1985).

Another non-systemic, protectant fungicide, the organochlorine chlorothalonil (2,4,5,6-tetrachloro-1,3-benzenedicarbonitrile), was registered for use on crops and ornamentals in 1970. It found widespread use as a broad-spectrum product against Oomycetes, Ascomycetes, Basidiomycetes and Fungi imperfecti on a wide variety of crops (Kelly 2012). Figure 2.20 depicts these compounds.

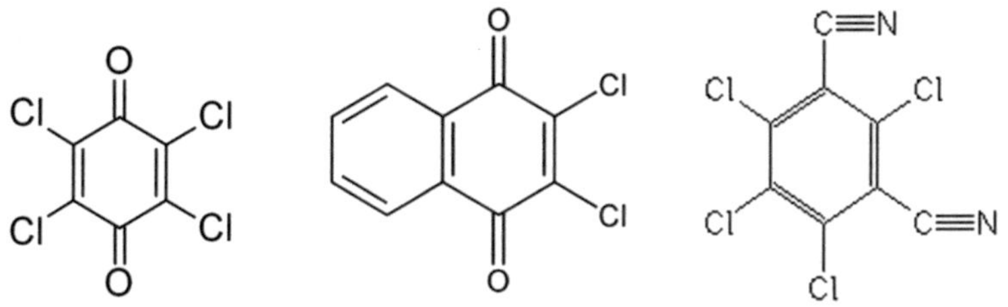

Fig. 2.20 Molecular structure of, from left to right, chloranil, dichlone and chlorothalonil, non-systemic, protectant fungicides

2.5.2   Insecticides

As with fungicides research into synthetic insecticides became a major goal in the 1930s, at this time, screens were developed for several easily reared species which allowed the many compounds of hitherto unknown biological activity to be tested. The system of synthesis and screening became well established allowing rapid advances in insecticide research (Casida and Quistad 1998). During this period, several major classes of insecticides were discovered and commercialised, including chlorinated hydrocarbons, organophosphates and methyl carbamates. Despite the widespread use of inorganic and botanical insecticides, such as lead arsenate or nicotine-based products, there were still many insect species that were poorly controlled. With the introduction of synthetic organic insecticides, pests could be better controlled and crop yields increased. For example, it became possible to grow economically sweet corn in the winter in southern Florida, which had an ideal climate for this. Similarly, economical control of onion thrips was not successful until the introduction of synthetic chemical insecticides which increased yield by 40–80%. Significant yield increases were also found for dates and apples. The area of field corn also grew dramatically in the 1950s after the introduction of inexpensive, broad-spectrum soil-applied synthetic organic insecticides which were less expensive than inorganic or botanical products (Gianessi 2009). Whilst significant yield increases have been obtained by the use of insecticides, it would appear that, in the USA at least, the percentage crop loss, not to be confused with absolute crop loss, has not decreased as much as might be expected. This can be attributed to several major factors in agricultural practice, such as pest resistance, absence of crop rotation, increase in crop monoculture and reduced crop diversity (Pimental et al. 1993).

2.5.2.1   Chlorinated Hydrocarbons

Both lindane (Fig. 2.21) and DDT were synthesised many years before their potential as insecticides was realised. After the discovery of the broad-spectrum insecticidal properties of lindane, gamma-1,2,3,4,5,6-hexachlorocyclohexane, in 1943, at Imperial Chemical Industries in the UK, it became widely used for control of both soil

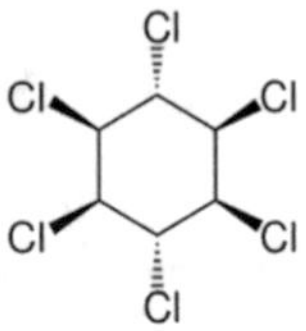

Fig. 2.21 Molecular
structure of lindane

dwelling and plant-eating insects on a large variety of crops (EXTOXNET 1996b). Lindane found particular use as a seed treatment against wireworms, which can cause significant stand losses and ultimately crop yield reductions, and other soil pests in a range of vegetable and field crops (Lange et al. 1950, 1951, 1953).

DDT (1,1,1-trichloro-2,2-bis(4-chlorophenyl)ethane) was developed as the first of the modern synthetic insecticides in the early 1940s. It was initially used with great effect to combat malaria, typhus and the other insect-borne human diseases among both military and civilian populations (Mishke et al. 1985; US EPA 2015). Later, DDT was introduced into agriculture and was hailed as a miracle product (Muir 2012) due to its properties of:

- Broad spectrum of activity;
- Persistence—did not need to be reapplied often;
- Practically insoluble in water—was not washed off by rain;
- Relatively inexpensive;
- Easy to apply.

In large-scale tests in California in 1945, it was shown to be extremely effective against a variety of insects on wide range of crops (Essig 1946). Further trials looked at the efficacy of DDT against, for example, mealy bugs on avocado trees (Ebeling and Pence 1947), citrus thrips (Ewart 1948), caterpillars on tomatoes (Michelbacher et al. 1950) and codling moth on walnuts (Ortega 1948), apples and pears (Borden 1947). By 1966, DDT was being suggested for the control of about 20 insects on cane fruit, 50 on vegetables and 50 on tree fruits, together with insects on nut trees and other food crops (FAO 1967). Derivatives of DDT were also prepared, Fig. 2.22, with methoxychlor (1,1'-(2,2,2-trichloro-1,1-ethanediyl)bis(4-methoxybenzene)), used against various arthropods commonly found on field crops, vegetables, fruits and stored grains, introduced in the USA in 1946 (USDA 2002) and dicofol (2,2,2-trichloro-1,1-bis(4-chlorophenyl)ethanol) introduced in 1957, for use against mites (PPDB 2018). Interestingly, whilst DDT is inactive against mites, and its use has been said to increase the mite population in, for example, apples (Baker 1952) and walnuts (Michelbacher and Bacon 1952); its close relative dicofol has little effect on insects but is an excellent miticide on a wide range of crops.

After the discovery of the significant insecticidal properties of lindane and DDT, other chlorinated hydrocarbons were synthesised and, on the basis that every new chlorinated hydrocarbon might be a potential DDT, screened for insecticidal activity. One of these, toxaphene, was produced by the chlorination of camphene and was claimed to be the most complex mixture of active ingredients ever marketed as

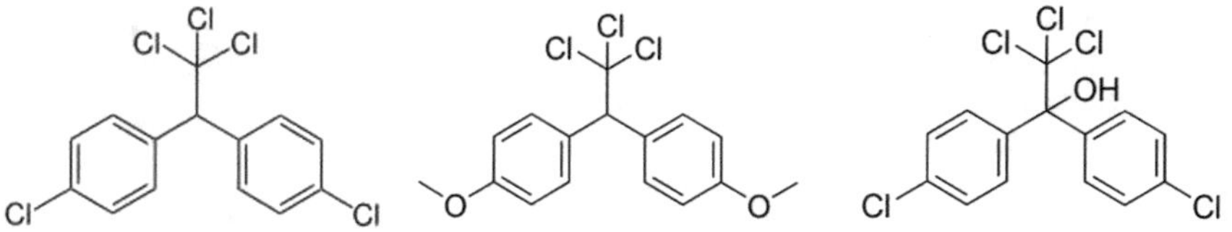

Fig. 2.22   Molecular structure of organochlorine insecticides derived from DDT, from left to right, DDT, methoxychlor and dicofol

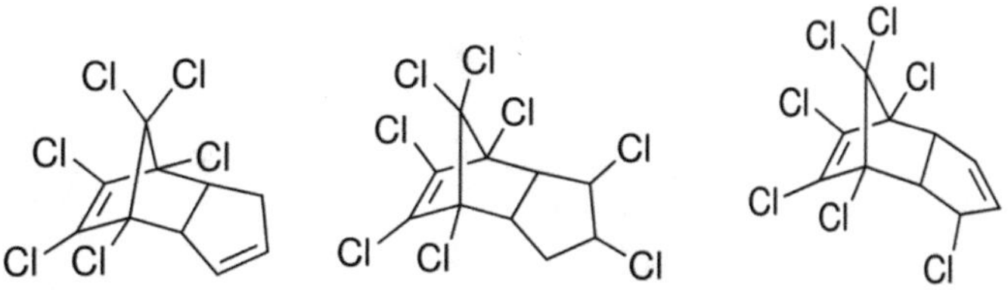

Fig. 2.23   Molecular structure of insecticides derived from hexachlorocyclopentadiene, from left to right chlordene, chlordane and heptachlor

a synthetic insecticide (Casida and Quistad 1998), containing over 200 different compounds (Saleh 1983; Kimmel et al. 2000).

In a research programme looking for new uses for cyclopentadiene, a by-product of synthetic rubber production, it was unexpectedly found that it reacted with hexachlorocyclopentadiene to give mono- and bis-adducts via the Diels–Alder reaction. These adducts were tested for insecticidal activity, and it was found that the mono-adduct, chlordene (4,5,6,7,8-hexachloro-3a,4,7,7a-tetrahydro-4,7-methano-1h-indene), did indeed have some activity. However, it was too volatile to compete with DDT as a persistent insecticide, but this could be overcome by chlorinating the reactive double bond to give the less volatile chlordane (1,3,4,7,8,9,10,10-octachlorotricyclo[5.2.1.0^{2,6}]dec-8-ene), (Brooks 1977). Discovered in 1945, chlordane was widely used on agricultural food crops such as vegetables, small grains, maize, oilseeds, potatoes, sugar cane, sugar beet, fruits and nuts (Roarck 1951). Technical chlordane also contains 10–20% heptachlor (1,5,7,8,9,10,10-heptachlorotricyclo[5.2.1.0^{2,6}]deca-3,8-diene), an insecticide in its own right with a close structural similarity to chlordane (IARC 2001), Fig. 2.23.

It was also found that cyclopentadiene reacted with acetylene to give bicyclo[2.2.1] hepta-2,5-diene (norbornadiene), which, when reacted with hexachloro-cyclopentadiene via a Diels–Alder reaction, gave the first preparation of aldrin ((1R,2R,3R,6S,7S,8S)-1,8,9,10,11,11-hexachlorotetracyclo[6.2.1.1^{3,6}.0^{2,7}]dodeca-4,9-diene) in 1948. In order to reduce the volatility of aldrin, the corresponding epoxide was made to give dieldrin ((1R,2S,3S,6R,7R,8S,9S,11R)-3,4,5,6,13,13-hexachloro-10-oxapentacyclo[6.3.1.1^{3,6}.0^{2,7}.0^{9,11}]tridec-4-ene), Fig. 2.24, which kept the significant insecticidal properties found with aldrin (Brooks 1977).

Another member of this family, endrin ((1R,2S,3R,6S,7R,8S,9S,11R)-3,4,5,6,13,13-hexachloro-10-oxapentacyclo[6.3.1.1^{3,6}.0^{2,7}.0^{9,11}]tridec-4-ene),

Fig. 2.24 Molecular structure of the chlorinated hydrocarbon insecticides aldrin (left) and dieldrin (right)

Fig. 2.25 Molecular structure of the chlorinated insecticides endrin (left) and endosulfan (right)

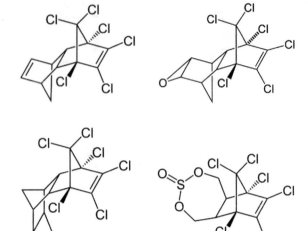

Fig. 2.25, was prepared by reacting hexachloronorbornadiene with cyclopentadiene to give isodrin and then forming the corresponding epoxide (Brooks 1977). Aldrin and dieldrin were used for soil treatment against various soil pests, as a seed treatment on grains, sugar beet, beans, leeks and onions and as a foliar application on various agricultural crops and fruits (FAO 1968). The main use of endrin was to control insects on cotton, but there was some use in controlling insect pests on rice, sugar cane, grain crops and sugar beet (FAO 1971). Another important insecticide in the cyclodiene family of insecticides is endosulfan (1,9,10,11,12,12-hexachloro-4,6-dioxa-5-thiatricyclo[7.2.1.0^{2,8}]dodec-10-ene 5-oxide), Fig. 2.25, formed by reacting hexachlorocyclopentadiene with *cis*-butene-1,4-diol, via a Diels–Alder reaction, and the product formed reacted with thionyl chloride to give endosulfan (IPCS 1984). It was used globally on a wide variety of food crops to control insects such as the Colorado potato beetle, flea beetle, cabbageworm, peachtree borer, and tarnished plant bug, as well as several species of aphid and leafhopper (IPCS 1984).

The introduction of the chlorinated hydrocarbon insecticides in the late 1940s and early 1950s brought significant benefits to agriculture, particularly in tropical and developing countries, where their general effectiveness, and being relatively inexpensive, allowed farmers to protect their crops in an efficient manner. For example, aldrin was the insecticide of choice in many countries to control the majority soil pests and thus gave significant yield increases in many crops (Snelson 1977):

- Rice—in southern Brazil, irrigated rice attacked by weevils lost up to 20% of the yield if not treated with aldrin, whilst upland rice lost up to 25% of the yield if attacked by termites and other soil pests.
- Maize—in Guatemala in land infested with wireworms and other soil pests, losses in the order of 40% occurred if the land was not treated with aldrin. In Mexico, up to 60% of the crop yield was lost if areas with a severe infestation of soil pests were not treated with either aldrin or heptachlor.

- Sugarcane—in most countries where sugarcane is grown, attack by indigenous soil pests could lead to a loss of up to 50% of potential yield unless treated with aldrin, benzene hexachloride or lindane.
- Sugar beet—in Chile and other South American countries, aldrin was essential to protect sugar beet over a sufficiently long period.
- Bananas—in countries where bananas are grown, aldrin, dieldrin or kelevan (a cyclodiene chlordecone derivative) was essential to protect the crop against the banana weevil borer.
- Potatoes and sweet potatoes—in some countries, particularly in the tropics, these important staple food crops cannot be grown economically without treatment with aldrin, dieldrin or chlordane against wireworms and other soil pests.

2.5.2.2   Organophosphates

As previously mentioned, Gerhard Schrader, working at I. G. Farbenindustrie, discovered insecticidal organophosphates in 1937 (Casida and Quistad 1998), following a German government initiative to lessen its reliance on food importation by emphasising the need for new insecticides to increase food production (Gilbert 2014). However, due to World War II, the organophosphates did not achieve their prominence as insecticides until the 1950s. The era of discovery of organophosphate insecticides was one of simple structures and ease of making many analogues for the selection of those derivatives with the most suitable properties (Casida and Quistad 1998), and between 1940 and the late 1960s, a significant number of organophosphate insecticides were developed. Organophosphate insecticides have two distinctive features; they are generally more toxic to vertebrates than other classes of insecticides, and most are chemically unstable or non-persistent. This last characteristic led to them being used in agriculture as substitutes for the more persistent organochlorines.

Organophosphates can be divided into three separate groups: aliphatic, phenyl and heterocyclic derivatives (Ware and Whiteacre 2004). Some commonly used aliphatic organophosphates are malathion, (diethyl 2-[(dimethoxyphosphorothioyl)sulfanyl]succinate), dimethoate (O,O-dimethyl S-[2-(methylamino)-2-oxoethyl] phosphorodithioate), dichlorvos (2,2-dichlorovinyl dimethyl phosphate) and mevinphos (2-methoxycarbonyl-1-methylvinyl dimethyl phosphate) Fig. 2.26; common phenyl derivatives include parathion (*O,O*-diethyl *O*-(4-nitrophenyl) phosphorothioate) and methyl-parathion (*O,O*-dimethyl *O*-(4-nitrophenyl)phosphorothioate), Fig. 2.27, whilst examples of heterocyclic derivatives are diazinon (*O,O*-diethyl *O*-(2-isopropyl-6-methyl-4-pyrimidinyl) phosphorothioate), azinophos-methyl (*O,O*-dimethyl S-[(4-oxo-1,2,3-benzotriazin-3(4H)-yl)methyl] phosphorodithioate), chlorpyrifos (*O,O*-diethyl *O*-(3,5,6-trichloro-2-pyridinyl) phosphorothioate and phosmet (S-[(1,3-dioxo-1,3-dihydro-2H-isoindol-2-yl)methyl] *O,O*-dimethyl phosphorodithioate), Fig. 2.28.

The organophosphate insecticides were used on a wide range of crops, some examples taken from US data are (Gianessi 1997):

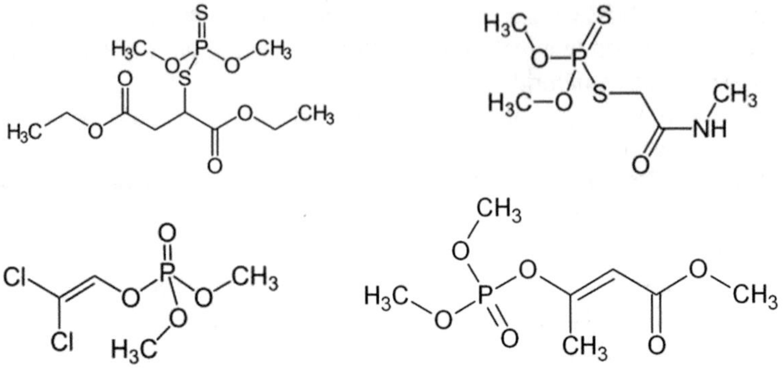

Fig. 2.26 Molecular structure of some aliphatic organophosphate insecticides, malathion (top left), dimethoate (top right), dichlorvos (bottom left) and mevinphos (bottom right)

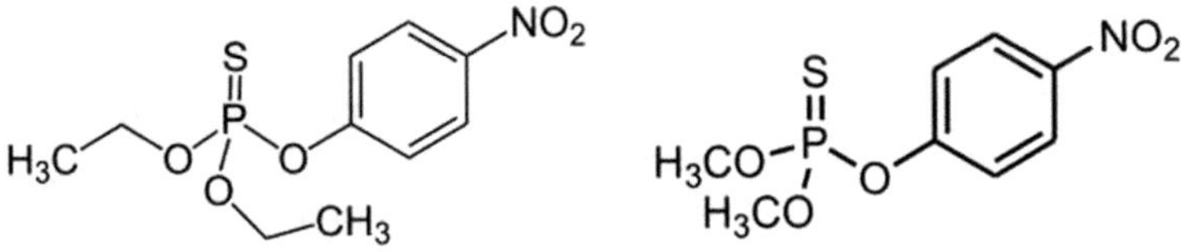

Fig. 2.27 Molecular structure of phenyl organophosphate insecticides parathion (left) and methyl-parathion (right)

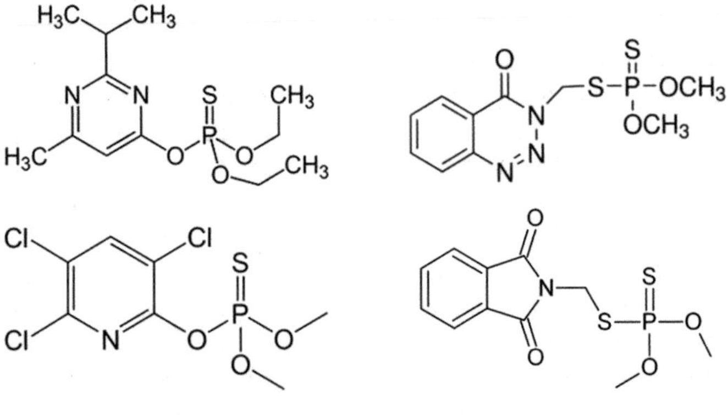

Fig. 2.28 Molecular structure of some heterocyclic organophosphate insecticides, diazinon (top left), azinphos-methyl (top right), chlorpyrifos (bottom left) and phosmet (bottom right)

- Apples—if codling moths were not controlled, crop losses of up to 50% could occur in one or two years, and this pest can be controlled with azinphos-methyl, phosmet methyl-parathion and chlorpyrifos.
- Avocados—the major pest of California avocados is greenhouse thrips which leave scars on the fruit resulting in losses of up to 50%; the thrips can be controlled with malathion.
- Blueberries—the blueberry maggot, from the blueberry fruit fly, is the major insect pest of blueberries in Maine. The larva feeds from inside the fruit leading to shrunken berries; the presence of infested fruit at harvest can result in the condemnation of whole fields of harvested fruit. The most widely used insecticides are azinphos-methyl and phosmet.
- Cherries—the most important insect pests in Michigan tart cherry orchards are cherry fruit fly and plum curculio, both of which directly damage the fruit. Azinphos-methyl, phosmet and chlorpyrifos are widely used against these pests because they give effective control of all the major insect species.
- Corn—corn rootworm larvae are the most damaging pest in corn production, resulting in yield losses of up to 55% due to root pruning and lodging. Chlorpyrifos and terbufos are widely used against both cutworms and rootworms.
- Dates—four species of beetles are pests of dates in California, and these have been known to cause crop losses of up to 75%; malathion will give control of these pests.
- Sugarbeet—the sugarbeet root maggot is the most destructive insect pest of sugarbeets in the USA, terbufos and chlorpyrifos can control this pest which, if left unchecked, can reduce yields by up to 42%.
- Sugarcane—the sugarcane borer reduces sugar yield by causing retarded growth and stunted stalks, with a resulting loss in plant weight. Before azinphos-methyl was introduced, about 13% of yield was lost in Louisiana despite the use of inorganic insecticides.
- Walnuts—codling moth can be a serious pest in walnuts; chlorpyrifos has been widely used to control this pest since it is effective and is less toxic to beneficial organisms in the orchard.

Several organophosphates also play a key role in reducing toxic levels of aflatoxins in food by controlling insects which spread the contaminant. Aflatoxins are produced by certain strains of *Aspergillus* fungi and are mutagenic, carcinogenic and acutely toxic to most animals and humans. Aflatoxins are linked with damage to almond and pistachio nut kernels caused by navel orangeworm larvae, and in peanuts damage to the pods caused by the lesser cornstalk borer, these provide a point of entry for the aflatoxin producing fungi (Gianessi 1997).

The mode of action of the organophosphate insecticides is to inhibit the normal breakdown of acetylcholine by blocking the acetylcholinesterase enzyme. The result is a build up of acetylcholine and continuous stimulation of the receptors of target cells. Research into alternative chemicals with an anticholinesterase activity led in the early 1950s to the development of the N-methyl carbamate family of insecticides. Thus, although the organophosphates and carbamates are distinct families of chemicals, they both have a similar mode of action (Fukuto 1990).

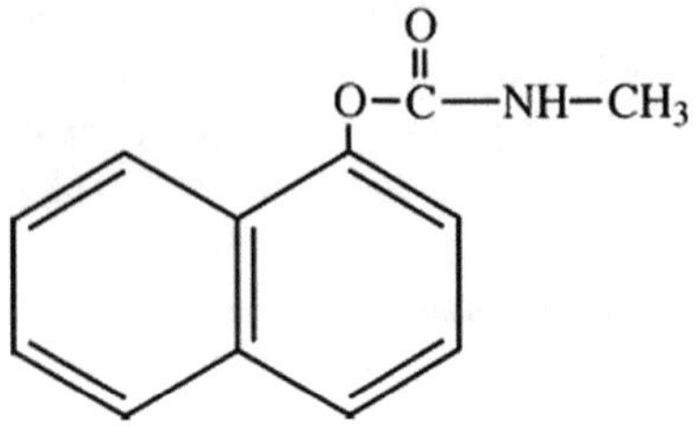

Fig. 2.29 Molecular
structure of carbaryl, the first
N-methyl carbamate
insecticide

$$O-\overset{\overset{\displaystyle O}{\|}}{C}-NH-CH_3$$

2.5.2.3   Methyl Carbamates

As mentioned in the previous section, the *N*-methyl carbamate insecticides came from the original tests on the botanical alkaloid physostigmine, a naturally occurring methyl carbamate, and follow-on work by Gysin at J. R. Geigy in the 1940s (Casida and Quistad 1998). Although, the *N*-methyl carbamates share a similar mode of action with the organophosphates, there is one important difference in that the *N*-methyl carbamyl-acetylcholinesterase combination dissociates more readily than the phosphoryl-acetylcholinesterase complex. This difference limits the duration of *N*-methyl carbamate poisonings and gives a greater difference between the onset of symptoms and the lethal dose compared to most organophosphate compounds (Roberts and Reigart 2016). This advantage meant that the use of *N*-methyl carbamates was often preferred to that of the organophosphates. The first *N*-methyl carbamate was carbaryl (1-naphthyl methylcarbamate) Fig. 2.29, introduced in 1956 and which over the course of eight years was used to control over 150 major pests. In addition, it could be used as a reliable substitute for chlorinated hydrocarbons in certain cases of resistance development (Back 1965). Under the Food Quality Protection Act (FQPA) of 1996 in the USA, carbaryl was found to be used on all major crops including fruit and nut trees, fruits, vegetables and grain crops, with the establishment of over 140 tolerances. Crops with the greatest amounts of carbaryl used annually included apples, pecans, grapes, alfalfa, oranges and corn (US EPA 2004).

Other carbamate insecticides introduced after carbaryl included carbofuran (2,2-dimethyl-2,3-dihydro-1-benzofuran-7-yl methylcarbamate), pirimicarb (2-(dimethylamino)-5,6-dimethyl-4-pyrimidinyl dimethylcarbamate) Fig. 2.30, methomyl (methyl (1E)-*N*-[(methylcarbamoyl)oxy]ethanimidothioate) and aldicarb (2-methyl-2-(methylthio)propanal *O*-(*N*-methylcarbamoyl)oxime, Fig. 2.31. Carbofuran was introduced in 1969 for control of soil and foliar pests of field crops, fruits and vegetables; pirimicarb was also introduced in 1969 for control of aphids, including species resistant to organophosphate insecticides, on a wide range of crops including cereals, sugar beet, potatoes, fruits and vegetables; methomyl was introduced in 1966 as a broad-spectrum insecticide for foliar use on field crops, fruits and vegetables; aldicarb was introduced in 1970 as a soil-applied product to control mites, nematodes and aphids and was widely used on potatoes, soybeans and peanuts (EXTOXNET 1996c).

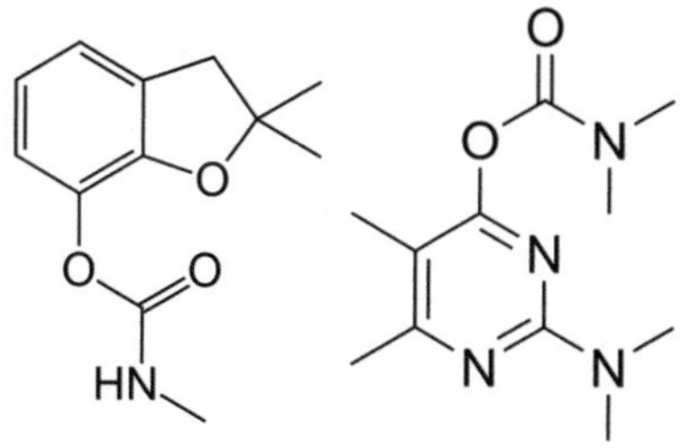

Fig. 2.30 Molecular structure of *N*-methyl carbamates carbofuran (left) and pirimicarb (right)

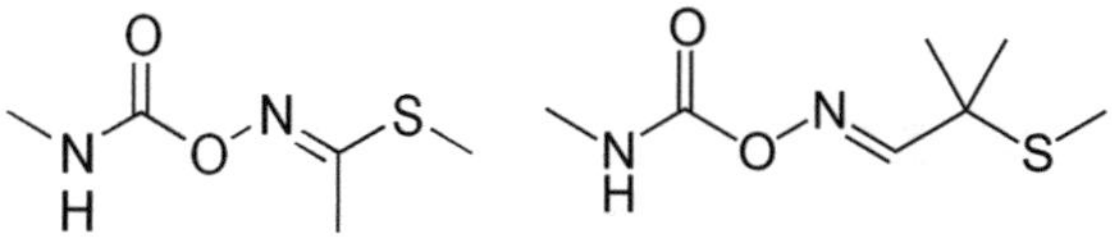

Fig. 2.31 Molecular structure of methyl carbamate insecticides methomyl (left) and aldicarb (right)

The importance of the *N*-methyl carbamate, and organophosphate, insecticides in maintaining crop yield was highlighted in a report from the Agricultural and Food Policy Center at Texas A&M University. This report calculated the loss in yield and increase in costs, if both families of insecticides were withdrawn from the US market. Even if alternative insecticides or other methods of pest control were used, there would still be significant yield losses, and increased costs, for some crops (Knutson and Smith 1999), see Table 2.1.

2.5.3   Herbicides

Crop yield losses would be very significant without the use of herbicides (Gianessi and Reigner 2007); overall, it has been estimated that globally the potential crop yield loss due to weeds is 34%, with wide variations for different crops and regions (Oerke 2006). For example, on average in North America, 52.1% of soybean yield would be lost with no herbicide weed control; even if best management practices were followed (Soltani et al. 2017), the corresponding value for the loss of corn yield was 50% (Soltani et al. 2016). However, crop yield losses could be even greater under low moisture conditions with losses of 84% for soybean and 96% for corn (MAFF Canada 2017). In a study, under field conditions in the UK, looking at the effect of weed species, cleavers (*Galium aparine* L.), wild oats (*Avena fatua* L.), field speedwell (*Veronica persica* Poir.), common chickweed (*Stellaria media* L. Cyrill.) and scented mayweed (*Matricaria recutita*) on beans, wheat, barley and oilseed rape, it was found that the highest grain yield loss was caused by cleavers at 33.9%, with smaller losses due to wild oats at 26.5%, chickweed at 26.3%, speedwell at 21.6% and mayweed at 8.8% (Karim 2002). Significant crop yield losses due to wild oats were

Table 2.1 Commodities and regions found to be most vulnerable to reduced yields and increased costs if organophosphates and carbamates were eliminated

Commodity	Regions/states most adversely affected	Yield reduction (%)	Variable cost increase (%)
Apples	Washington[a]	38	66
Carrots	Texas/Washington	25/20	3/20
Corn	Southeast	5	9
Cotton	Southwest/Southern Plains	21/19	27/28
Grapes	California (table)	32	32
Oranges	California	15	14
Peaches	Georgia	17	19
Peanuts	Virginia/North Carolina	17	15
Potatoes	Central States	5	13
Rice	Gulf Coast	15	17
Grain Sorghum	Southern Plains	14	13
Soybeans	Delta	8	15
Tomatoes	Florida (Fresh)	21	20
Wheat	Southeast	5	3

[a]Only region studied

also found in trials with wheat and barley under various conditions (Scursoni et al. 2011; Fahad et al. 2015). Studies with field peas in Canada indicated that pea yield losses of up to 70% occurred after a full season of weed competition (Harker et al. 2001). Similar crop yield losses can be seen in the rice crop of sub-Saharan Africa where losses due to uncontrolled weeds were estimated to be between 28 and 74% in transplanted lowland fields and 28–89% in direct-seeded lowland rice (Makokha et al. 2016). Interestingly, yield losses also occur in orchard fruit production due to the effect of weeds, with yield reductions of 16–49% being reported compared to herbicide-treated orchards (Derr 2001).

Up until, about 1940 weeds were generally controlled either by cultivation techniques, including hand weeding, or by chemical means with inorganic salts which had been in use since the late 1800s with varying degrees of success. As an example, field bindweed (*C. arvensis*) was recognised as a serious weed pest in parts of the USA early in the twentieth century, and studies using various inorganic chemicals, sodium chlorate, sodium arsenite, sodium hypochlorite, zinc chloride and copper sulphate to control it were carried out. Of the chemicals used, sodium arsenite and zinc chloride killed the tops of plants but regrowth soon appeared after each treatment and only sodium chlorate gave promising results (Latshaw and Zahnley 1927). Bindweed can cause significant yield losses if uncontrolled, see Table 2.2, and studies continued into its eradication, focussing on sodium chlorate, but also using sodium chloride and carbon bisulphide, the latter being used as a fumigant (Timmons 1941).

Table 2.2 Effect of
bindweed on the yield of
crops at Hays, Kansas,
1936–1940

Crop	Yield reduction (%)
Barley	32
Oats	26
Rye	20
Wheat	42
Corn	67
Kafir[a]	85
Milo[b]	89

Average yields, 1936–1940, comparing infested and non-infested
fields
[a]Kafir corn, *Sorghum vulgare* var. *caffrorum*
[b]Milo, *Sorghum vulgare* var. *subglabrescens*

Trials continued with sodium chlorate and from 1945 with the newly introduced
2,4-D, which did not give the complete answer to bindweed eradication but had the
advantage of giving greater reductions early in the eradication programme, compared
to intensive cultivation or use of competitive crops (Phillips 1961). Despite not being
the complete answer for the control of bindweed 2,4-D was rapidly accepted for the
control of other broad leaved weeds in cereals. In the USA, 5,466,000 lb of 2,4-D
was produced in 1946 and by 1950 production exceeded 14,000,000 lb (Peterson
1967). The discovery of 2,4-D and the other chlorophenoxyacetic acids 2,4,5-T and
MCPA, which could kill many weeds selectively in small and relatively inexpensive
quantities, was said to have initiated an agricultural revolution and modern weed
science (Troyer 2001). In general, the phenoxy acid herbicides were so effective and
economical for selectively controlling broadleaf weeds in grass crops that they put
selective weed control in the "public spotlight" worldwide. As stated in the report of
the United States Department of Agriculture National Agricultural Pesticide Impact
Assessment Program (NAPIAP) in 1996 (Burnside 1996)—*"More effective weed
control has been a major factor in increasing crop yields in developed countries
worldwide since the introduction of 2,4-D and subsequent selective herbicides"*.
The discovery of 2,4-D and the resultant publicity triggered a keen interest in weed
research, and gave rise to the development of chemical synthesis and screening programmes for the discovery of new herbicides that had a broader spectrum of activity,
were generally more economical and safer to use, compared to those available at the
time (Timmons 1970). Following on from the phenoxy acid herbicides, in the period
1940 to the 1960s, several important new families of herbicides were introduced
on to the market, such as the phenylurea, triazine, dinitroaniline, bipyridinium and
chloroacetamide herbicides.

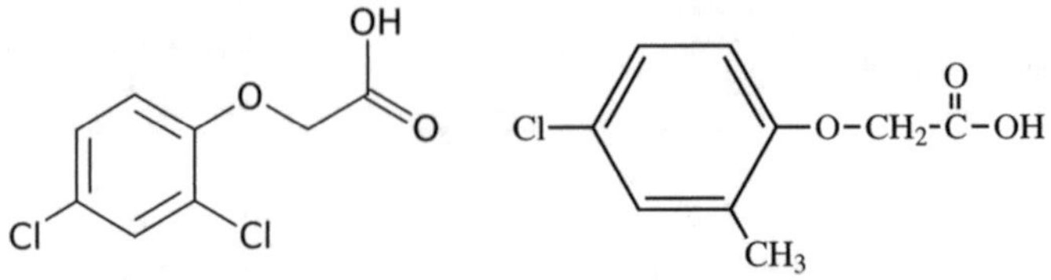

Fig. 2.32 Molecular structure of phenoxyacetic acids 2,4-D (left) and MCPA (right)

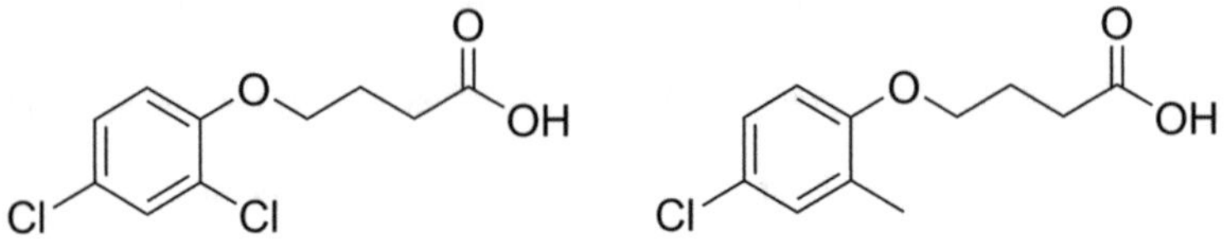

Fig. 2.33 Molecular structures of phenoxybutyric acids 2,4-DB (left) and MCPB (right)

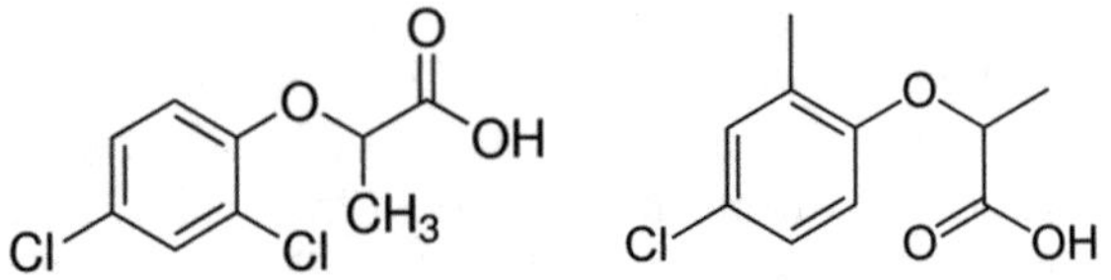

Fig. 2.34 Molecular structures of phenoxypropionic acids, from left to right, 2,4-DP (dichlorprop) and MCPP (mecoprop)

2.5.3.1 Phenoxy Acid Compounds

Phenoxy acid herbicides are selective for broadleaf weeds, notably against those in cereal crops. They work by mimicking indole acetic acid, an auxin plant growth regulator, thereby producing rapid uncontrolled growth lethally damaging the weeds. The phenoxy acid group of herbicides is comprised of three separate sub-groups; phenoxyacetic, e.g. 2,4-D ((2,4-dichlorophenoxy)acetic acid) and MCPA ((4-chloro-2-methylphenoxy)acetic acid), Fig. 2.32, phenoxybutyric, e.g. 2,4-DB (4-(2,4-dichlorophenoxy)butanoic acid) and MCPB (4-(4-chloro-2-methylphenoxy)butanoic acid), Fig. 2.33 and the phenoxypropionic chemicals such as 2,4-DP (2-(2,4-dichlorophenoxy)propanoic acid) and MCPP (2-(4-chloro-2-methylphenoxy)propanoic acid), Fig. 2.34.

The phenoxy acid herbicides were first marketed from the mid-1940s to the early 1960s. As previously mentioned, the first phenoxy acid herbicide was 2,4-D, first introduced in 1945, and widely used on a variety of small grain, fruit, nut and vegetable crops for the control of broadleaf weeds and woody plants (Peterson et al. 2016). Overall, the phenoxy acid herbicides had a significant impact on agriculture and on securing an economically sustainable food supply. A study carried out in the USA in 1996 calculated that if the phenoxy acid herbicides were banned farmers would suffer an annual loss of 1.868 billion dollars due to yield loss and increased

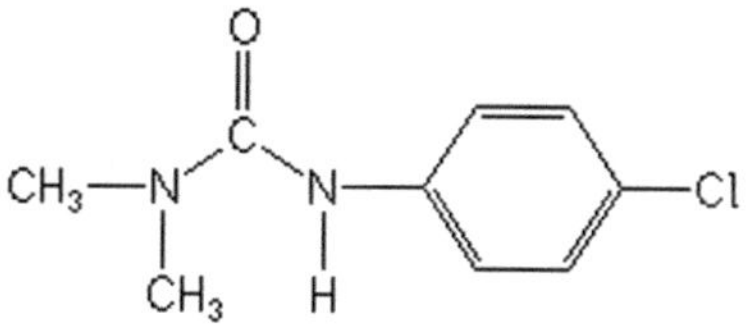

Fig. 2.35 Molecular structure of the first phenylurea herbicide, monuron

costs for alternative herbicides in over 65 crops and many non-cropland situations (Burnside 1996). In a 2006 study in Canada, it was estimated that if the phenoxy acid herbicides, 2,4-D, MCPA and MCPP, were withdrawn completely for use on wheat and barley, there would be an increase in costs of 338 million dollars. This increase being due to increased cost of alternative herbicides and lower yields, as the herbicides used would be less effective. This increase in costs can be considered to be conservative as the loss in yield was estimated as 0.5–3.5%, whereas other studies have estimated yield losses of 15–37% (RIAS Inc. 2006).

2.5.3.2 Phenylurea Compounds

Phenylurea derivatives were used extensively as herbicides for the selective control of germinating grasses and broad-leaved weeds in many crops, including fruit trees, berries, asparagus, sugarcane and especially cereals. They are typically photosystem II inhibitors and concentrate in the leaves at the site of photosynthesis. They then block the electron transport system causing a build up of destructive high energy products which destroy chlorophyll and ultimately the leaf tissue (Cudney 1996). The first phenylurea herbicide, monuron (3-(4-chlorophenyl)-1,1-dimethylurea), was reported in 1951 (Bucha and Todd 1951), Fig. 2.35.

Because the urea molecule is easily accessible to multiple substitutions, thousands of urea derivatives were screened as potential selective herbicides (Kudsk and Streiberg 2003) and monuron was followed by up to 21 other substituted phenylurea herbicides, including diuron (3-(3,4-dichlorophenyl)-1,1-dimethylurea), linuron (3-(3,4-dichlorophenyl)-1-methoxy-1-methylurea), isoproturon (3-(4-isopropylphenyl)-1,1-dimethylurea) and chlortoluron (3-(3-chloro-4-methylphenyl)-1,1-dimethylurea) (Liu 2010), Fig. 2.36.

In Europe, a weed of concern in cereal crops is black-grass (*Alopecurus myosuroides*) which can cause significant crop yield losses if not controlled. For example, 25 black-grass plants per square metre can cause losses, on average, of between less than 5–25%, whilst 100 plants can cause losses of between 5 and 50%, depending on conditions (Moss 2013). The phenylurea herbicides isoproturon and chlortoluron were found to give excellent control of black-grass, thus mitigating the potential yield losses (Orson et al. 2001). In Pakistan, where wheat is the staple crop, a study using isoproturon was reported to give an increase in grain yield of 76% when compared to a "weedy" check plot (Fahad et al. 2013). In sugar cane, a

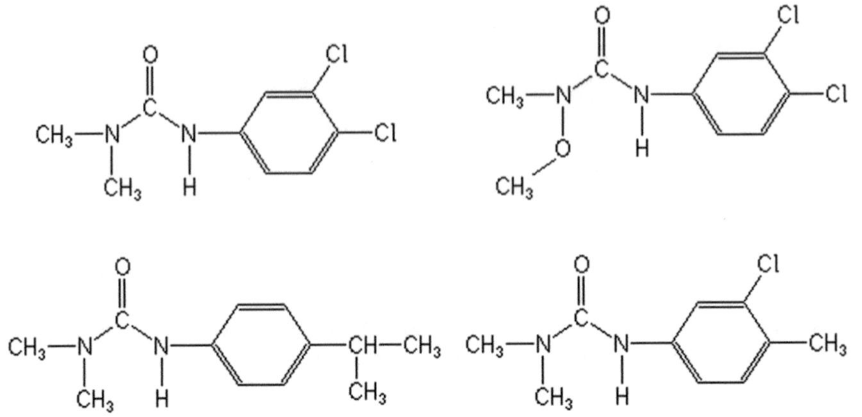

Fig. 2.36 Molecular structure of some phenylurea herbicides diuron (top left), linuron (top right), isoproturon (bottom left) and chlortoluron (bottom right)

pre-emergence treatment with diuron also gave an increased yield of 17% of sugar cane per hectare (Cisneros 1970). Phenylurea herbicides rapidly had an important role in weed control and were used globally on a wide range of crops, as can be seen, for example, from the large number of residue tolerances granted in the USA (US EPA 2003).

2.5.3.3 Triazine Compounds

The first triazine was prepared in 1952, and triazine herbicides were subsequently registered for use in more than 100 countries on a great number of crops (LeBaron et al. 2008). Like the phenylurea herbicides, they are photosystem II inhibitors. The triazine family of herbicides can be divided into three separate groups, chlorotriazines, methoxytriazines and methylthiotriazines, each of which vary in terms of selectivity and use. The most important triazines were atrazine (6-chloro-N-ethyl-N'-isopropyl-1,3,5-triazine-2,4-diamine) and simazine (6-chloro-N,N'-diethyl-1,3,5-triazine-2,4-diamine), Fig. 2.37, both members of the chlorotriazine group of herbicides. Atrazine became the most widely used triazine with corn as its major crop use (Mitchell 2014; US EPA 2006a). Simazine was registered for use on a variety of food and feed crops including fruit and nut crops, as well as corn (US EPA 2006b). Several studies have been carried out in the USA on the potential crop yield loss if atrazine was not used but was replaced by alternative herbicides. The results indicated potential crop yield losses of up to 6% and a rise in the price of corn or a reduction in farmers' incomes (Ackerman 2007). Atrazine and simazine can be used on corn and sweet corn in the USA, and crop yield losses in this crop, after replacing these two herbicides with alternatives, have been estimated at about 20%. This figure is higher than for corn since sweet corn has generally a greater sensitivity to weed populations, and there are

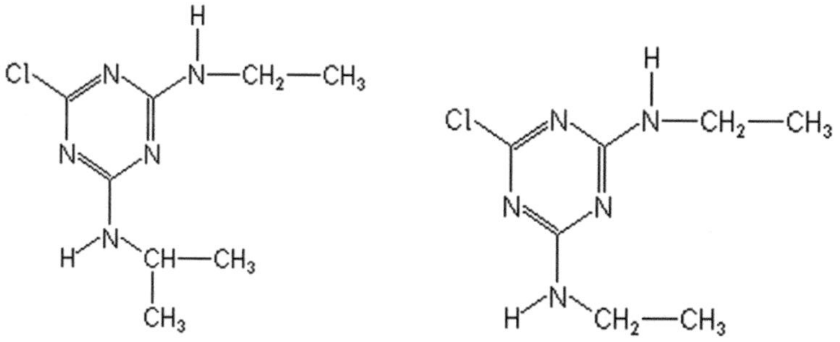

Fig. 2.37 Molecular structure of the triazine herbicides atrazine (left) and simazine (right)

fewer alternative herbicides available. Atrazine is also used in sugarcane, and crop yield losses have been estimated to be up to 25% if it was not available (Mitchell 2011). The use of simazine for weed control in strawberries has also been investigated, and it was shown that crop yields could be increased more than threefold, compared to untreated control plots, with applications 6 and 12 weeks after planting (Collins and Everett 1965). In grapes, simazine treatment gave a 33 and 95% increase in yield, compared to untreated controls, in two successive years, this discrepancy being explained by differences in rainfall between the two years (Sanguankeo et al. 2009). Simazine treatment of apple plots gave a 46% increase in crop yield compared to trees grown areas of mowed grass (Derr 2001).

2.5.3.4   Dinitroaniline Compounds

Dinitroaniline compounds had been researched as dye intermediates many years before some members of this chemical family were discovered to be useful as selective herbicides. Of the potential isomers, it was found that 2,6-dinitroanilines were more effective than either the 2,3- or 2,4-dinitroanilines, with the first member of the substituted 2,6-dinitroanilines being reported in 1960 (Zimdahl 2010). Unlike the phenylurea and triazine herbicides, the dinitroanilines are not photosystem II inhibitors, but act by inhibiting mitosis (cell division), thereby limiting normal root growth (Vaughn and Lehnen 1991). The first successful dinitroaniline commercialised was trifluralin (2,6-dinitro-N,N-dipropyl-4-(trifluoromethyl)aniline) which was registered in the USA in 1963 as a selective pre-emergent herbicide and was used on a large variety of food crops, most notably broccoli, cabbage, onions, a number of leafy green vegetables, beans, tomatoes, potatoes, wheat, soybeans, sugar beet and sugar cane (Williams et al. 2011). Trifluralin was followed by several more dinitroaniline herbicides, including oryzalin (4-(dipropylamino)-3,5-dinitrobenzenesulfonamide) and pendimethalin (3,4-dimethyl-2,6-dinitro-N-(3-pentanyl)aniline), Fig. 2.38, indicating the wide range of substitutions that were tried.

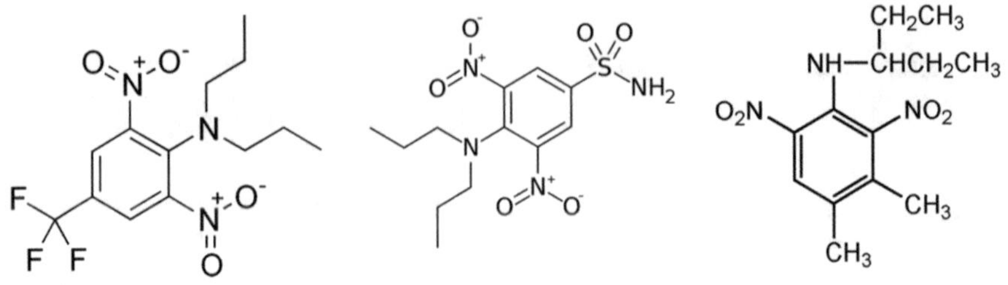

Fig. 2.38 Molecular structure of the dinitroaniline herbicides, from left to right, trifluralin, oryzalin and pendimethalin

Significant yield increases for rapeseed of more than 150% have been reported after the use of dinitroaniline herbicides; however, the actual increase depended on the weed pressure, climatic conditions, timing of application and placement in soil (Friesen and Bowen 1973; Chow 1976). The use of trifluralin in soybean gave an increase in yield of 54% compared to the untreated control plot (Movahed-pour et al. 2013), whilst preplant incorporation and/or pre-emergence application of pendimethalin in peanuts generally gave significant yield increases, in some cases more than double that from untreated plots (Prostko et al. 2001). Preplant incorporation of trifluralin in sesame gave an average crop yield increase of between 134 and 159% over a 5 year trial in the USA, whilst pendimethalin gave average increases of between 124 and 136% (Langham et al. 2007). When used in corn, pendimethalin gave yield increases which varied between 17 and 99% (Janak and Grichar 2016).

2.5.3.5 Bipyridinium Compounds

The discovery of the bipyridinium herbicides is a good example of the benefit of screening readily available chemicals to test for possible pesticidal effects. These compounds were first recognised as herbicides in 1954 following screening by Imperial Chemical Industries in the UK, although paraquat was first synthesised in 1882 and had been used as an oxidation–reduction indicator, under the name of methyl viologen, since 1932 (Bromilow 2003). Although related bipyridinium compounds are also active, only diquat (6,7-dihydrodipyrido[1,2-a:2′,1′-c]pyrazinediium) and paraquat, (1,1′-dimethyl-4,4′-bipyridinium), Fig. 2.39, were successfully commercialised. Initially, the non-selective properties of these herbicides, i.e. rapid action, lack of effect on roots and rhizomes and its rapid deactivation by strong adsorption to soil, could be advantageous in areas where weeds needed to be eliminated prior to replanting or reseeding and in control of weeds between rows in orchards, etc. Their mode of action, which requires both light and oxygen, involves disrupting Photo System I by diverting electrons from the iron–sulphur centres to form, for example, reduced paraquat species which react with oxygen to ultimately give highly reactive radicals which attack membranes, thereby inducing breakdown of cells in the sprayed green tissue (Bromilow 2003), causing the foliage to dry out. Apart from

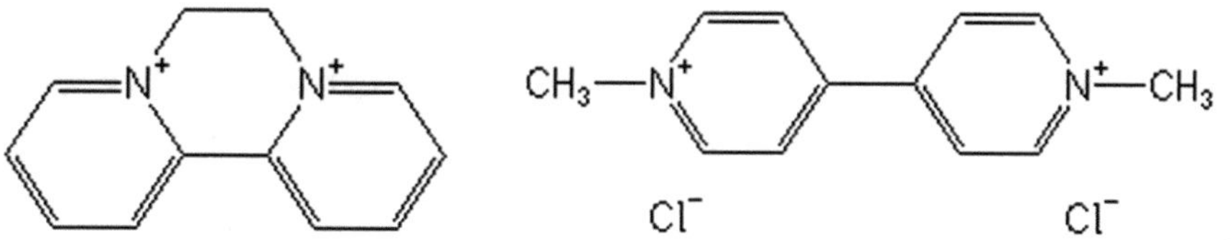

Fig. 2.39 Molecular structure of the bipyridinium herbicides diquat and paraquat

the weed killing utility, this property has been successfully used for preharvest desiccation in a wide range of crops. For example, diquat has been successfully used for preharvest desiccation to allow easier harvesting in peas, beans, oilseed rape and potatoes (Syngenta 2016). Paraquat has been widely used as a herbicide in almost 100 countries and by an estimated 25 million farmers, mainly on small farms. In 2013, it was the world's second largest herbicide in sales terms. A review of studies on paraquat in Vietnam, the Philippines and China shows that its use can result in significant increases in crop yields, for example, 18% for maize and 13% for tea in Vietnam, whilst in China, increases of 8, 4 and 3.8–7.1% were obtained for wheat, rice and oilseed rape, respectively (Shoham 2013).

2.5.3.6 Chloroacetamide Compounds

The chloroacetamide family of herbicides originated from a research programme initiated by Monsanto in 1952 to look for compounds that would be active against grasses. After preliminary screening of compounds under greenhouse conditions, two areas of chemistry were chosen for further research, based on the effect on grasses and the possibilities for further chemical modification. One of these two areas led to the chloroacetamide class of herbicides (Hamm 1974). The first of the chloroacetamide herbicides to be commercialised, in 1956, was allidochlor (*N,N*-diallyl-2-chloroacetamide), a pre-emergence herbicide which was recommended for the control of grasses. This was followed by propachlor (2-chloro-*N*-isopropyl-*N*-phenylacetamide) in 1964, butachlor (*N*-(butoxymethyl)-2-chloro-*N*-(2,6-diethylphenyl)acetamide in 1968 and alachlor (2-chloro-*N*-(2,6-diethylphenyl)-*N*-(methoxymethyl)acetamide) in 1969, Fig. 2.40; later additions to the family were metolachlor (2-chloro-*N*-(2-ethyl-6-methylphenyl)-*N*-(1-methoxy-2-propanyl)acetamide) and acetochlor (2-chloro-*N*-(ethoxymethyl)-*N*-(2-ethyl-6-methylphenyl)acetamide), Fig. 2.41.

The chloracetamide family of herbicides contains residual herbicides that share a common mode of action for the control of more than 30 grasses and broadleaf weeds particularly in corn, soya and rice. They inhibit and distort early seedling growth, which appears to be due to inhibition of cell elongation and cell division (Fuerst 1987). The utility of alachlor in reducing crop yield losses has been demonstrated with the control of yellow nutsedge (*Cyperus esculentus* L.) in corn. This can be particularly troublesome weed, with yield decreases of up to 41% registered in a heavy infestation

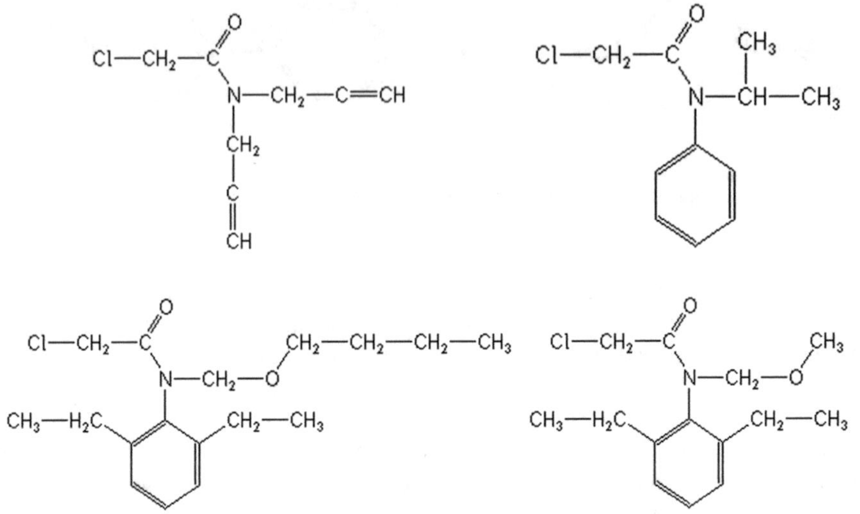

Fig. 2.40 Molecular structure of the chloroacetamide herbicides allidochlor (top left), propachlor (top right), butachlor (bottom left) and alachlor (bottom right)

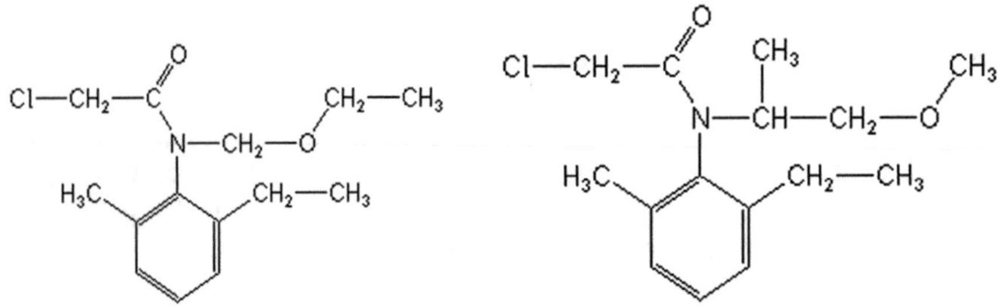

Fig. 2.41 Molecular structure of the chloroacetamide herbicides metolachlor (left) and acetochlor (right)

(initially infested with 1200 tubers m^{-2}), compared to hand weeded plots. However, a single preplant-incorporated treatment with alachlor prevented yield reductions from yellow nutsedge competition (Stoller et al. 1979). In soybean, sicklepod (*Cassia obtusifolia* L.) can also be a problem and it was demonstrated that the use of alachlor gave significant increases in crop yield compared to untreated plots (Adcock and Banks 1991). Soybean can be grown in rotation with rice as part of an integrated weed management system; in this case, red rice (*Oryza sativa* L.) and *Echinochloa* spp. can cause yield losses where the soybean is grown after the rice crop. In a study carried out over 3 years, it was found that alachlor and metolachlor applied preplant-incorporated or pre-emergence, although not completely controlling red rice, gave increased yields compared to non-treated controls, for example, with metolachlor in 1992 the yield was 2860 kg ha^{-1} compared to 970 kg ha^{-1} for the untreated control

Fig. 2.42 Molecular
structure of dimethenamid

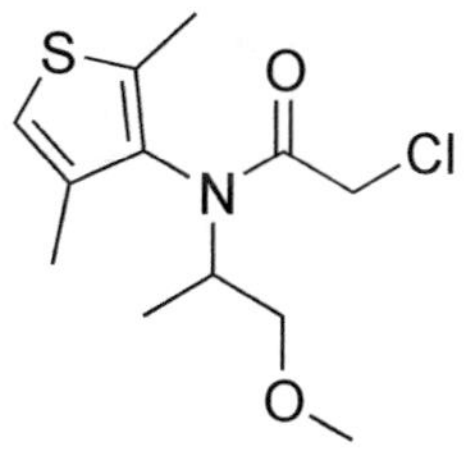

(Noldin et al. 1998). Red rice is major problem wherever rice is grown; it is very competitive with rice and difficult to control as both belong to the same species and share many morphological and physiological characteristics. However, after treatment of rice with the chloroacetamide herbicides, alachlor, dimethenamid 2-chloro-N-(2,4-dimethyl-3-thienyl)-N-(1-methoxy-2-propanyl)acetamide), Fig. 2.42, and metolachlor or acetochlor, 15 days before sowing, it was demonstrated that significant yield increases were obtained compared to control plots; for example, in 1998 the rice yield was 8440 kg ha^{-1} for acetochlor treated plots compared to just 200 kg ha^{-1} for control plots (Eleftherohorinos and Dhima 2002).

2.6   The Battle Continues

It was not until the nineteenth century that great strides were made in crop protection, particularly for the control of fungal diseases and insect pests with chemicals. The large number of inorganic chemicals used came from the growth in mining and the earlier advances in inorganic chemistry, whilst the use of dyestuffs and aromatic chemicals followed the development of organic chemistry in the latter half of the nineteenth century (Achilladelis et al. 2013). As previously mentioned, in the USA using insecticides on the apple crop protected 50–75% of the crop against larvae of the codling moth, whilst for plum and other small fruits lack of treatment could result in a complete loss of yield of sound fruit (Marlatt 1903).

Prior to 1940s, fruit and vegetable crops were routinely treated against fungal problems with inorganic chemicals, for example, sulphur, lime sulphur, copper and Bordeaux mixture. Bordeaux mixture, for example, proved effective against fungal attacks in controlling grape downy mildew, potato late blight and numerous other diseases. By the early 1900s, spraying Bordeaux mixture on potatoes was common practice in North America and Europe. In Vermont, for example, a yield increase of 64% was obtained as a result of late blight control in potatoes. Bordeaux mixture also reduced cranberry rots by 50% (Gianessi and Reigner 2006). The value of Bordeaux mixture in crop protection was demonstrated in Germany in 1916 when a major late blight epidemic on potatoes went untreated, due to a lack of copper which was needed for producing shell casings and wire. As a result, and as happened in Ireland in 1845–46, potatoes rotted in the fields and 700,000 German civilians died from starvation in the resulting famine (Gianessi and Reigner 2005).

However, chemical control of weeds was much less successful. Several inorganic chemicals, working as contact herbicides, including copper salts, sodium nitrate and ferrous sulphate were available, but the use of herbicides was not widespread. Treatment required large quantities of the chemicals, making treatments expensive, and they seldom worked effectively or consistently (Peterson 1967). In the absence of chemical control, weeds could significantly reduce crop yields, and it was calculated that moderately weeded areas could yield 40–50% more than non-weeded crops. However, since labour for weed removal was cheap and readily available, with substantial numbers of women and children involved (Braseley 2000), farmers were not generally interested in using herbicides to limit the impact of weeds on a crop.

Until the 1930s, studies on pest control continued to be predominantly carried out with chemicals used in the latter part of the nineteenth century, with a focus on more effective and consistent treatments. Indeed, even up to the early 1940s, farmers still relied substantially on tried and tested remedies and significant crop losses due to pests and disease were not unknown. However, there was increasing interest in the possibility of using various organic chemicals to control insects, weeds and fungal diseases. Companies were becoming more research oriented, and with an increasing interest in organic chemistry, many thousands of organic compounds was screened for potential pesticidal properties.

For herbicides, 2,4-D led the revolution in selective weed control, which before 1945 relied on cultivation techniques, including hand weeding, or on inorganic salts which had been used since the late 1800s with varying degrees of success. At the time of the introduction of new herbicide classes, which coincided with a reduction in available labour and increased mechanisation, farmers were ready for improved methods of selective, chemical weed control (Zimdahl 2010). There was rapid adoption of synthetic herbicides, and, for example, by 1962 about 100 herbicides in 6000 formulations were marketed in the USA (Gianessi and Reigner 2007). With improved mechanisation and the use of synthetic chemical herbicides in the USA, it was estimated that in 1920 one farmer could feed 8 people, in 1947 sixteen people and in 1977 fifty people (Adler et al. 1977).

With the significant increase in yields that could be attributed to the introduction of synthetic chemical pesticides, consumers and most policy-makers were not overly concerned about the potential risks to health or the environment. Food was cheaper because of the new chemical formulations, and with the new pesticides, there were no documented cases of people dying or being seriously hurt by their "normal" use (Ganzel 2007). There were some cases of harm from misuse of the chemicals, but the new pesticides seemed generally rather safe, especially compared to the forms of arsenic that had poisoned people in the 1920s and 1930s (Reinhardt and Ganzel 2003). However, things were about to change with the publication of Rachel Carson's book "Silent Spring" (Carson 2002), which exposed the dangers and environmental impacts caused by the indiscriminate spraying of pesticides, notably DDT. Carson's view (Beyl 1992) was that *"we have allowed these chemicals to be used with little or no advance investigation of their effects on soil, water, wildlife, and man him-*

self". The impact of "Silent Spring" was international and served as an important stimulus for the environmental movement in the USA and ultimately in the creation of the Environmental Protection Agency (EPA) in December 1970 (Zimdahl 2015). As these new pesticides were used commercially, resistance to them developed, particularly with fungicides and insecticides. An extreme example of the problem of insecticide resistance is that of the Colorado potato beetle (*Leptinotarsa decemlineata*) which developed resistance to 52 different insecticides, belonging to all major insecticide classes (Alyokhim et al. 2008). To combat this problem, pesticide resistance management programmes were put in place. These included alternating pesticides with different modes of action, which required an ongoing synthesis programme for new chemical families, particularly those targeting different active sites. Resistance problems, together with the increasingly stringent regulations introduced by the EPA and other regulatory bodies to protect human health and the environment, led agrochemical companies to look for new chemistries giving less toxic and more environmentally friendly products. Thus, a greater emphasis was put on finding chemicals which could be used at lower rates, were more rapidly degraded in the environment and had lower toxicity to non-target species. During this period, with changes in both market and regulatory environments, there was an eagerness for product novelty across all agrochemical sectors. For example, this can be seen with the phenomenal success of contemporary chemistries such as the sulfonylurea herbicides, the strobilurin fungicides and the neonicotinoid insecticide classes (Cole et al. 2000; Nauen and Bretschneider 2002). Finally, in the late 1990s, genetically modified crops were introduced which were resistant to insects, for example, Bt corn in 1995 (Tabashnik et al. 2009), and tolerance to herbicides such as glyphosate with soybeans and canola commercialised in 1996 and 1997, respectively (Sourcewatch 2012).

2.7 Conclusions

Ever since mankind changed from a hunter/gatherer way of life to that of a farmer, it has been necessary to combat the pests that threaten the food supply. Early techniques for pest control included religion or folk magic, but hand removal of weeds and insects would have proved to be more reliable. Chemical techniques used readily available raw materials such as smokes, minerals, oils and plant extracts known to have pesticidal activity. In the nineteenth and early twentieth centuries, chemicals, based on inorganic materials such as sulphur and copper, or organic chemicals from coal tar and the dyestuff industry became widely used. Even up to the early 1940s, farmers still relied substantially on tried and tested remedies from earlier periods and significant crop losses due to pests and disease were not unknown. During the so-called Golden Age of Pesticides in the 1940s to the 1960s, there was increasing interest in organic chemistry and thousands of compounds were routinely screened for their pesticidal properties. These new synthetic pesticides, such as the phenoxy acid herbicides, the dithiocarbamate fungicides and the *N*-methyl carbamate insecticides marked, at least in terms of their selective activity, a quantum leap in the progress of pest control. Nevertheless, the battle against the agricultural pests was not won and increasing resistance meant that new pesticides with new modes of action were needed. In addition, stricter regulatory controls required new chemistries giving less toxic and more environmentally friendly products, giving rise to contemporary chemistries such as the sulfonylurea herbicides, the strobilurin fungicides and the neonicotinoid insecticides. As part of the ongoing battle against pests to the increased resistance of pests and increasing resistance to pesticides, various crops were introduced which in themselves were resistant to insects, for example, Bt corn, and crops that are tolerant to non-selective herbicides such as glyphosate with soybeans and canola.

References

Achilladelis B, Schwarzkopf A, Cines M (1987) A study of inoovation in the pesticide industry: analysis of the innovation record of an industrial sector. Res Policy 16(2–4):175–212

Achilladelis B, Schwarzkopf A, Cines M (2013) A study of innovation in the pesticide industry: analysis of the innovation record of an industrial sector. In: Freeman C (ed) Output measurement in science and technology: essays in honour of Yvan Fabian. Elsevier Science Publishers B.V., Amsterdam

Ackerman F (2007) The economics of atrazine. J Occup Environ Health 13:441–449

Adcock TE, Banks PA (1991) Effects of pre-emergence herbicides on the competitiveness of selected weeds. Weed Sci 39:54–56

Adler EF, Wright WL, Klingman GC (1977) Development of the American herbicide industry. ACS symposium series—pesticide chemistry in the 20th Century (Chap. 3), vol 37, pp 39–55

Agropages (2015) Pests generated losses of US$ 250 billion in the World, Agronews, 13 Jan

Alyokhim A, Baker M, Monta-Sanchez D, Dively G, Grafius E (2008) Colorado potato beetle resistance to insecticides. Am J Potato Res 85:395–413

Andes JO, Epps JM (1956) Evaluation of fungicides for control of fruit diseases. Ag. Expt. Station, University of Tennessee, Bulletin No. 254

APS (2019) American Phytopathological Society, APS History, Forming the American Phytopathological Society. http://www.apsnet.org/about/history/Pages/default.aspx

Armelagos GT, Goodman AH, Jacobs KH (1991) The origins of agriculture: population growth during a period of declining health. Popul Environ 13:9–22

Attard E, Cuschieri A (2009) Cytotoxicity of Cucurbitacin E extracted from *Ecballium elaterium* in vitro. J Nat Remedies 4:137–144

Back RC (1965) Carbamate insecticides: significant developments in eight years with Sevin insecticide. J Agric Food Chem 13:198–199

Bais HP, Walker TS, Kennan AJ, Stermitz FR, Vivanco JM (2003) Structure-dependent phytotoxicity of catechins and other flavonoids: flavonoid conversions by cell-free protein extracts of *Centaurea maculosa* (spotted knapweed) roots. J Agric Food Chem 51:897–901

Baker H (1952) Spider mites, insects and DDT. The yearbook of agriculture, USDA, Washington DC, USA, pp 562–567

Baker MJ (2008) A structural model of the transition to agriculture. J Econ Growth 13:257–292

Baldwin RFW (2005) Public perceptions of urban pest management and the toxicity of fatty acid salts to cockroaches. A dissertation presented to the Graduate School of the University of Florida in Partial Fulfilment of the Requirements for the Degree of Doctor of Philosophy, University of Florida

Ball WE, French OC (1935) Sulfuric acid for control of weeds. University of California, College of Agriculture, Bulletin 596

Barss HP, Andre F (1945) Reducing Losses from Diseases and Pests, Report on Experiment Stations 1944, USDA, Agricultural Research Administration, Washington DC, USA

Bayer Seed Growth (2014) 100 years of innovation in seed treatment. Bayer CropScience, Monheim am Rhein, Germany. http://www.seedgrowth.bayer.com/~/media/SeedGrowth/Bayer100YearsFlash/index.ashx?iframe=true&width=995&height=700

Bedford CW, Winkelman HA (1923) Systematic survey of rubber chemistry. The Chemical Catalog Company Inc., New York

Beinhart EG (1951) Production and use of nicotine, United States Department of Agriculture, Crops in Peace and War, The Yearbook of Agriculture 1950–1951, pp 773–779

Bellwood P, Oxenham M (2008) The expansion of farming societies and the role of the neolithic demographic transition. In: Bocquet-Appel J-P, Bar-Yosef O (eds) The neolithic demographic transition and its Consequences. Springer Science + Business Media B.V., The Netherlands. ISBN 978-1-4020-8538-3

Bender B (1978) Gatherer-hunter to farmer: a social perspective. World Archaeol 10:204–222

Bermúdez-Torres K, Herrera JM, Brito RF, Wink M, Legal L (2009) Activity of quinolizidine alkaloids from three Mexican *Lupinus* against the lepidopteran crop pest *Spodoptera frugiperda*. Biocontrol 54:459–466

Bettinger R, Richerson P, Boyd R (2009) Constraints on the development of agriculture. Curr Anthropol 50:627–631

Bettiol W (1999) Effectiveness of cow's milk against zucchini squash powdery mildew (*Sphaerotheca fuliginea*) in greenhouse conditions. Crop Prot 18:489–492

Beyl CA (1992) Rachel carson, silent spring and the environmental movement. HortTechnology 2:272–275

Biswas K, Chattopadhay L, Banerjee RK, Bandyopadhay U (2002) Biological activities and medicinal properties of neem (*Azadirachta indica*). Curr Sci 82:1336–1345

Boote KJ, Jones JW, Mishoe JW, Berger RD (1983) Coupling growth simulators to predict yield reductions. Phytopathology 73:1581–1587

Borden AE (1947) Control of coddling moth with DDT spray on apples and pears: good in investigational work. Calif Agric 1:3–4

Boschin G, Annicchiarico P, Resta D, d'Agstina Arnold A (2008) Quinolizidine alkaloids in seeds of lupin genotypes of different origins. J Agric Food Chem 56:3657–3663

Bourcart E (1913) Insecticides, fungicides and weed killers: a practical manual on the diseases of plants and their remedies, for the use of manufacturing chemists, agriculturalists, arboriculturalists and horticulturalists. Translated from the French by Grant D, Scott, Greenwood and Son, London, UK

Brandes GA (1953) The history and development of the ethylene bisdithiocarbamate fungicides. Am J Potato Res 30:137–140

Braseley P (2000) Weed and pest control. In: Collins EJT and Thirsk J (eds) The agrarian history of England and Wales, vol VII, 1850–1914. Cambridge University Press, Cambridge. ISBN 0521 32926 4

Brent KJ (2003) Fungicides, an overview. In: Plimmer JR, Gammon DW, Ragsdale NA (eds) Encyclopedia of agrochemicals. Wiley Online Library. Online ISBN 9780471263630

Bromilow RH (2003) Paraquat and sustainable agriculture. Pest Manag Sci 60:340–349

Brooks GT (1977) Chlorinated insecticides: retrospect and prospect, pesticide chemistry in the 20th Century. In: Plimmer JR (ed) ACS Symposium Series No. 37. ACS, Washington DC, USA, pp 1–20

Brown JG, Streets RB (1928) Sulphuric acid spray: a practical means for the control of weeds. University of Arizona, College of Agriculture, Bulletin No. 128

Bucha HC, Todd CW (1951) 3-(p-Chlorophenyl)-1,1-Dimethylurea—a new herbicide. Science 114:493–494

Bulger JW (1928) Studies on elemental sulfur as a soil insecticide. Ohio J Sci 28:1–42

Bunsupa S, Yamasaki M, Saito K (2012) Quinolizidine Alkaloid Biosynthesis: Recent Advances and Future Prospects. Front Plant Sci 3:1–7

Burnside OC (1996) Biologic and economic assessment of benefits from use of phenoxy herbicides in the United States. Special NAPIAP Report, No. 1-PA-96, Washington D.C., USA

Buttress FA, Dennis RWG (1947) The early history of cereal seed treatment in England. Agric Hist 21:93–103

Carson R (2002) Silent spring. 40th Anniversary Edition, Houghton Mifflin Harcourt, New York, USA. ISBN 0618249060

Case HCM, Mosher ML (1932) Farm practices that pay. University of Illinois, College of Agriculture and Agricultural Experiment Station, Circular 389

Casida JE, Quistad GB (1998) Golden age of insecticide research: past, present or future. Annu Rev Entomol 43:1–16

Cates HR (1917) The weed problem in American agriculture. Yearbook of the Dept. Agriculture, USDA, pp 205–216

Chow PNP (1976) Dinitroaniline herbicides for grassy weed control in rapeseed. Can J Plant Sci 56:705–713

Cisneros J (1970) Estudio Sobre el Incremento de la Produccion de la Cana de Azucar en el Ingenio de Los Mochis, Mediante el Uso de Herbicidas (A study on yield increase of sugar cane using herbicides at Los Mochis estate). Bol Azucar Mex 246:12–19

Clark G (1991) Yields per acre in English agriculture, 1250–1860: evidence from labour inputs. Econ Hist Rev New Ser 44:445–460

Cloyd RA (2013) Lime-sulfur: a broad-spectrum pesticide. Garden & Greenhouse, April

Cohen MN, Crane-Kramer G (2007) Ancient health: skeletal indicators of agricultural and economic intensification. University Press of Florida, Gainesville. ISBN 978-0-8130-3082-1

Cole D, Pallett K, Rodgers M (2000) Discovering new modes of action for herbicides and the impact of genomics. Pest Outlook 11:223–229

Collins WB, Everett CF (1965) Simazine for weed control in strawberries in Eastern Canada. Can J Plant Sci 45:541–547

Copper Development Association Inc. (2019) Uses of copper: agricultural uses. http://www.copper.org/resources/properties/compounds/agricultural.html

Cornell University (1985) Pesticide management education program. Chemical Profile 2/85, Dichlone (Phygon, Quintar)

Cornell University (1986) Pesticide management education program. Chemical Fact Sheet 3/86, Captan

Crosby DG (1995) Environmental fate of pyrethrins. In: Casida JE, Quistad GB (eds) Pyrethrum flowers: production, chemistry, toxicology and uses. Oxford University Press, New York, pp 177–178. ISBN 10:0195082109

Cudney DE (1996) Why herbicides are selective, California Exotic Pest Council, Symposium Proceedings, San Diego, California, USA. http://www.cal-ipc.org/symposia/archive/pdf/1996_symposium_proceedings1827.pdf

De Groot I (2004) Protection of stored grains and pulses, Agrodok 18, p 27, Agromisa Foundation, Wageningen, The Netherlands. ISBN 90-77073-49-3

de Saulieu G, Testart A (2015) Innovation, food storage and the origin of agriculture. Environ Archaeol 20:314–320

Deng Z, Qin L, Gao Y, Weisskopf AR, Fuller DQ (2015) From early domesticated rice of the middle Yangtze Basin to millet, rice and wheat agriculture: Archaeobotanical macro-remains from Baligang, Nanyang Basin, Central China (6700–500 BC). PLoS 10(10):e0139885

Derr JF (2001) Biological assessment of herbicide use in apple production 1. Background and current use estimates. HortTechnology 11:1–19

Diener UL, Eden G, Carlton CC (1955) Control of leaf spot and strawberry weevil on trailing blackberries. Ag. Expt. Station, Alabama Polytechnic Institute, Leaflet No. 46

Dow GK, Olewiler N, Reed C (2005) The transition to agriculture: climate reversals, population density and technical change. SSRN Electron J. https://doi.org/10.2139/ssrn.698342

Duke SO, Cantrell CL, Meepagala KM, Wedge DE, Tabanca N, Schrader KK (2010) Natural toxins in pest management. Toxins (Basel) 2:1943–1962

DuPont Digest (1948) Rubber accelerators lead the way to new agricultural fungicides. The Michigan Technic, Jan 1948

Ebeling W, Pence R J (1947) The mealy bug problem on newly top-grafted avocado trees. California Avocado Society, Yearbook 32, pp 44–45

Eddins AH (1947) New fungicides. Proc Florida State Hort Soc 60:124–127

Egli T, Sturm E (1981) Bacterial diseases and their control, 6.2.1 Chemistry of bactericidal compounds. In: Wegler R (ed) Chemie der Pflanzenschutz- und Schädlingsbekämpfungsmittel. Springer, Berlin

Eleftherohorinos IG, Dhima KV (2002) Red rice (*Oryza sativa*) control in rice (*O. sativa*) with preemergence and postemergence herbicides. Weed Technol 16:537–540

Emsley J (1985) Whatever happened to arsenic? New Scientist, 19–26 Dec. http://johnemsley.com/articles/new_scientist/ns_arsenic.html

Essig EO (1946) Investigations with DDT and other new insecticides in 1945, Circular 365, University of California, College of Agriculture, Agricultural Experiment Station, Berkeley, California, USA

Evans WH (1896) Copper sulphate and germination: Treatment of seed with copper sulphate to prevent the attacks by Fungi. USDA, Dept. of Vegetable Physiology and Pathology, Bulletin No. 10

Evans LT (1980) The natural history of crop yield. Am Sci 68:388–397

Ewart WH (1948) Citrus thrips control with DDT: investigated in two Coachella Valley groves. Calif Agric 2:11–16

EXTOXNET (1996a) Pesticide information profiles, Captan

EXTOXNET (1996b) Pesticide information profiles, Lindane

EXTOXNET (1996c) Pesticide information profiles, Aldicarb

Fahad S, Nie L, Rahman A, Chen C, Wu C, Saud S, Huang J (2013) Comparative efficacy of different herbicides for weed management and yield attributes in wheat. Am J Plant Sci 4:1241–1245

Fahad S, Hussain S, Chauhan BS, Saud S, Wu C, Hassan S, Tanveer M, Jan A, Huang J (2015) Weed growth and crop yield loss in wheat as influenced by row spacing and weed emergence times. Crop Prot 71:101–108

FAO (1967) JMPR—evaluation of some pesticide residues in food. DDT

FAO (1968) JMPR—evaluations of some pesticide residues in food. Dieldrin

FAO (1971) JMPR—evaluations of some pesticide residues in food. Endrin

FAO (2018) The state of food security and nutrition in the world 2018; Building Climate Resilience for Food Security and Nutrition Food and Agriculture Organisation of the United Nations, Rome, Italy

Fay IW (1919) The chemistry of the coal tar dyes. Pub. D. Van Nostrand Company, New York, USA

Ferguson C (1992) Bayer crop protection—the first 100 years 1892–1992. Pestic Outlook 3:32–34

Fisher DF (1921) Control of apple powdery mildew. Farmers' Bulletin No. 1120, USDA, Washington D.C., USA

Freed VH (1980) Weed science: the emergence of a vital technology. Weed Sci 28:621–625

Friesen HA, Bowen KE (1973) Factors affecting the control of wild oats in rapeseed with trifluralin. Can J Plant Sci 53:109–205

Fuerst EP (1987) Understanding the mode of action of the chloroacetamide and thiocarbamate herbicides. Weed Technol 1:270–277

Fukuto TR (1990) Mechanism of action of organophosphate and carbamate insecticides. Environ Health Perspect 87:245–254

Fusetto R, O'Hair RAJ (2015) Nicotine as an insecticide in Australia: a short history. Chemistry in Australia, 18–21 Oct

Ganzel W (2007) Farming in the 1950s & 60s, silent spring & the environmental movement. Wessels Living History Farm, York, Nebraska, USA

Gasquez J (2015) Désherbant, Herbicide: Un Peu d'Histoire, Les Mots de l'Agronomie, 1 Apr, INRA, Paris, France. http://mots-agronomie.inra.fr/mots-agronomie.fr/index.php/D%C3% A9sherbant,_herbicide_:_un_peu_d%27histoire

Gianessi LP (1997) The uses and benefits of organophosphate and carbamate insecticides in U.S. crop production. National Center for Food and Agricultural Policy, Washington DC, USA

Gianessi LP (2009) The value of insecticides in U.S. crop production. Crop Protection Research Institute, CropLife Foundation, Washington DC, USA

Gianessi LP, Reigner N (2005) The value of fungicides in US crop production. CropLife Foundation, Crop Protection Research Institute

Gianessi LP, Reigner N (2006) The value of fungicides in US crop production. Outlooks Pest Manag 17:209–213

Gianessi LP, Reigner NP (2007) The value of herbicides in U.S. Crop Prod Weed Technol 21:559–566

Gignouxa CR, Hennb BM, Mountain JL (2011) Rapid, global demographic expansions after the origins of agriculture. Proc Natl Acad Sci USA 108:6044–6049

Gilbert SG (2014) Chemical Weapons, interwar years, Toxipedia. http://www.toxipedia.org/display/ toxipedia/Chemical+Weapons

Goldsworthy MC, Green EL, Smith MA (1943) Fungicidal and phytocidal properties of some metal dialkyl dithiocarbamates. J Agric Res 66:277–291

Goyal P (2003) Ancient crop protection practices: their relevance today. History of Indian Science and Technology. http://www.indianscience.org/essays/t_es_goyal_crop.shtml

Gray GP (1916) Phenolic insecticides and fungicides, Bulletin No. 269. University of California, Berkeley, California, USA

Guiteras JM (1909) Mercuric chloride as an insecticide. Public Health Rep (1896–1970) 24:1859–1861

Gullino ML, Tinivella F, Garibaldi A, Kemmitt GM, Bacci L, Sheppard B (2010) Mancozeb: past present and future. Plant Dis 94:1076–1087

Guo Z, Vangapandu S, Sindelar RW, Walker LA, Sindelar RD (2009) Biologically active quassinoids and their chemistry: potential leads for drug design. Curr Med Chem 4:285–308

Hamm PC (1974) Discovery, development and current status of the chloroacetamide herbicides. Weed Sci 22:541–545

Harker KN, Blackshaw RE, Clayton GW (2001) Timing weed removal in field Pea. Weed Technol 15:277–283

Haskell RJ, Doolittle SP (1942) Vegetable seed treatments. Farmers' Bulletin No. 1862, USDA, Washington D.C., USA

Haverkort AJ, Struik PC, Visser RGF, Jacobsen E (2009) Applied technology to combat late blight in potatoes caused by *Phytophthora infestans*. Potato Res 52:249–264

Haywood JK (1902) Insecticides and fungicides: chemical composition and effectiveness of certain preparations. United States Department of Agriculture, Farmers' Bulletin No. 146

Hester WF (1943) Fungicidal composition. United States Patent Office, US Patent 2,317,765

Hirst KK (2017) Oasis theory—did climate change cause the invention of agriculture? ThoughtCo.com, 6 Mar. https://www.thoughtco.com/k-kris-hirst-166730

Hirst KK (2018) The eight founder crops and the origins of agriculture. ThoughtCo.com, October 19th, https://www.thoughtco.com/founder-crops-origins-of-agriculture-171203

Holland R (2002) A history of tar distillation at Crew's Hole, Bristol, SCI Lecture Papers Series, LPS 123/2002

Holm FA, Johnson EN (2009) The history of herbicide use for weed management on the prairies. Prairie Soils Crop J Weeds Herbic Manag 2:1–11

Holmes E (1958) The role of industrial research and development in weed control in Europe. Weeds 6:245–250

Horsfall JG (1975) Fungi and fungicides: the story of a nonconformist. Ann Rev Phytopathol 13:1–14

Huang HT, Yang P (1987) The ancient cultured citrus ant. BioScience 37:665–671

Hublin J-J, Ben-Ncer A, Bailey SE, Friedline SE, Neubauer S, Skinner MM, Bergmann I, Le Cabec A, Benazzi S, Harvati K, Gunz P (2017) New fossils from Jebel Irhoud, Morocco, and the Pan-African origin of *Homo sapiens*. Nature 546:289–292

IARC (2001) On the evaluation of carcinogenic risks to humans. Some Thyrotropic Agents, Chlordane and Heptachlor, IARC Monograph, vol 79, pp 411–492

IARI (2010) Achievements in Neem Research, Indian Agricultural Research Institute, Indian Council of Agricultural Research, Department of Agricultural Research and Education, Ministry of Agriculture & Farmers Welfare, Government of India, New Delhi, India

IPCS (1984) Environmental health criteria 40, Endosulfan, WHO Geneva, Switzerland

Janak TW, Grichar WJ (2016) Weed control in corn (*Zea mays* L.) as influenced by pre-emergence herbicides. Int J Agron 2016:1–9

Janakat SM, Hammad F (2013) Chemical composition of amurca generated from Jordanian Olive Oil. Nutr Food Sci 3:186

Janek J (2008) History of Horticulture, Roman Agricultural History; Purdue University. http://www.hort.purdue.edu/newcrop/Hort_306/text/lec18.pdf

Jarman WM, Ballschmiter K (2012) From coal to DDT: the history of the development of the pesticide DDT from synthetic dyes till silent spring. Endeavour 36(4):131–142

Jentsch P (n.d.) Historical perspectives on apple production: fruit tree pest management. Regulation and New Insecticidal Chemistries, Dept. Entomol, Cornell University, Hudson Valley Lab, Highland, New York 12528. http://web.entomology.cornell.edu/jentsch/assets/historical-perspectives-on-apple-production.pdf

Johnson GF (1935) The early history of copper fungicides. Agric Hist 9:67–79

Johnson SB, Lambert D (2010) Common scab diseases of potatoes. University of Maine, Co-operative Extension, Bulletin #2440

Karim SMR (2002) Competitive ability of different weed species. Pak J Agron 1:116–118

Kaushik N, Gurudev Singh B, Tomar UK, Naik SN, Vir S, Bisla SS, Sharma KK, Banerjee SK, Thakar P (2007) Regional and habitat variability in azadirachtin content of Indian neem (*Azadirachta indica* A. Jusieu). Curr Sci 92:1400–1406

Kellerman WA, Swingle WT (1890) Preliminary experiments with fungicides for stinking smut of wheat. Kansas State Agricultural College, Bulletin No. 12, Aug 1890. http://www.ksre.k-state.edu/historicpublications/pubs/SB012.PDF

Kelly RL (1992) Mobility/sedentism: concepts, archaeological measures and effects. Ann Rev Anthropol 21:43–66

Kelly D (2012) Chlorothalonil biological and use profile. Syngenta Presentation, pp 1–46. http://www.cdpr.ca.gov/docs/emon/surfwtr/presentations/syngenta_presentation_2012.pdf

Kelton JA, Price AJ (2009) Weed science and management. In: Verheye WH (ed) Soils, plant growth and crop production. Encyclopedia of life support systems (EOLSS), developed under the auspices of the UNESCO. EOLSS Publishers, Oxford

Kenawy MA, Abdel-Hamid YM (2015) Insects in ancient (pharaonic) Egypt: a review of Fauna, their mythological and religious significance and associated diseases. Egypt Acad J Biol Sci 8:15–32

Kimmel L, Coelhan M, Leupold G, Vetter W, Parlar H (2000) FTIR specroscopic characterization of chlorinated camphenes and bornenesin technical toxaphene. Environ Sci Technol 34:3041–3045

Kislev ME, Weiss E, Hartmann A (2004) Impetus for sowing and the beginning of agriculture: ground collecting of wild cereals. Proc Natl Acad Sci 101:2692–2694

Kittleson AR (1951) Parasiticidal compounds containing the $NSCCl_3$ group. United States Patent Office, US Patent 2,553,770

Kittleson AR (1952) A new class of organic fungicides. Science 115:84–86

Klittich CJ (2008) Milestones in fungicide discovery: chemistry that changed agriculture. Plant Health Progress, vol 9, Apr 2008

Knorr LC, Denmark HA, Burnett HC (1968) Occurrence of Brevipalpus mites, leprosis and false leprosis on citrus in Florida. Fla Entomol 51:11–17

Knutson R, Smith EG (1999) Impacts of eliminating organophosphates and carbamates from crop production. AFPC Policy Working Paper 99-2, Agricultural and Food Policy Center, Department of Agricultural Economics, Texas A&M University, Texas, USA

Koul O, Walia S, Dhaliwal GS (2008) Essential oils as green pesticides: potential and constraints. Biopestic Int 4:63–84

Kraemer H (1906) Dilute sulphuric acid as a fungicide. Proc Am Phil Soc 45:157–163

Kudsk P, Streiberg JC (2003) Herbicides—a two edged sword. Weed Res 43:90–102

Kureel RS, Kishore R, Dutt D (2009) Neem: a tree borne oilseed, National Oilseeds & Vegetable Oils Development Board (Ministry of Agriculture, Govt. of India), Gurgaon-122015, India

Laforge FB, Haller HL, Smith LE (1933) The determination of the structure of rotenone. Chem Rev 18:181–213

Lange WH, Leach LD, Carlson EC (1950) Lindane for wireworm control. Calif Agric 4:5–6

Lange WH, Carlson EC, Corrin WR (1951) Seed treatments for the control of seed-corn maggot in Northern California. J Econ Entomol 44:202–208

Lange WH, Carlson EC, Leach LD (1953) Pest control by seed treatment. Calif Agric 7:7–8

Langham DR, Grichar J, Dotray P (2007) Review of herbicide research on sesame (*Sesamum indicum* L.), Version 1, American Sesame Growers Association

Latham KJ (2013) Human health and the neolithic revolution: an overview of impacts of the agricultural transition on oral health, epidemiology, and the human body. Nebraska Anthropologist, University of Nebraska, Lincoln, Nebraska

Latshaw WL, Zahnley JW (1927) Experiments with sodium chlorate and other chemicals as herbicides for field bindweed. J. Agric Res 35:757–767

LeBaron HM, McFarland JE, Burnside OC (2008) The triazine herbicides: a milestone in the development of weed control technology (Chap. 1). In: The triazine herbicides: 50 years of revolutionizing agriculture. Elsevier B.V., Amsterdam, pp 1–12

Leonard KJ, Szabo LJ (2005) Stem rust of small grains and grasses caused by *Puccinia graminis*. Mol Plant Pathol 6:99–111

Lipnick RL, Muir DCG (2001) History of persistent, bioaccumulative and toxic chemicals (Chap. 1). ACS Symposium Series 772, pp 1–12

Liu J (2010) Phenylurea herbicides (Chap. 80). In: Krieger R (ed) Hayes' handbook of pesticide toxicology, 3rd edn., pp 1725–1731. ISBN 978-0-12-374367-1

MacMillan HG, Christensen A (1927) A study of potato seed treatment for rhizoctonia control. University of Wyoming, Agricultural Experiment Station, Bulletin No. 152

MAFF Canada (2017) Agronomy guide for field crops, Pub. 811, Section 13. In: Brown C (ed) Weed control, crop yield losses due to weeds, impact of soil moisture on weed competitiveness. MAFF, Ontario, Canada. ISBN 978-1-4606-9021-5

Makokha D, Irakiza R, Malombe I, Le Bourgeois T, Rodenburg J (2016) Dualistic roles and management of non-cultivated plants in lowland rice systems of East Africa, South African. J. Bot. 108:321–330

Market Research.com (2015) Global Pyrethrin Market 2015–2019. https://www.marketresearch.com/Infiniti-Research-Limited-v2680/Global-Pyrethrin-9358950/

Marlatt CL (1894) Important insecticides: directions for their preparation and use. United States Department of Agriculture, Farmers' Bulletin No. 19

Marlatt CL (1903) Important insecticides: directions for their preparation and use. United States Department of Agriculture, Farmers' Bulletin No. 127

Matson PA, Parton WJ, Power AG, Swift MJ (1997) Agricultural intensification and ecosystem properties. Science 227:504–509

McCallan SEA (1930) Studies on fungicides, III the solvent action of spore excretions and other agencies on protective copper fungicides, Cornell experiment station, Ithaca, New York. Memoir 128:25–79

McDougall I, Brown FH, Fleagle JG (2005) Stratigraphic placement and age of modern humans from Kibish, Ethiopia. Nature 433:733–736

McIndoo NE (1943) Insecticidal uses of nicotine and tobacco: a condensed summary of the literature 1690–1934. United States Department of Agriculture, Agricultural Research Administration, Bureau of Entomology and Quarantine, E597, Washington, May 1943

McIndoo NE, Sievers AF (1917) Quassia extract as a contact insecticide. J Agric Res 10:497–531

McLaughlin Gormley King Company (2010) Pyrethrum: nature's insecticide, about pyrethrum. https://pyrethrum.com/About_Pyrethrum/History.html

Melber J, Kielhorn J, Mangelsdorf I (2004) Coal tar creosote. Concise International Chemical Assessment Document 62

Meyer-Thurow G (1982) The industrialization of invention: a case study from the German chemical industry. Isis 73:363–381

Michelbacher AE, Bacon OG (1952) Spider mites on walnuts. Calif Agric 6:4–15

Michelbacher AE, Middlekauf WW, Akesson NB (1950) Tomato insect studies: DDD and DDT in three year tests with chlorinated hydrocarbons. Calif Agric 4:11–12

Mishke T, Brunetti K, Acosta V, Weaver D, Brown M (1985) Agricultural sources of DDT Residues in California's Environment. A Report Prepared in Response to House Resolution N[o.] 53 (1984), California Dept. of Food and Agriculture, Sacramento, California, USA

Mitchell PD (2011) Economic assessment of the benefits of Chloro-s-Triazine Herbicides to U.S. Corn, Sorghum and Sugarcane Producers, University of Madison-Wisconsin, Dept. of Agriculture and Applied Economics, Staff Paper No. 564

Mitchell PD (2014) Market level assessment of the economic benefits of atrazine in the United States. Pest Manag Sci 70(11):1684–1696

Moechnig MJ, Stoltenberg DE, Boerboom CM, Binnir LK (2013) Variations in corn yield losses due to weed competition. University of Wisconsin-Extension, Madison, USA. https://extension.soils.wisc.edu/wcmc/variations-in-corn-yield-losses-due-to-weed-competition/

Monaco TJ, Weller SC, Ashton FM (2002) Weed science: principles and practices. In: Introduction to weed science, weed impacts, 4th edn. Wiley, New York, pp 6–9. ISBN 0-471-37051-7

Morgan ED (2007) Azadirachtin, a scientific gold mine. Bioorg Med Chem 17:4096–4105

Morone JG, Woodhouse EJ (1986) Averting catastrophe: strategies for regulating risky technologies. In: Toxic chemicals. University of California Press, Berkeley

Moss S (2013) Black-grass (*Alopecurus myosuroides*), Rothamsted Technical Publication—revised 2013, Rothamsted Research, Rothamsted, UK

Moss SR, Storkey J, Cussans JW, Perryman SAM, Hewitt MV (2004) The Broadbalk long-term experiment at rothamsted: what has it told us about weeds? Weed Sci 52:864–873

Movahedpour F, Nassab ADM, Shakiba MR, Amini S, Aharizad S (2013) Weed interference on soybean performance by using integrated weed management and empirical model. Int Res J Appl Basic Sci 4:118–124

Muir P (2012) History of pesticide use: DDT case study. Agricultural Pesticides, Human Impact on Ecosystems, Oregon State University

Mukula J, Ruuttunen E (1969) Chemical weed control in Finland in 1887–1965. Seria Agricultura No. 31, Supp 1, Ann Agric Fenniae 8:1–43

Müller U (2002) Chemical crop protection research. Methods and challenges. Pure Appl Chem 74:2241–2246

Mummert A, Esche E, Robinson J, Armelagos GJ (2011) Structure and robusticity during the agricultural transition: evidence from the bioarchaeological record. Econ Human Biol 9:284–301

National Academy of Sciences (1968) Classification and chemistry of herbicides (Chap. 10). In: Principles of plant and animal pest control, vol 2. Weed Control, Pub. No. 1597, pp 164–165

National Academy of Sciences (2000) History of pest control (Chap. 1). In: The future role of pesticides in US Agriculture. National Academy Press, Washington, D.C., USA, pp 20–32. ISBN 0-390-06526-7

Nauen R, Bretschneider T (2002) New modes of action of insecticides. Pest Outlook 241–245

Nietzki R (1892) Chemistry of the organic dyestuffs. Pub. Gurney & Jackson, London

Noldin JA, Chandler JM, Mccauley GN, Sij JW Jr (1998) Red rice (*Oryza sativa*) and *Echinochloa* spp. control in texas gulf coast soybean (Glycine max). Weed Technol 12:677–683

Oerke E-C (2006) Crop losses to pests. J Agric Sci 144:31–43

Ogbuewu IP, Odoemenam VU, Obikaonu HO, Opara MN, Emenaiom OO, Uchegbu MC, Okoli IC, Esonu BO, Iloeje MU (2011) The growing importance of neem (*Azadirachta indica* A Juss) in agriculture, industry, medicine and environment: a review. Res J Med Plants 5:230–245

Ohta H (2013) Historical development of pesticides in Japan, Center of the History of Japanese Technology, Survey Reports on the Systemization of Technologies, No. 18, March 2013, National Museum of Nature and Science of Japan

Olsson O (2001) The rise of neolithic agriculture. Working Paper in Economics No. 57, Dept. of Economics, Gothenburg University, Gothenburg, Sweden

Orson J, Thomas M, Kudsk P (2001) Impact of generic herbicides on current and future weed problems and crop management. Proc BCPC Conf Weeds 1:123–132

Ortega JC (1948) Codling moth in walnuts: Southern California studies of varying methods of DDT application. Calif Agric 2:4–14

Pammel LH (1911) Weeds of the farm and garden. Pub. Kegan Paul, Trench, Trübner & Co. Ltd., London

Pan L, Feng X, Zhang H (2017) Dissipation and residues of pyrethrins in leaf lettuce under greenhouse and open field conditions. Int J Environ Res Public Health 14:Art.822

Papworth DS (1958) Practical experience with the control of ants in Britain. Ann Appl Biol 46:106–111

Pascall-Villalobos MJ, Miras AR (1999) Anti-insect activity of plant extracts from the wild flora in Southeastern Spain. Biochem Syst Ecol 27:1–10

Perkins JH (1982) A new technology: the introduction of insecticides (Chap. 1). In: Insects, experts and the insecticide crisis. Plenum Press, New York. ISBN-13: 978-1-4684-4000-3

Peryea FJ (1998) Historical use of lead arsenate insecticides, resulting soil contamination and implications for soil remediation. In: Proceedings, 16th world congress of soil science, Montpellier, France, 20–26

Peterson GE (1967) The discovery and development of 2,4-D. Agric Hist 41:243–254

Peterson MA, McMaster SA, Riechers DE, Skelton J, Stahlam PW (2016) 2,4-D past, present and future: a review. Weed Technol 30:303–345

Petruzzello M (2019) 18 food crops developed in the Americas, Encyclopaedia Britannica. https://www.britannica.com/list/18-food-crops-developed-in-the-americas

Pharaonic Egypt (2000) Pests in ancient Egypt: extermination and remedies, insects, rodents, birds. http://webcache.googleusercontent.com/search?q=cache, http://www.reshafim.org.il/ad/egypt/timelines/topics/pests.htm

Phillips WM (1961) Control of field bindweed by cultural and chemical methods. Agricultural Research Service, U.S. Department of Agriculture, Technical Bulletin No. 1249, Washington D.C., USA

Pierce NB (1900) Peach leaf curl: its nature and treatment. US Department of Agriculture, Department of Vegetable Physiology and Pathology, Bulletin No. 20

Pimental D (2009) Pesticides and pest control. In: Rajinder P, Dhawan A (eds) Integrated pest management: innovation-development process, vol 1. Springer, Netherlands, pp 83–87. ISBN 978-1-4020-8992-3

Pimental D, McLaughlin L, Zepp A, Lakitan B, Kraus T, Kleinman P, Vancini F, Roach WJ, Graap E, Keeton WS, Selig G (1993) Environmental and economic effects of reducing pesticide use in agriculture. Agric Ecosyst Environ 46:273–288

Pokorny R (1941) New compounds, some chlorophenoxyacetic acids. J Am Chem Soc 63:1768

Potter MF (2008) The history of bed bug management. Pest Control Technology Magazine, Aug 2008

PPDB (2018) University of Hertfordshire, Pesticides Properties Database, Dicofol. https://sitem.herts.ac.uk/aeru/ppdb/en/Reports/223.htm

Price TD, Bar-Yosef O (2011) The origins of agriculture: new data, new ideas, an introduction to supplement 4. Curr Anthropol 52(Suppl 4):S163–S174

Prostko EP, Johnson WC III, Millinix BG Jr (2001) Annual grass control with preplant incorporated and pre-emergence applications of ethalfluralin and pendimethalin in Peanut (*Arachis hypogaea*). Weed Technol 15:36–41

Raman VK, Radcliffe EB (1992) Pest biology, damage and distribution in the potato crop (Chap. 11). In: Harris PM (ed) Part 2, Insect Pests, vol 1. Springer-Science + Business Media, B.V., p 485

Reilly D (1951) Salts, acids and alkalis in the 19th century: a comparison between advances in France, England and Germany. Isis 42:287–296

Reinhardt C, Ganzel W (2003) Farming in the 1930s, Backlash—100,000 Guinea Pigs & the FDA, Wessels Living History Farm, York, Nebraska, USA

Reynolds JL (1915)The fruit industry, the benefits of spraying. Colonist 77:2

Food and Agricultural Materials Inspection Centre (FAMIC) Japan, What are Agricultural Chemicals? The History of Agricultural Chemicals. https://www.acis.famic.go.jp/eng/chishiki/01.htm

RIAS Inc. (2006) Assessment of the economic and related benefits to Canada of phenoxy. Herbicides, RIAS Inc., Toronto

Richardson HD (1847) The pests of the farm, with instructions for their extirpation. Pub. J. McGlashan, Dublin, Eire

Richardson HD (1852) The pests of the farm which annoy the American farmer with directions for their destruction. Pub. A.O. Moore & Co., New York

Richardson HW (1997) Copper fungicides/bactericides (Chap. 5). In: Handbook of copper compounds and applications. CRC Press, Boca Raton. ISBN 9781482277463

Richerson PJ, Boyd R, Bettinger RL (2001) Was agriculture impossible during the pleistocene but mandatory during the holocene? A climate change hypothesis. Am Antiq 66:387–411

Roarck RC (1951) A Digest of information on Chlordane, E817, Agricultural Research Administration, Bureau of Entomology and Plant Quarantine, United States Department of Agriculture

Roberts JW (1936a) Recent developments in fungicides: spray materials. Bot Rev 2:586–600

Roberts JW (1936b) Recent developments in fungicides II: spray materials 1936–1944. Bot Rev 12:538–547

Roberts JR, Reigart JR (2016) N-methyl carbamates, recognition and management of pesticide poisonings, 6th edn, Section II—Insecticides (Chap. 6), US EPA, Washington D.C., USA, p 56

Robertshaw P (1998) Medieval household pest control, the medieval world view and *Ars Magica*, Natural History of the Middle Ages. http://www.granta.demon.co.uk/arsm/jg/pest.html

Royal Society of Chemistry (2019) Periodic table, arsenic, history. http://www.rsc.org/periodic-table/element/33/arsenic

Rudrappa T, Bonsall J, Bais HP (2007) Root-secreted allelochemical in the noxious weed *Phragmites australis* deploys a reactive Oxygen species response and microtubule assembly disruption to execute rhizotoxicity. J Chem Ecol 33:1898–1918

Ruehle GD (1944) New fungicides for potatoes and tomatoes. Proc Florida State Hort Soc 57:201–206

Russell PE (2006) The development of commercial disease control. Plant Pathol 55:585–594

Saleh MA (1983) Capillary gas chromatography-electron impact and chemical ionization mass spectrometry of toxaphene. J Agric Food Chem 31:748–751

Sanguankeo PP, Leon RG, Malone J (2009) Impact of weed management practices on grapevine growth and yield components. Weed Sci 57:103–107

Sauls JW (1999) False spider mite damage, texas citrus & subtropical fruits. Horticultural Sciences Department, Texas A&M University, Texas 77843–2133, USA

Schooley T, Weaven MJ, Mullins D, Eick M (2008) The history of lead arsenate use in apple production: comparison of its impact in Virginia with other states. J Pestic Saf Educ 10:22–53

Scursoni JA, Martín A, Catanzaro MP, Quiroga J, Goldar F (2011) Evaluation of post-emergence herbicides for the control of wild oat (*Avena fatua L.*) in Wheat and Barley in Argentina. Crop Prot 30:18–23

Secoy DM, Smith AE (1977) Superstition and social practices against agricultural pests. Environ Rev 2:2–18

Secoy DM, Smith AE (1983) Lineage of lime sulfur as an insecticide and fungicide. Bull Entomol Soc Am 29:18–23

Semple EC (1928) Ancient mediterranean history: part I. Agric Hist 2:61–98

Sharma V, Malik AK, Aulakh JS (2003) Thiram: degradation, application and analytical methods. J Environ Monit 5:17–23

Shaw H (1946) Some uses of DDT in agriculture. Nature 157:285–287

Shearer PW (2001) Management of San Jose Scale for Orchard Crops, Mid-Columbia Agricultural Research & Extension Center, University of Oregon, Oregon, USA. http://extension.oregonstate.edu/umatilla/mf/sites/default/files/Peter_Shearer_OR_Blue_Mountain_09.pdf

Shiva V, Hollabhar R (2008) Piracy by Patents; The Case of the Neem Tree, Published by: YANAC (Scribd), 19 Nov 2008. http://www.scribd.com/doc/8156919/Piracy-by-Patents-the-Case-of-the-Neem-tree-VANDANA-SHIVA-and-RADHA-HOLLA-BHAR

Shoham J (2013) Quantifying the economic and environmental benefits of paraquat. Outlooks Pest Manag 24:64–69

Smart NA (1968) Use and residues of mercury compounds in agriculture. Residue Rev/Rückstands-Ber 23:1–36

Smith LE (1937) The use of phenothiazine as an insecticide. US Department of Agriculture, Bureau of Entomology and Plant Quarantine, E-399

Smith VL (1975) The primitive hunter culture, pleistocene extinction and the rise of agriculture. J Polit Econ 83:727–756

Smith AE, Secoy DM (1975) Forerunners of pesticides in classical Greece and Rome. J Agric Food Chem 23:1050–1055

Smith AE, Secoy DM (1976a) A compendium of inorganic substances used in European pest control before 1850. J Agric Food Chem 24:1180–1186

Smith AE, Secoy DM (1976b) Salt as a pesticide, manure and seed steep. Agric Hist 50:506–516

Snelson JT (1977) The importance of chlorinated hydrocarbons in world agriculture. Ecotoxicol Environ Saf 1:17–30

Soltani N, Dille JA, Burke IC, Everman W (2016) Potential corn yield losses from weeds in North America. Weed Technol 30:979–984

Soltani N, Dille JA, Burke IC, Everman W (2017) Perspectives on potential soybean yield losses from weeds in North America. Weed Technol 31:1–7

Sourcewatch (2012) History of roundup ready crops, sourcewatch. Center for Media and Democracy

Stoller EW, Wax LM, Slife FW (1979) Yellow Nutsedge (*Cyperus esculentus*) competition and control in corn (*Zea mays*). Weed Sci 27:32–37

Streibig JC (2003) Assessment of herbicide effects, short history of chemical control. European Weed Research Society, Education and Training, pp 4–7

Syngenta (2016) Reglone® desiccant herbicide. Syngenta Canada Inc., Guelph, Ontario, Canada

Tabashnik BE, van Rensburg JB, Carrière Y (2009) Field-evolved insect resistance to Bt crops—definition, theory and data. J Econ Entomol 102:2011

Texas A&M (2019) Landscape IPM, Botanical Insecticides, Texas A&M University—Department of Entomology, College Station, Texas, USA

Thacker JRM (2002) An introduction to arthropod pest control (Chap. 1). In: A brief history of arthropod pest control. Cambridge University Press, Cambridge. ISBN 0-521-56106X

Thompson NF, Robbins WW (1926) Methods of eradicating the common barberry (*Berberis Vulgaris* L.), Bulletin No. 1451, United States Department of Agriculture, Washington D.C., USA

Timmons FL (1941) Results of bindweed control experiments at the fort hays branch station, Hays, Kansas, 1935 to 1940. Bulletin 296, Kansas State College of Agriculture and Applied Science, Manhattan, Kansas, USA

Timmons FL (1970) A history of weed control in the United States and Canada. Weed Sci 18:294–307

Tisdale WH, Williams I (1934) Disinfectant, United States Patent Office, US Patent 1,972,961. http://www.freepatentsonline.com/1972961.pdf

Torkey HM, Abou-Yousef HM, Abdel Azeiz AZ, Farid HEA (2009) Insecticidal Effect of Cucurbitacin E Glycoside Isolated from *Citrullus colocynthis* against *Aphis craccivora*. Aust J Basic Appl Sci 3:4060–4066

Tozzi A (1998) A brief history of the development of piperonyl butoxide as an insecticide synergist (Chap. 1). In: Jones DG (ed) Piperonyl butoxide: the insecticide synergist. Academic Press. ISBN 0-12-286975-3

Travis AS (1990) Perkin's mauve: ancestor of the organic chemical industry. Technol Cult 31:51–82

Troyer JR (2001) In the beginning: the multiple discovery of the first hormone herbicides. Weed Sci 49:290–297

Turner RS (2005) After the Famine: Plant Pathology, *Phytophthora infestans*, and the Late Blight of Potatoes, 1845–1960. Hist Stud Phys Biol Sci 35:341–370

Tweedy BG (1981) Inorganic sulfur as a fungicide. Res Rev 78:43–68

Upawansa GK, Wagachchi R (2018) Activating all powers in Sri Lankan Agriculture, Living Heritage Trust, Sri Lanka. http://goviya.com/activating-powers.htm

US EPA (2003) United States Environmental Protection Agency, Reregistration Eligibility Decision (RED) for Diuron, 30 Sept, pp 85–91

US EPA (2004) Interim reregistration eligibility decision for carbaryl. United States EPA, Washington D.C., USA

US EPA (2006a) United States environmental protection agency, Interim Reregistration Eligibility Decision (IRED) for Atrazine, 6 Apr, pp 79–84

US EPA (2006b) United States environmental protection agency, Reregistration Eligibility Decision (RED) for Simazine, 6 Apr, pp 46–49

US EPA (2015) US Environmental Protection Agency, DDT—a brief history and status, Development of DDT, US EPA, 5 Nov

US Department of Health and Human Services, Agency for Toxic Substances and Disease Registry (2002) Toxicological Profile for Methoxychlor, Production, Chapter 5 Import/Export, Use and Disposal, pp 143–146

Vaughn KC, Lehnen LP Jr (1991) Mitotic disrupter herbicides. Weed Sci 39:450–457

Veitch GE, Beckmann E, Burke BJ, Boyer A, Maslen SL, Ley SV (2007) Synthesis of azadirachtin: a long but successful journey. Angew Chem Int Ed 46:7629–7632

Verma LR (1998) Indigenous technology knowledge for watershed management in Upper North-West Himalayas of India. Negri A, Sharma PN (eds) PWMTA Field Document No.15, FAO Document Repository

Vivian DL, Acree F Jr (1941) A second list of organic sulfur compounds used as insecticides. US Bureau of Entomology and Plant Quarantine, Division of Insecticide Investigations, E-539

Wardle RA, Buckle P (1923) Principles of insect control (Chap. VI). In: Insecticides. University Press, Manchester, p 81

Ware GW, Whiteacre DM (2004) An introduction to insecticides, 4th edn. MeisterPro Information Resources, Ohio, USA, pp 61–69

Watson J (2004) The significance of Mr. Richard Buckley's exploding trousers: reflections on an aspect of technological change in New Zealand dairy farming between the world wars. Agric Hist 78:346–360

Weber GF (1931) Spraying and dusting cucumbers for control of downy mildew from 1925–1930. University of Florida, Agricultural Experiment Station, Bulletin No. 230

Weisdorf J (2005) From foraging to farming, explaining the neolithic revolution. J Econ Surv 19:561–586

Weisdorf J (2009) Why did the first farmers toil? Human metabolism and the origins of agriculture. Eur Rev Econ Hist 13:157–172

Weiss E, Zohary D (2011) The neolithic southwest asian founder crops: their biology and archaeobotany. Current Anthropol 52(S4):S237–S254

Westgate WA, Raynor RN (1940) A new selective spray for the control of certain weeds. University of California, College of Agriculture, Agricultural Experiment Station, Berkeley, California, USA, Bulletin 634

White RP (1928) Potato experiments for the control of rhizoctonia, scab and blackleg: 1922–1927. Technical Bulletin No.24, Kansas State Agricultural College, Kansas, USA

Williams JS, Cooper RM (2004) The oldest fungicide and newest phytoalexin—a reappraisal of the fungitoxicity of elemental sulphur. Plant Pathol 53:263–279

Williams MM, Thoreby E, Mergel M (2011) Trifluralin in toxipedia (Connecting Science and People)

Wink M (1994) Biological activities and potential application of Lupin Alkaloids, Advances in Lupin Research, Neves-Martin JM, Beirao da Costa ML (eds). ISA Press, Lisbon

Wink M, Meisner C, Witte L (1995) Patterns of quinolizidine alkaloids in 56 species of the genus *Lupinus*. Phytochemistry 38:139–153

Woolman HM, Humphrey HH (1924) Summary of the literature on bunt, or stinking smut, of wheat. United States Dept. of Agriculture, Dept. Bulletin No. 1210

World Bank (2018) World Bank Data, Cereal Yield 2010–2014. https://data.worldbank.org/indicator/AG.YLD.CREL.KG

Zadoks JC (2013) Crop protection in medieval agriculture. Sidestone Press, Leiden. ISBN 978-90-8890-187-4

Zimdahl RL (1969) The etymology of herbicide. Weed Sci 17:137–139

Zimdahl RL (2010) Development of herbicides after 1945 (Chap. 6). In: A history of weed science in the United States. Elsevier B.V., Amsterdam. ISBN 978-0-12-381495-1

Zimdahl RL (2015) DDT: an insecticide (Chap. 7). In: Six chemicals that changed agriculture. Academic Press, Elsevier B.V., Amsterdam, pp 115–133

Zucconi LM (2007) Medicine and religion in ancient Egypt. Relig Compass 1:26–37

Chapter 3
Semiochemicals for Integrated Pest Management

Maria C. Blassioli-Moraes, Raúl A. Laumann, Mirian F. F. Michereff and Miguel Borges

Abstract In Brazil, implementation of integrated pest management in the mid-1970s until the mid-1990s allowed to develop one of the most robust tropical agriculture systems in the world. However, at the beginning of this century, the intensification of the no-tillage cultivation system combined with multiple crops cultivated in a rotation system provided food and hosts for insects throughout the year. These two factors have been responsible for provoking pest outbreaks. In order to overcome these pest outbreaks, farmers started applying huge amounts of pesticides to arable crops. The excess of pesticides, climate changes and more restrictive laws concerning insecticide use combining with the high costs of developing new synthetic molecules, and taking into account the increase in the world's population, have put pressure on all food production sectors to develop more sustainable tools for controlling pests. In this aspect, in the last years, scientists have put effort to develop new technologies based on semiochemicals aiming to provide more sustainable, with less cost pest control methods to farmers. In this chapter, the principles of semiochemical use for monitoring and controlling pests as well as the way in which these natural molecules work are presented and discussed.

Keywords Pheromones · Natural enemies · Sustainable agriculture · Control pests · Arable crops

3.1 Semiochemicals: Definition and Uses

3.1.1 Introduction

During the 1960s, the Green Revolution led to agricultural intensification, which allowed a significant growth in food production. For instance, the cereal production in 1961 was 8.7 million tonnes per year and in 2016 was around 2.7 billion tonnes

M. C. Blassioli-Moraes (✉) · R. A. Laumann · M. F. F. Michereff · M. Borges
Brazilian Agricultural Research Corporation, National Research Center for Genetic Resources and Biotechnology—Laboratory for Semiochemicals, Parque Estação Biológica, W5 Asa Norte, Brasilia-DF CEP 70770-917, Brazil
e-mail: carolina.blassioli@embrapa.br

© Springer Nature Switzerland AG 2019
S. Vaz Jr. (ed.), *Sustainable Agrochemistry*,
https://doi.org/10.1007/978-3-030-17891-8_3

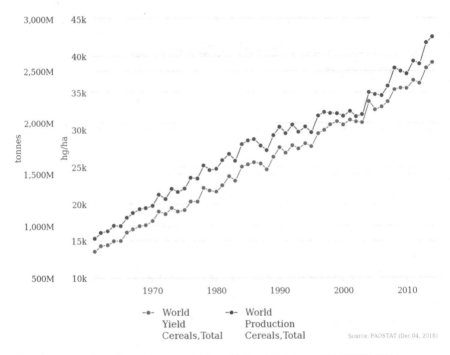

Fig. 3.1 Increase in total cereal production from 1961 to 2016. *Source* FAOSTAT (2018)

(Fig. 3.1). The Green Revolution enhanced food production that allowed the world's population to be fed but with costs. The huge areas for growing arable crops with high productivity involved higher uses of fertilizers, pesticides and water, which presented negative effects to the environment such as pesticides and fertilizer residues that can negatively affect non-target organisms, causing water depletion and/or pollution.

In the last century in Brazil, the use of pesticides was very high at the beginning of the 1960s, but in the mid-1970s, the implementation of integrated pest management (IPM) approach using different tools, such as baculovirus for Lepidoptera larvae and egg parasitoids for stink bugs, led to a 50% reduction in pesticide application (Moscardi et al. 1999; Panizzi 2013; Bortolotto et al. 2015). However, since the start of this century, the intensification in use of the no-tillage cultivation system combined with multiple crops cultivated in a rotation system, provides food and hosts for insects all year. These two factors are believed to be responsible for inducing pest outbreaks, including pests that were previously considered secondary pests, such as stink bugs *Dichelops melacanthus* (Hemiptera: Pentatomidae), *Piezodorus guildinii* (Hemiptera: Pentatomidae) and *Euschistus heros* (Hemiptera: Pentatomidae). These pests are found mainly in soybean but also in cotton and maize (Moscardi et al. 1999; Panizzi 2013; Bortolotto et al. 2015). In order to overcome this pest's outbreaks, the farmers returned to applying huge amounts of pesticides to arable crops (Panizzi 2013; Bortolotto et al. 2015; Bombardi 2017). A recent study (Aquino et al. 2018) described a survey about the stink bug community and its natural enemies in nine different soybean regions in Brazil. After comparing this study's data with data from the 1990s, the authors concluded that the very low level of parasitoid and predator communities observed nowadays were due to overuse of pesticide application, which had replaced the IPM used in the 1990s.

Therefore, new methods for monitoring and controlling pests, which should be sustainable and have less negative impact on the environment, are necessary for farmers' use. In this context, semiochemicals use have become a key IPM approach for agricultural systems.

3.1.2   Semiochemicals: Definitions

Communication among living beings involves exchange of information between an organism that emits a signal (emitter/transmitter) and the one that receives it (recipient) and then shows a physiological and/or behavioural change. The signal can be sonorous, visual or chemical. Chemical communication in insects and plants probably is the most important signal used to locate partners for mating, hosts, food, perception of dangerous situations and repelling enemies. Chemical signals have some advantages compared to visual and vibrational signals; these types of signals can be transmitted in the dark, are independent of visual contact, can be perceived from distances ranging from short to long distances and/or transmitted after contact (Price 1997; Bradbury and Vehrencamp 2011).

Chemical communication between organisms is mediated by chemical signals called semiochemicals, which can act in intra- or inter-specific interactions. When communication through chemical compounds occurs within the same species, the semiochemical is called pheromone (Nordlund and Lewis 1976; Borges and Blassioli-Moraes 2017; Weber et al. 2018). Pheromones are divided into different classes depending on how they work: (1) trail; (2) oviposition; (3) alarm; (4) marking; (5) aggregation; and (6) sex. When the semiochemicals are related to heterospecific interactions, they are called allelochemicals, and these are classified as allomones, kairomones and synomones depending on the organisms that receive and benefit from this interaction. The convergent ladybird beetle, *Hippodamia convergens* (Coleoptera: Coccinellidae) produces the compound alkylmethoxypyrazine that presents a negative effect on predators and functions as an allomone (Wheeler and Cardé 2014). Other examples of allomones are soybean plants with high amino acids levels that cause a delay in the development of the aphid *Aphis glycines* (Hemiptera: Aphididae) (Chiozza et al. 2010). Kairomones, which are beneficial to the recipient and have a negative effect on the emitting organism, can be exemplified by volatile chemical release from plants that attract herbivores as observed in chemical communication between cotton plants and the boll weevil, *Anthonomus grandis* (Coleoptera: Curculionidae) (Magalhães et al. 2012, 2016). Other examples of kairomones are the compounds released by host eggs that attract egg parasitoids as observed in the association between *E. heros*' eggs and the egg parasitoid *Telenomus*

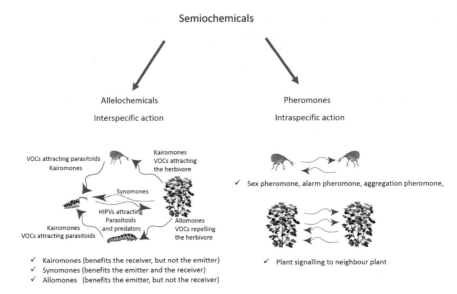

Fig. 3.2 Semiochemical classification based on interactions with plants and insects. VOCs = volatiles organic compounds; HIPV = herbivore-induced plants volatiles

podisi (Hymenoptera: Platygastridae) (Tognon et al. 2014; Michereff et al. 2016). The synomones that benefit both transmitter and recipient can be observed in tri-trophic interactions. Herbivore-inflicted injuries to plants can induce release of volatile compounds, which are used by their natural enemies as a cue for locating their prey/hosts, so the plant benefits from herbivore elimination and the natural enemies' capability for finding their host (Blassioli-Moraes et al. 2013; Michereff et al. 2011, 2018; Dias et al. 2016) (Fig. 3.2).

Semiochemical classification can change, depending on the interaction involved and their modes of action; for example, the compound (*E*)-β-farnesene (Eβf), released by a transgenic plant of *Arabidopsis thaliana* (Brassicales: Brassicaceae), repels the herbivore *Myzus persicae* (Hemiptera:Aphididae), acts as an allomone and attracts the parasitoid *Diaeretiella rapae* (Hymenoptera: Braconidae) by acting as a synomone (Beale et al. 2006). *Euschistus heros*' sex pheromone, methyl-2,6,10-trimethyltridecanoate, attracts conspecific females for mating, but it is also perceived by the egg parasitoid *T. podisi* and *Trissolcus urichi* (Hymenoptera: Platygastridae) then acting as a kairomone (Borges et al. 2011). Maize plants produce enhanced indole levels when injured by *S. frugiperda* (Lepidoptera: Noctuidae) larvae, and the higher indole levels repel (allomone) the parasitoid *Microplitis rufiventris* (Hymenoptera: Braconidae) (D'Alessandro et al. 2006), but also induce priming defence mechanisms in neighbouring maize plants (alarm pheromone) (Erb et al. 2015).

3.1.2.1 An Overview of Insect Pheromones

There are recent reviews about chemical communication in stink bugs (Borges and Blassioli-Moraes 2017; Weber et al. 2018) and about the chemical communication in insects associated with agricultural plants (Schulz 2005).

Pheromones are classified depending on their effects on insect behaviour: (1) trail pheromones are used by social insects to signal to conspecifics about food and shelter, indicating the size of a tunnel to be excavated and recruiting workers to perform this task; (2) oviposition pheromones have been observed in hematophagous mosquitoes and have the function of attracting pregnant females to oviposition in an environment already selected by individuals of the same species, increasing the number of larvae, and reducing predation risk; (3) alarm pheromones provide an alarm signal as to the presence of a predator, parasitoid, or a threat by eliciting two distinct behaviours, including recruitment of more individuals to the site of the disturbance with a subsequent attack on the aggressor as observed in bees, ants and termites or escape through the rapid dispersion of insects, which occurs in aphids, stink bugs, and Coleoptera; (4) marking pheromones are found in ants and bees and are used to mark the territory where the nest is located, indicating the entry of the nest and differentiating the nests of different colonies; in fruit flies and parasitoids these pheromones are used to indicate that the site is already being colonized, thus preventing competition between individuals; (5) aggregation pheromones are produced by males or females, act in attraction of both genders, and could favour protection, reproduction, and feeding in addition to the combination of all of these factors for the species that use it; and (6) sex pheromones are defined as the odour produced by a male or female that stimulates the change in the opposite sex's behaviour, leading to attraction and closeness of both for mating proposes (Borges et al. 1987; Cook et al. 2007; Weber et al. 2018).

In terms of IPM, sex and aggregation pheromones are the most used for controlling and monitoring pests in the fields (Witzgall et al. 2010; Borges et al. 2011).

3.1.2.2 Sex and Aggregation Pheromones

Sex and aggregation pheromones are small organic molecules with a C_8–C_{26} linear chain, saturated or containing double bonds. They can present a branched chain and can have different chemical functional groups such as esters, aldehydes, ketones, epoxides and lactones. In general, sex pheromones in Lepidoptera are produced by females and consist of at least two components. Most parts of the molecules can be divided into two major groups: (1) compounds of group I present linear carbon chains ranging from C_{10} to C_{18} with terminal functional groups (alcohols, acetates, or aldehydes) and double bonds that range from 0 to 3; and (2) compounds from group II have longer carbon chains from C_{17} to C_{23}, and they may or may not present double bonds and/or their epoxide forms. The specificity of the pheromone blends is guaranteed by multicomponent blends, different lengths of the carbon chain and numbers and positions of the double bonds.

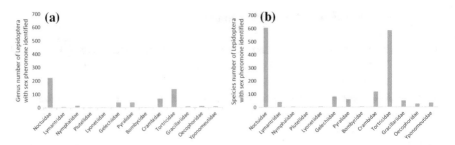

Fig. 3.3 Number of Lepidoptera **a** genus and **b** species with identified sex pheromones from the main families related to agricultural plants

Due to the economic importance, sex pheromones from several Lepidoptera species have been identified. A search in the Pherobase (2018) indicates that 1617 species have had their sex pheromones elucidated. The families Noctuidae and Tortricidae are the most studied (Fig. 3.3a, b). A survey in the Pherobase website shows that there are at least 228 pheromones on the market to be used for mass control or mating disruption. In Brazil, there are still a few moth pheromones on the market that are available to the farmers; only the pheromones of *Spodoptera frugiperda*, *Anticarsia gemmatalis* (Lepidoptera: Noctuidae), *Grapholita molesta* (Lepidoptera:Tortricidae), *Neoleucinodes elegantalis* (Lepidoptera:Crambidae), *Leucoptera coffeella* (Lepidoptera:Lyonetiidae), *Bonagota cranaodes* (Lepidoptera:Tortricidae), *Cydia pomonella* (Lepidoptera:Tortricidae), *Tuta absoluta* (Lepidoptera:Gelechiidae), *Ephestia cautela* (Lepidoptera:Pyralidae), *Ephestia elutella* (Lepidoptera:Pyralidae), *Plodia interpunctella* (Lepidoptera:Pyralidae), *Pseudoplusia inclundes* (Lepidoptera:Noctuidae) and *Helicoverpa armigera* (Lepidoptera:Noctuidae) can be acquired on the Brazilian market.

Several Lepidoptera species such as *Bombyx mori* (Lepidoptera:Bombycidae), the silkworm moth, produce pheromones from Group I. This moth was the first to have its pheromone identified by the Germany research Butenandt et al. (1959); the pheromone was called bombikol. In order to identify this molecule, it was used an astonishing number of 500,000 female individuals; nowadays just few insects are necessary for chemical structural elucidation (Butenandt et al. 1959). Bombikol or $(10E,12Z)$-10,12-hexadecadienol presents a linear chain with 16 carbons, *trans* and *cis* double bonds in the C_{10} and C_{12}, and an alcohol functional group on the end of the carbon chain. This compound can be represented in shortened form as $E10,Z12$-16:OH. Other examples of lepidopteran that are important agricultural pests and produce group I pheromones are the complex of *Spodoptera* sp.: *S. frugiperda*, *S. cosmioides* and *S. eridania; H. armigera* and *H. zea, Agrotis* sp. (Lepidoptera: Noctuidae), *Elasmopalpus lignosselus* (Lepidoptera: Pyralidae), *C. pomonella*, and *G. molesta* (Fig. 3.4).

For *S. frugiperda*, a detailed study was conducted with different populations around the world, and the researchers were able to confirm that sex pheromones are species-specific but also present specificity within the same species that are geo-

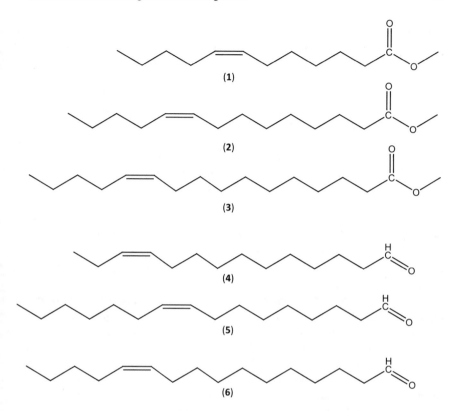

Fig. 3.4 Structures of Lepidoptera pheromone Type I. Compounds **1** (7Z)-7-dodecenyl acetate (Z7-12:OAc), **2** (9Z)-9-tetradecenyl acetate (Z9-14:OAc), **3** (11Z)-11-hexadecenyl acetate (Z11-16:OAc) the three components present in commercial lures to capture *S. frugiperda* in Brazil and Costa Rica. Compounds **4** (11Z)-11-tetradecenal (Z11-14:O), **5** (9Z)-9-hexadecenal (Z9-16:O) and **6** (11Z)-11-hexadecenal (Z11-16:O) are the three pheromone components present in *Helicoverpa armigera* female glands that are attractive to conspecific males

graphically isolated or separated by other factors. *S. frugiperda* maize and rice strain produce a different blend of sex pheromones with the same components but with different ratios between them (Andrade et al. 2000; Groot et al. 2008). Seven different compounds with potentials to work as sex pheromones were identified in virgin female gland extracts of the maize *S. frugiperda* Brazilian strain (Batista-Ferreira et al. 2006). The compounds are (7Z)-7-dodecenyl acetate (Z7-12:OAc), (7E)-7-dodecenyl acetate (E7-12:OAc), (9Z)-9-dodecenyl acetate (Z9-12:OAc), (9Z)-9-tetradecenyl acetate (Z9-14:OAc), (10Z)-10-tetradecenyl acetate (Z10-14:OAc), (11Z)-11-tetradecenyl acetate (Z11-16:OAc), and (11Z)-11-hexadecenyl acetate (Z11-16:OAc) (Batista-Fereira et al. 2006). In this study, the authors reported that blends containing only three components: Z9-14:OAc, Z7-12:OAc, and Z11-16:OAc were efficient to attract *S. frugiperda* males, and E7-12:OAc increases this attraction. The other components did not present synergic effect to the blend. The commercial pheromone of *S. frugiperda* that is available on Brazilian market contains a blend of the three components: (1) Z9-14:OAc; (2) Z7-12:OAc; and (3) Z11-16:OAc. The commercial lure used in Costa Rica for *S. frugiperda* also presents these three components in the same ratio of the Brazilian lure (Andrade et al. 2000) (Fig. 3.4).

A fewer number of moths that are pests to agriculture plants produce pheromones from Group II. The first pheromone of type II identified in 1981 in *Estigmene acrea* (Lepidoptera: Arctiidae) moth was (3Z,6Z)-*cis*-9,10-epoxy-3,6-heneicosadiene (Hill and Roelofs 1981). Another important agricultural pest that produces type II pheromone is *A. gemmatalis*, the soybean pest in Brazil, which produces a sex pheromone with two components, (3Z,6Z,9Z)-3,6,9-eicosatriene (Z3Z6Z9-20:Hy) and (3Z,6Z,9Z)-3,6,9-heneicosatriene (Z3Z6Z9-21:Hy).

Coleoptera is the second order of insects with the largest number of identified sex pheromones. This insect's sex pheromone consists of compounds of terpene origin, generating linear monoterpenes or having cyclic chains and by compounds with one or more functional group (epoxies, alcohols and/or ketones) with short linear carbon chains, derived from small units of acetogenins and propionates, originating from alcohols, ketones, and branched hydrocarbons (Francke and Dettner 2005).

Most of Coleoptera's pheromones that have been identified to date are aggregation pheromones; in this case, the pheromone is produced by one of the genders and attracts both genders, and in some cases also immature stages, but there are examples of both genders producing the sexual pheromone. The males of the cotton pest, *A. grandis*, produce four compounds used as aggregation pheromones: (1) two alcohols *(Z)*-isopropenyl-1-methylcyclobutaneethanol and (Z)-2-(3,3-Dimethyl)-cyclohexylideneethanol and (2) two aldehydes, (Z)-2-(3,3-Dimethyl)-cyclohexylideneethanol and (E)-(3,3-Dimethyl)-cyclohexylideneacetaldehyde (Francke and Dettner 2005) (Fig. 3.5). This pheromone attracts males and females and has been commercialized around the world in order to monitor and control *A. grandis* in cotton crops. The aggregation pheromones of *Cosmopolitus sordidus* (Coleoptera: Curculionidae) and *Sternechus subsignatus* (Coleoptera: Curculionidae) are also terpene derivative, *whereas Rhynchophorus palmarum* (Coleoptera: Curculionidae) produce as pheromone a simple alcohol, the compound (2E,4S)-methyl-2-hepten-4-ol (Fig. 3.5).

Over the last 20 years, Hemiptera, mainly the stink bugs (Pentatomidae), have become severe pests to wheat, soybean, maize, cotton, and other crops, and because of this, studies aiming to understand the chemical communication of these species received higher attention. Forty-four different species of stink bugs had their sex pheromones identified. The first stink bug that had its sex pheromone identified was *Nezara viridula* (Hemiptera: Pentatomidae) from a Neartic population (Pavis and Malosse 1986) and from a Brazilian population in 1987 (Baker et al. 1987; Borges et al. 1987). This species produces sex pheromones consisting of two terpenoid derivatives, *cis* and *trans*-(Z)-bisabolene epoxides (Pavis and Malosse 1986; Baker et al. 1987; Borges et al. 1987). The sex pheromone of the brown soybean stink bug, *E. heros*, was identified by Aldrich et al. (1994), and it consists of three methyl esters: (1) methyl-2,6,10-trimethyldodecanoate; (2) methyl-2,6,10-trimethyltridecanoate; and (3) (2E,4Z)-2,4-methyl-decadienoate. To date, the studies

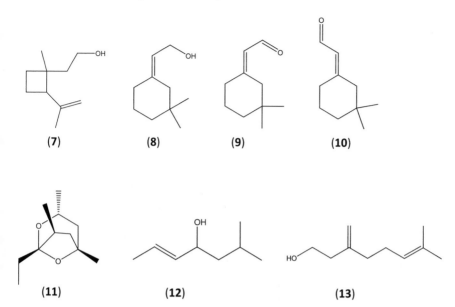

(7) **(8)** **(9)** **(10)**

(11) **(12)** **(13)**

Fig. 3.5 Structures of pheromone compounds for Coleoptera bugs. Compounds **7** *cis*-isopropenyl-1-methylcyclobutaneethanol (grandlure I), **8** (Z)-2-(3,3-dimethyl)-cyclohexylideneethanol (grandlure II), **9** (Z)-2-(3,3-dimethyl)-cyclohexylideneethanol (grandlure III) and **10** (E)-(3,3-dimethyl)-cyclohexylideneacetaldehyde (grandlure IV). The four compounds are the aggregation pheromone of *A. grandis*. Compound **11** is sordidin, (1S,3R,5R,7n)-1-ethyl-3,5,7-trimethyl-2,8-dioxa-biciclo[3.2.1]octane, the aggregation pheromone of *Cosmopolitus sordidus*, **12** is ryncopherol, (2E,4S)-methyl-2-hepten-4-ol, the aggregation pheromone of *Rhynchophorus palmarum* and **13** isogeraniol, 7-methyl-3-methyleneoct-6-en-1-ol, aggregation pheromone of *Sternechus subsignatus*

conducted with these 44 species have shown that the sex pheromone of Pentatomidae is present as two major classes of compounds: (1) terpenoids and/or (2) methyl esters. The stink bug *Piezodorus guildinii* produces as sex pheromone the sesquiterpene 7R-sesquiphellandrene. The rice stink bugs, *Oebalus poecilus* (Hemiptera: Pentatomidae) and *Tibraca limbativentris* (Hemiptera: Pentatomidae), produce different stereoisomers of the sesquiterpene zingiberenol (Oliveira et al. 2013; Blassioli-Moraes et al. 2018). *Chinavia* sp. (Hemiptera: Pentatomidae) and *N. viridula* produce as sex pheromones sesquiterpenes *cis*-(Z)- and *trans*-(Z)-bisabolene epoxides. The specificity of these species is guaranteed by the different ratio between these two components (Blassioli-Moraes et al. 2012) (Fig. 3.6). There are a few stink bugs-derived pheromones on the market; the sex pheromone of the brown stink bug is commercially available in addition to those from other Neartic species such as *Euchistus* sp., *Plautia stali* (Hemiptera: Pentatomidae), *Piezodorus hybneri* (Hemiptera: Pentatomidae), *Halyomorpha halys* (Hemiptera: Pentatomidae) and *N. viridula*.

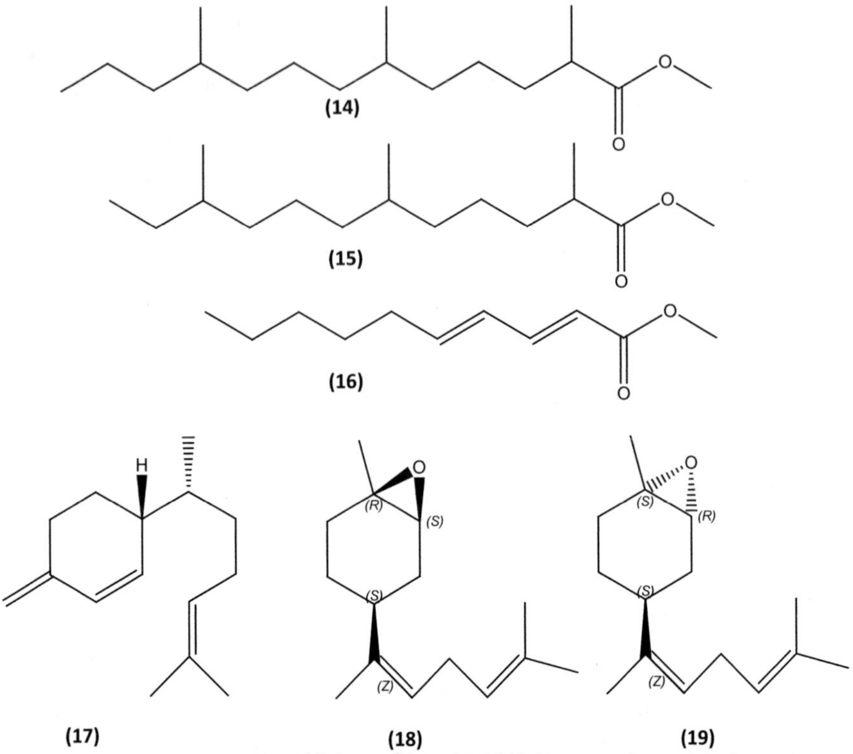

Fig. 3.6 Structures of pheromone compounds for Pentatomidae bugs. Compounds **14** methyl-2,6,10-trimethyltridecanoate, **15** methyl-2,6,10-trimehyldodecanoate, and **16** (2E,4Z)-2,4-methyl-decadienoate are produced by *Euschistus heros* males, and compound **14** is attractive to females. **17** (7R)-sesquiphellandre, the sex pheromone produced by males of *Piezodorus guildinii*, and **18** cis-(Z)- and **19** trans-(Z)-bisabolene epoxides are components of *Chinavia* sp. and *Nezara viridula* sex pheromones; for these species, the pheromone specificity is due to the different ratio between these two compounds

3.1.2.3 Alarm Pheromones

Alarm pheromones, also known as defensive compounds, are produced by different insect orders. They have been identified, and the roles of these chemicals have been demonstrated, mainly in the hymenopteran, coleopteran and hemipteran orders. Alarm pheromones are defined as chemical compounds produced and released by an organism that elicits an alert in other individuals of the same species. In Coleoptera, the alarm pheromone of *Alphitobius diaperinus* (Coleoptera: Tenebrionidae) is composed of three benzoquinones: (1) 1,4 -benzoquinone; (2) 2-methyl-1,4-benzoquinone; and (3) 2-ethyl-1,4-benzoquinone. Recently, Hassemer et al. (2015) showed that a blend of these three compounds has a role as an alarm pheromone, repelling both genders. There is a huge chemical diversity in the compound structures produced by Coleoptera, and these compounds were studied in insects divided into 95 different families (Pherobase). The defensive compounds range from simple amino acid-derived carboxylic acids (such as such as tiglic and crotonic acids) to more complex molecules such as sesquiterpenes, steroids, linear hydrocarbons with double bonds or methyl branches, methyl, ethyl, or propyl esters, quinones derivatives, and others (Francke and Dettner 2005).

Stink bugs are famous because of their smell; this smell is due mainly to the aldehydes ((*E*)-octenal, (*E*)-nonanal and (*E*)-2-decenal) produced and released by these insects when they are disturbed. Nymphs and adults produce species-specific blends of defensive compounds (Pareja et al. 2009). These compounds are produced in the abdominal dorsal glands in nymphs and in the metathoracic glands in adults (Borges and Blassioli-Moraes 2017; Weber et al. 2018). The defensive compounds in Pentatomidae consist of (*E*)-2-alkenals, 4-oxo-(*E*)-2-alkenals and linear hydrocarbons, mostly tridecane and undecane. It has been shown that only genus *Edessa* sp. produced high levels of *n*-undecane. All other genus of stink bugs studied produce and release the hydrocarbon, *n*-tridecane, when disturbed. The function of these hydrocarbons remains to be elucidated. Borges and Aldrich (1992) linked the repellent effects of (*E*)-2-alkenals and 4-oxo-(*E*)-2-alkenals to the nymph stage of different stink bug species.

The alarm pheromone of several species of aphids is sesquiterpene (*E*)-β-farnesene (Eβf) that was first identified in 1973 in *Myzus persicae* (Homoptera: Aphididae) (Edwards et al. 1973). This pheromone is released by different aphid species when they are disturbed and has a series of roles aimed at aphid protection. The released Eβf causes other aphids in the vicinity to stop feeding and move away (Hardie et al. 1999) and also influences the number of aphid progeny that are alate (Kunert et al. 2005), attracts predators (Al Abassi et al. 2000), and increases parasitoids' foraging behaviours (Du et al. 1998). A new approach, which has not yet been successful, is to use genetically modified (GM) plants to produce and emit insect pheromones in order to protect them from herbivores. GM wheat plants were designed to produce and release Eβf with the aim of decreasing the aphid population in wheat plants (Bruce et al. 2015); this will be discussed in more detail in the IPM and Semiochemicals section.

3.1.2.4 Semiochemicals from Plants

Plants are sessile organisms but present a very sophisticated defence system that mainly involves the use of secondary metabolism-originating semiochemicals in order to defend themselves against herbivores, microorganisms, and other biotic stress. It is known that plants produce more than 200,000 secondary metabolites (Pichersky and Gang 2000; Bino et al. 2004). The main volatile organic compounds involved in plant defence are terpenoids, phenylpropanoid/benzenoids compounds and fatty acid derivatives (Dudareva et al. 2013). The plant defence can be constitutive or induced by insect injury and/or microorganism infestation.

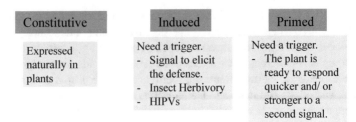

Fig. 3.7 Classification of plant defences

The constitutive defence mechanism generally protects plants from direct insect or microorganism attack; this defence includes morphological and structural characteristics and also constitutive chemicals such as benzoxazinoids and phenolic compounds that can have a negative effect on feeding insect development. In general, the defence is triggered by an external factor such as an herbivore- or microorganism-induced attack; and plants subject to an insect herbivore attack, undergo enhanced production of volatile and non-volatile compounds, which are involved with their defence. These compounds can decrease the herbivore's performance (direct defence) (Bruce and Pickett 2011) or their population after attack (indirect defence) and can attract the natural enemy of the herbivores (Michereff et al. 2011, 2018; Blassioli-Moraes et al. 2016). In addition, herbivore-induced plant volatiles (HIPVs) can act at the first trophic level, protecting the plants against weeds or signalling neighbouring plants within the herbivore's vicinity (Engelberth et al. 2004; Erb et al. 2015; Borges et al. 2017). The induced defence can be activated directly by the stressor (as described above) but also can be primed. In primed defence, the factor stressing consists of a less intense stress, such HIPVs or a short-term herbivore injury that makes the plants activate their molecular defence system, but does not trigger direct phenological changes. In order to induce phenological changes, the primed plants need to undergo a second stress, and it is expected that the plant defence would involve faster, more intense, and earlier activation compared to plants that did not have previous contact with the stress (Engelberth et al. 2004; Erb et al. 2015; Borges et al. 2017) (Fig. 3.7).

Plant semiochemicals related with its defence can act locally, being released in the area in which the injury occurred and protect this area through emission of toxic compounds against the herbivore (Frost et al. 2007). Alternatively, plant semiochemicals can act systemically. Semiochemicals are transported or produced in the non-damaged areas of the plant as observed in lima beans *Phaseolus lunalus* L. (Fabales: Fabaceae) (Heil and Bueno 2007), poplar saplings *Populus deltoids* (Magnoliopsida: Malpighiales) x *Populus nigra* (Magnoliopsida: Malpighiales) (Frost et al. 2008) and the highbush blueberry *Vaccinium corymbosum* L. (Ericales: Ericaceae) (Rodriguez-Saona et al. 2009).

The main volatiles involved in plant-insect and plant-plant communication are terpenoids and compounds that are derived from the shikimic acid pathway. Several agricultural plants, including soybean, maize, cotton, tomato, potato, rice, and wheat, have been reported to enhance their volatile production when damaged by herbivores (Blassioli-Moraes et al. 2013, 2016). Michereff et al. (2011) reported that the sesquiterpene (E,E)-α-farnesene and the ester methyl salicylate are the main compounds produced by soybean plants when undergoing herbivore-induced attack by the stink bug, *E. heros*. These compounds are involved in attraction of the egg parasitoid *T. podisi*, which is the major egg parasitoid of this stink bug (Michereff et al. 2013). The plant's defence responses to herbivores depends on the type of feeding insect (Magalhães et al. 2012) and plant species. Different genotypes can have different responses to the same herbivore. For instance, Michereff et al. (2011) showed that the soybean cultivars, Dowling and IAC 100, had enhanced production of volatile chemicals after herbivory injury provoked by the brown stink bug and these volatiles attracted the egg parasitoid *T. podisi*. Whereas the cultivar Silvania did not show an enhanced volatile production and did not have activation of its indirect plant defence mechanisms. A similar result was obtained when six different maize genotypes were evaluated; only one genotype, Sintetico Spodoptera, showed activation of its indirect defences when herbivory injured by *S. frugiperda* larvae, enhancement of volatile chemical production, and attraction of the egg parasitoid *Telenomus remus* (Hymenoptera: Platygastridae) (Michereff et al. 2018). The maize plants' herbivore-induced damage by *S. frugiperda* larvae was shown to cause production of higher levels of the volatiles compounds such as (E)-β-ocimene, linalool, indole, (E)-4,8-dimethyl-1,3,7-nonatriene (DMNT), (E)-β-farnesene, cyclosativene, and $(E$-$E)$-4,8,12-trimethyltrideca-1,3,7,11-tetraene (TMTT) (Michereff et al. 2018).

The non-volatile plants' semiochemicals have different roles in the nature; they can protect the plants against herbivores or insects can sequester these chemicals when they are feeding in order to protect them against predators. The classical example is the monarch butterfly *Danaus plexippus* (Lepidoptera: Nymphalidae), which feeds on leaves of *Asclepia scurassavica* (Gentianales: Asclepiadaceae) and sequesters and stores cardenolides, which makes them unpalatable to birds (Trigo 2000; Petschenka and Agrawal 2015). The non-volatile plants' semiochemicals are derived from the same biosynthetic pathways as volatiles compounds, which is the melavonate pathway that generates diterpenes, iridoids glycosides and cardenolides. Amino acids provide cyanogenic glycosides, glucosinolates, pyrrolizidine alkaloids, and the shikimic acid pathway provides many aromatic compounds such as benzoxazinoids, furonocouramins, phenolics and others. Several cereals such as wheat, maize and barley produce benzoxazinoids compounds that are stored as non-toxic glycosides, and when the plants suffer herbivore infestation, these compounds undergo hydrolytic release in the aglycone form, which is toxic to insects. The aphid *Rhopalosiphum padi* (Hemiptera: Aphididae) induces the plants to release DIMBOA (2,4-dihydroxy-7-methoxy-1,4-benzoxazin-3-one) when feeding on wheat plants thus increasing the toxicity of the plant against the herbivores (Gordon-Weeks et al. 2010). Hydroxamic acids also are involved in the maize plant defence against a complex of *Spodoptera* sp. (such as *S. exigua* and *S. littoralis*), but *S. frugiperda* is capable of detoxifying these compounds (Wouters et al. 2014). It was shown that benzoxazinoids did not negatively affect its development, and studies have also shown that the presence of these compounds in maize plants stimulates *S. frugiperda* to feeding on the plants (Róstas 2007).

Plants naturally produce a series of chemical compounds that can be used for crop protection. Plant defence mechanisms can be enhanced through conventional breeding or GM techniques, not only for yield and human nutrition, but also for more natural resistance to pests. The plants can have their metabolism altered to produce and release insect sex or alarm pheromones or even produce higher level of compounds from their secondary metabolism that is involved with their defence. In the next sections, we discuss the use of semiochemicals, insect pheromones and GM plants that are engineered to enhance pest defence and the use of companion plants for naturally increasing a plant's defence.

3.2 Integrated Pest Management: Definition and Applicability

3.2.1 Introduction

Since the last half of the last century, IPM has evolved as the principal and more rational strategy for addressing insects and other organisms that threaten human beings' well-being. This strategy was built and developed from the concepts of integrated control (Stern et al. 1959) and pest management (Geier 1966). These two concepts were developed as a response to the intensive and non-rational use of pesticides that was conducted to address problems as resistance, resurgence of primary pests, onset of secondary pest, environmental pollution and toxicity to humans as described in the seminal work of Carson (1962). Since the IPM strategy was developed, many definitions of it have been proposed. The Food and Agricultural Organization of the United Nations (FAO) committee recognize IPM as "Integrated Pest Management (IPM) means the careful consideration of all available pest control techniques and subsequent integration of appropriate measures that discourage the development of pest populations and keep pesticides and other interventions to levels that are economically justified and reduce or minimize risks to human health and the environment. IPM emphasizes the growth of a healthy crop with the least possible disruption to agroecosystems and encourages natural pest control mechanisms" (FAO 2018). In this way, IPM prioritizes integrative and harmonious use of control tactics and rational decision rules based on ecological principles and economic and social considerations in order to prioritise more adequate control tactics and maintain pest populations below economically injurious levels (Kogan 1998).

The practice of IPM requires the application of basic principles (Dent 2000):

- Monitoring: survey studies to identify the pest and establish the population levels or density of a pest. Sequential monitoring allows for the establishment of trends and patterns concerning pest populations and determines whether control levels are or will be achieved.

- Action levels: pest population size used to determine the moment when remedial action is necessary for human health, economic or aesthetic reasons.
- Prevention: Preventive measures for maintaining the pest population levels below control threshold.
- Control tactics: Centred principally around the ecological or natural process.
- Evaluation: Periodical evaluation for determining the efficiency and success of the program.

The broad adoption of IPM in different agronomic systems and regions of the world has been combined with increasing research and technical development of other pest control tactics such as biological controls and semiochemical-based techniques. Recently, a new approach considering the new technologies of pest control in a synergistic combination with traditional control techniques was proposed (Stenberg 2017). In this approach, an IPM pyramid, in which the pyramid base is formed by abiotic actions (including mechanical, physical, cultural practices, and others), is applied during crop installation and maintained until harvest. The pyramid's mid-section is formed by ecology-based tactics as intrinsic heritable plant-resistance, indirect plant defences and plant vaccinations such as plant defence induction, biorational synthetic volatiles, botanical diversity and biological control (Stenberg 2017). Finally, the top of the pyramid includes chemical pesticides. Interestingly, this conceptual framework pays particular attention to the interactions of ecology-based tactics, three of which are based on semiochemical use.

3.2.2   Possibilities of Uses of Semiochemicals in IPM

3.2.2.1   Pheromones

Pheromones have been the most prevalent semiochemicals used in IPM because of the power of pheromones to attract partners of pheromone transmitters; they are very useful tools for monitoring and even controlling insects. Notwithstanding, in the last 30 years other semiochemicals, mainly plant volatiles, have become relevant.

As described above, monitoring is the central point of any IPM strategy. In general, in agricultural systems, monitoring can be defined as any system or technique used to determine the identity, number, density, spatial location, and/or level of pest damage in the field (Dent 2000; Lampson et al. 2013). For this type of monitoring, an efficient sampling technique is required. Insect sampling could be performed using direct (direct counting insects) or indirect (estimation of population size without direct counting of insects, for example, estimations of leaf damage level) techniques (Dent 2000). Sampling for direct insect counting could be performed actively, direct counting insects on plants or capturing insects with nets, beat sheets, or mechanical aspirators. On the other hand, direct monitoring can be performed passively by the use of traps (Grootaert et al. 2010). Traps baited with semiochemicals, mainly pheromones, have been a very efficient worldwide tool for insect monitoring since the 1970s in different crops, fruit, and forest agroecosystems (Baker and Heath 2005; Witzgall et al. 2010). Pheromone traps (containing aggregation pheromones) have been especially useful for moth (Lepidoptera) monitoring but also for Coleoptera, especially bark beetles (Curculionidae: Scolytinae) (Bakke 1991; Wermelinger 2004) and stink bugs (Pentatomidae) (Borges et al. 2011; Tillman and Cotrell 2017). In the last several years, automated traps are able to automatically count insects (Beroza 2002); these data can be transmitted to remote data processing stations by various methods, including wireless systems. Long-term automated monitoring also gives growers valuable information regarding trends in field pest population densities in order to predict insect migrations, estimate insect infestations on a regional scale and determine their spatial distribution (Laumann et al. 2017).

Another use of a pheromone trap includes the reduction of population levels via the mass trap technique, which aims to catch a significant proportion of the population in pheromone-baited traps. The practical use of this technique requires intensive research to determine the best trap densities for capturing a large number of insects, which results in a reduction in the pest population increase (Baker 2009). Between the aspects that need to be considered are pest densities, competitiveness of traps with insects releasing pheromone, ecology, target pest behaviour and costs. This technique has been used to target Lepidoptera, Coleoptera, Diptera, and Homoptera in crops, orchards, and forests (Tewari et al. 2014a, b). A general view shows that this technique is more efficient in situations in which the insect populations are at low levels and are more or less isolated (El-Sayed et al. 2006).

An alternative to this approach is termed the "attract and kill tactic", that consists of attracting insects with pheromones and use a toxic substance to eliminate them (El-Sayed et al. 2009). One interesting example of this tactic is the control of the cotton boll weevil, *A. grandis*. An efficient mass trapping technique used during the 1980s encouraged the development of a boll weevil bait stick (BWBS) (McKibben et al. 1990). This device consists of polyvinyl chloride (PVC) stakes containing the aggregation pheromone, Grandlure®, and an insecticide at very low dose (30 mg). The BWBS proved to be more effective than pheromone traps in a semi-field test and subsequently, their efficiency was demonstrated in different field tests in the USA and Central America (Smith et al. 1995). With this technique and use of a mean of 14 traps ha^{-1}, a successful eradication program was performed in the US (Ridgway et al. 1990), and it has been estimated that use of ~250,000 ha in South America (Smith 1998) would be adequate. A new device that substituted synthetic insecticides with glue, called the grandlure-and-glue tube (GGT), was shown to be more effective than BWBS both during the growing and off-seasons. In this case, insects are killed via arrestment in glue. Because of the absence of insecticide in the formulation, this device also has the advantage of being an efficient tool for organic growers (Neves et al. 2018).

Another approach to the attract and kill strategy is incorporation of ento-mopathogenic microorganisms in pheromone lure devices. An interesting case was reported with the banana weevil, *C. sordidus*. An attract and kill device was developed using pellets of soybean or palm stearin fats blended with the ento-mopathogenic fungus *Beauveria bassiana* (Hypocreales: Cordycipitaceae) sensu lato and the aggregation pheromone sordidin (Cosmolure®). This formulation promoted auto-dissemination because the manipulation of insect behaviour using artificial devices can contaminate some individuals that subsequently infect others (Lopes et al. 2014). Similar approaches have been developed for others beetles such as *Carpophilus lugubris* (Coleoptera: Nitidulidae) (Vega et al. 1995; Dowd and Vega 2003), *Cylas formicarius* (Coleoptera: Curculionidae (Yasuda 1999) and *Ips typographus* (Coleoptera: Curculionidae) (Kreutz et al. 2004).

Mating disruption with pheromones consists of the artificial release of an adequate pheromone amount (in most cases synthetic pheromones) in crops in order to avoid the successful mate searching behaviour of insects, generally males. Three main mechanisms could explain these phenomena: (1) competing between natural pheromone released by insects (female or males) with the artificial dispensers; (2) camouflage of natural pheromone plume by pheromone concentrate homogenization in the field or (3) sensory habituation or adaptation of recipients (Byers 2007; Miller et al. 2010). The use of this tactic requires two principal technical developments that include the capacity to produce large-scale syntheses of pheromonal components and the development of specific dispensers for controlled release, which guarantees pheromonal concentrate homogenization in the field. This dispenser uses different material (most common are rubber septa and other plastic polymers) that can be applied by hand, mechanically or via mechanical spraying with time and dose automation. In the last years, nano-formulations have also been developed (Hellmann et al. 2011; Bhagat et al. 2013). Mating disruption has proven to be a very efficient technique for controlling mainly lepidopteran pests; and has been extended mainly to orchards, vineyards, and some crops as cotton and rice in order to control lepidopteran (Witzgall et al. 2010).

3.2.2.2 Plant Volatiles

The relevance of insect pheromones for IPM in the last 30 years of exploration of the amazing diversity of chemical compounds from plants offers new possibilities for developing monitoring and pest control tactics. Some plant volatiles have synergistic effects when incorporated into pheromone traps (reviewed in Witzgall et al. 2010; Tewari et al. 2014a, b). Synergistic effects have been shown for green leaf volatiles and some terpenoids in Coleoptera, Lepidoptera, and Diptera (fruit flies). One interesting case concerned the cotton boll weevil (*A. grandis*). For this insect, synergistic effects of cotton green leaf volatiles (GLVs) ((*E*)-2-hexen-1-ol, (*Z*)-3-hexen-1-ol, or 1-hexanol) and aggregation pheromones were shown by Dickens (1989). Subsequently, similar synergistic effects with the aggregation pheromone was demonstrated by Magalhães et al. (2016, 2018), when six cotton volatiles, including (*R*)-linalool, (*E*)-4,8-dimethyl-1,3,7-nonatriene (DMNT), methyl salicylate, β-caryophyllene, geranylacetone and (3*E*,7*E*)-4,8,12-trimethyltrideca-1,3,7,11-tetraene (TMTT), were used in the same ratios produced by cotton plants during reproductive stages, attracting both males and females. Synergistic effects could contribute to an increase in the efficiency of pheromone traps and consequently in the precision and accuracy of population-level estimates.

The use of plant volatiles has also been proposal for monitoring natural enemies. Using herbivory induced plant volatiles (HIPVs) and/or floral volatiles or their mixtures and sticky traps, Jones et al. (2016) showed that population densities and phenology of natural enemies such as Hymenoptera and Diptera parasitoids and some predators (Chrysopidae and Syrphidae) could be efficiently estimated in walnut, pear, and apple orchards. The information related to natural enemy densities and temporal and spatial distributions can be an efficient tool for integrating biological control in IPM programmes (Jones et al. 2016).

Since discovering that injury caused by herbivores induces plants to produce indirect defence mechanisms, researchers have tried to use these semiochemicals for natural enemy manipulation with the aim of increasing biological control. A large volume of information has shown that different groups of natural enemies as predators (examples are ladybirds, predatory stink bugs) or parasitoids can be attracted by semiochemical applications. For example, with respect to arable crops, more than 50 HIPVs for natural enemies' manipulation have been used (Blassioli-Moraes et al. 2013). Use of plant volatiles for natural enemy attraction can be achieved by application of synthetic compounds or via environmental manipulation (Blassioli-Moraes et al. 2013).

3.2.2.3 Application of Synthetic Semiochemicals

For synthetic compound application, two strategies are being developed. The first uses synthetic compounds, either individually or in a mixture, and generally uses slow-release dispensers to act directly on insect behaviour, whereas the second uses chemical application of synthetic compounds or phytohormones as sprays to induce a plant's chemical defence, which also plays a role in the attraction of natural enemies or repelling herbivores (Blassioli-Moraes et al. 2013; Simpson et al. 2011a; Michereff et al. 2016). These tactics have been tested in different agricultural systems. Strong attraction and redistribution of insects have been shown in different situations (Simpson et al. 2011a, b; Braasch and Kaplan 2012; Vieira et al. 2013, 2014). Commercial use of plant volatiles for natural enemy management is reduced, but new use tactics (described in the next section) could improve their efficiency and stimulate their use in agroecosystems.

3.2.2.4 Indirect Application of Semiochemicals

Indirect application of semiochemicals can be achieved or used in combination with environmental manipulation (Landis et al. 2000; Gurr et al. 2004). One strategy proposes using plants as a food reward for natural enemies. The "attract and reward strategy" considers that attraction of natural enemies into a treated area would be insufficient to increase biological control when natural enemies do not find other resources as hosts, food, or shelter in those selected areas. Therefore, response to a semiochemical without a satisfactory oviposition site or attack on a host/prey would result in a weak response or worse. In the worst-case scenario, the pest would not find the host, and the parasitoid/predator would then lose or change its capability to respond to this cue. Therefore, the presence of additional resources such as an adequate food supply could indirectly improve the actions of natural enemies, positively influencing their longevity, fecundity, locomotion, and host-searching capabilities. The potential usefulness of this strategy was proven in maize fields using several HIPVs formulations and buckwheat (*Fagopyrum esculentum* [Caryophyllales: Polygonaceae]) as the reward (Simpson et al. 2011b). The presence of plants acting as a source of nectar, increased the abundance of Eulophidae and Encyrtidae parasitoids and predators in maize plots. The recruitment and retention of parasitoids in the attract and reward plots significantly reduced the density of the herbivores *Helicoverpa* spp., and plant damage was lower in the reward-based plots.

In the attract and reward strategy, host volatiles (such as sex pheromones) with proven efficient actions on natural enemies (Borges et al. 1998, 1999) could be used as attractive semiochemicals. Another perspective is the use of the natural enemies' own pheromones as sex pheromones of parasitoids (Steiner et al. 2006; Steiner and Ruther 2009), aggregation pheromones of predators such as predacious stink bugs *Podisus* spp. (Pentatomidae) (Sant'Ana et al. 1997; Aldrich et al. 1997), or combinations of HIPVs and sex pheromones (Jones et al. 2011).

A push–pull system is an additional strategy to use plant semiochemicals to control pests. In this strategy, one type of semiochemical could be used to repel herbivorous insects from crops (push) and another semiochemical could be used to attract natural enemies of the herbivorous (pull). This could be achieved by direct application of synthetic semiochemicals, but the main advantage of push–pull systems compared to other techniques is use crop, inter-crop, cover, or surrounded plants as sources of natural semiochemicals (Khan et al. 2014).

In this way, this strategy has a wide spectrum of action and is not centred exclusively on behavioural manipulation of natural enemies. Push-pull strategies involve the behavioural manipulation of pests and natural enemies using different semiochemical stimuli. Several push-pull systems have been proposed (Cook et al. 2007; Khan et al. 2014), but the only one currently in practice by farmers was developed in Africa for maize intercropped with a plant that is repellent to the stemborer, *Desmodium uncinatum* (Leguminosa: Fabaceae), and an attractive plant, Napier grass, *Pennisetum purpureum* (Poales: Poaceae). The latter is cultivated as a trap crop around the maize field, while *D. uncinatum* is cultivated between the maize plants. The volatile semiochemicals emitted by the *D. uncinatum* plants act by repelling the gravid female moths, which are attracted by the volatiles emitted by the Napier grass. There was a higher predation rate of eggs and larvae by natural enemies compared with monoculture maize plots (Midega et al. 2016).

Finally, the possibility of increasing plant semiochemical release has been considered over the last several years. One of these possibilities is to explore the priming effect (Kim and Felton 2013; Hilker et al. 2016). The other possibility explored in the last years is production of plants with increasing capacity to release HIPVs or other useful semiochemicals. This could be evaluated using traditional selective breeding or GM crops. Crop plants have been selected for environmental resistance and yield, but in the process of selection, some defence mechanisms could be reduced or lost. Therefore, by the strategies discussed previously, this capacity could be reintroduced. No selective breeding has been explored until now, but the knowledge of high resistance to herbivores by landraces opens new opportunities. For example, some landraces of maize were shown to produce higher quantities of HIPVs, were more attractive to egg parasitoids, and induced priming in other maize genotypes (Borges et al. 2017; Michereff et al. 2018).

On the other hand, transgenic *A. thaliana* plants have been obtained by insertion of the sesquiterpene synthase gene from maize in order to increase the production of (E)-β-farnesene $(E\beta f)$, (E)-α-bergamotene, and others sesquiterpenes. Transformed plants were more attractive to the larvae parasitoid *Cotesia marginiventris* (Hymenoptera: Braconidae) (Schnee et al. 2006). Recently, a hexaploid variety of wheat was transformed to release $E\beta f$, the alarm pheromone of aphids (Bruce et al. 2015). The GM wheat in laboratory experiments was able to repel cereal aphids and attract the natural enemy. However, the field experiments did not reproduce the same results, and there were no differences in the numbers of aphids or parasitoids in areas with GM and non-GM wheat crops. The authors speculated that the failure of GM wheat in the field occurred because these plants continuously release $E\beta f$; therefore, the GM wheat loses its function as an alarm pheromone against aphids. Therefore, the timing of semiochemical release from a plant needs to mimic what actually happens in nature.

3.2.3 Extension of Semiochemical Use in IPM

3.2.3.1 General Trends of Use and Successful Cases

The use of semiochemicals in agriculture has shown an explosive increase in the last 10 years. It was estimated that $\approx$22 million has have been used for semiochemical-related insect monitoring and population control (Witzgall et al. 2010). World semiochemical markets were estimated at U$S 540 million in 2016 with an annual increase of 17% in the period from 2011 to 2016 (Dunham 2018). It is estimated that for 2022, the market could reach U$S 4.23 billion (Market and Market 2018).

In 2015, the regional market shares were 30% for Europe, 27% for Asia-Pacific, 25% for the USA and Canada, 14% for Latin America and 4% for the rest of the world (Dunham 2018). These figures also reflect the use of IPM-associated semiochemical application in different regions of the world. This increase in semiochemical use reflects the effectiveness of semiochemical-based products to be included in IPM. Witzgall et al. (2010) reviewed the use of Lepidoptera pheromones for mating disruption in Europe and showed that in Mezzocorona vineyards in Trento-Italy, the use of this control tactic reduced the insecticide application over a period of 10 years from 5 kg ha^{-1} to less than 0.5 kg; during the same period, insect damage was reduced from ~5 to 0.3%. *C. pomonella* mating disruption programs have also been extremely efficient and are normally used in Europe, the US and South America (Witzgall et al. 2010). This phenomenon appears to have resulted from high quantities of semiochemical production; it was estimated that in 2010, the total quantity of this pheromone produced was 25,000 kg for codling moth control on more than 200,000 ha (Witzgal et al. 2010).

Another example of an extremely successful use of semiochemicals is the mass-trap program for the American palm weevil (*R. palmarum*) in Costa Rica. The insect is a serious pest affecting oil and coconut palms, principally as the vector of the nematode, *Rhadinaphelenchus cocophilus* (Aphelenchida: Parasitaphelenchidae). It causes a severe disease in palm trees. This programme uses traps baited with the sex pheromone of the species in combination with insecticides. Over a 10-year period, the attract and kill system was able to practically eradicate the disease from the country because only 50 trees were identified with the nematode infestation after the program was implemented (Oehlschlager et al. 2002).

3.3 Use of Semiochemicals in Brazil: A Case Study

In Brazil, the use of semiochemicals in agriculture has been increasing year by year. This could be a consequence of the increasing capacity of Brazilian researchers to isolate, identify and test semiochemicals in the field to be incorporated in IPM programs. For example, in the period from 1960 to 2008, Brazilian researchers identified pheromones of 110 insect species (Bergmann et al. 2009). A search of the ISI Web of Science for the period from 2009 to 2018 used the keywords pheromone identification and Brazil, and it was shown that during this period, more than 19 new pheromones were isolated and identified by Brazilian researchers.

Despite the lack of organized information in the country, 47 commercial products based on semiochemicals by 8 companies and for 23 target pests have been officially registered (Agrofit 2018). On the official website of Brazilian Institute of Environment, (IBAMA), which is the official environmental agency, a description of more than 800 kg of commercialized semiochemicals within the country in 2017 is given (IBAMA 2017).

The advances in commercial semiochemical use for IPM are not widespread despite their successes. One interesting case is the use of the sex pheromone of the moth *C. pomonella* in an eradication program used for monitoring and identification of this moth in areas infested with this insect. The early identification of the areas infested with this moth contributed to the total eradication of this pest from commercial orchards. Nowadays, *C. pomonella* is considered an invasive pest, has been eradicated from Brazil, and pheromones are used to check for its possible reintroduction (Kovaleski and Mumford 2007). In the case of monitoring to determine population densities, an example is the brown sink bug's sex pheromone, which can be used for monitoring it in soybean fields. Borges et al. (2011) showed that the pheromone of this stink bug is more efficient for its monitoring than the technique commonly used for monitoring (the beating cloth). The pheromone is placed in the field every 100 m in plastic bottle traps (Borges et al. 1998) and can be distributed only on the edges of the crop without the need to place it inside the plantations (Borges et al. 2011). In Rio Grande do Sul State, the pheromone of the moths *G. molesta* (Padilha et al. 2018) and (Lepidoptera: Pyralidae) are used for their control via the technique of mating disruption in apple and vineyards, respectively. For *Cryptoblabes* gindiella (Lepidoptera: Pyralidae), the mating disruption technique yielded a drastic population density reduction (proximal to 0 infestation) (Oliveira et al. 2014).

3.4 Conclusions

Nowadays, most semiochemical use in agriculture is deployed for pest monitoring. Semiochemicals are natural molecules, that when applied in very tiny amounts, nanograms or less, present any adverse effect of non-target insects, including beneficial insects and pollinators, decrease the risk of pest insect resistance, the molecules do not persist in the nature, they are easily degraded when exposed to air, therefore do not pollute the environment. Using synthetic pheromones in large areas is not an easy task due to the difficulties of setting traps and visiting them, to develop good release devices, and they still have high costs. However, in the near future, with the development of new technologies to deliver semiochemicals in the field and to transmit the insect caught information in real-time using wireless systems, as discussed above, this tool may become more accessible to farmers. The use of modified plants that release the semiochemicals in all areas or the use of companion plants can be promising for successful application of semiochemicals for plant defence. The new generation of GMO plants releasing toxins against herbivores or volatiles to attract natural enemies or repel the herbivore could trigger the new semiochemical biosynthesis only when the plant is under attack. In this case, plants will not produce the semiochemical continuously, thus avoiding development of insect resistance to these semiochemicals. The advances in the Internet-of-things development new types of multi-sensors, combined to wireless monitoring trap system together with the knowledge of plant defence and the chemical communication of plant-plant and plant-insects are essential for developing more sustainable agriculture to be applied worldwide in the upcoming decades.

References

Agrofit (2018) Agrotóxicos registrados no AGROFIT. http://agrofit.agricultura.gov.br/agrofit_cons/principal_agrofit_cons. Accessed 12 Oct 2018

Al Abassi SAL, Birkett MA, Pettersson J et al (2000) Response of the seven-spot ladybird to an aphid alarm pheromone and an alarm pheromone inhibitor is mediated by paired olfactory cells. J Chem Ecol 26:1765–1771

Aldrich JR, Oliver JE, Lusby WR et al (1994) Identification of male-specific volatiles from Neartic and Neotropical stink bugs (Heteroptera: Pentatomidae). J Chem Ecol 20:1103–1111

Aldrich JR, Zanuncio JC, Vilela EF et al (1997) Field tests of predaceous pentatomid pheromones and semiochemistry of *Podisus* and *Supputius* species (Heteroptera: Pentatomidae: Asopinae). An Soc Entomol Brasil 26:1–14

Andrade R, Rodrigues C, Oehlschager C (2000) Optimization of a pheromone lure for *Spodoptera frugiperda* (Smith) in central America. J Braz Chem Soc 11:609–613

Aquino MFS, Sujii ER, Borges M et al (2018) Diversity of stink bug adults and their parasitoids in soybean crops in Brazil: influence of a latitudinal gradient and insecticide application intensity. Environ Entomol. Nyv 174. https://doi.org/10.1093/ee/nvy174

Bhagat D, Samanta SK, Bhattacharya S (2013) Efficient management of fruit pests by pheromone nanogels. Sci Rep 3:1–8

Baker R, Borges M, Cooke NG et al (1987) Identification and synthesis of (Z) (1'S,3'R,4'S)–2–(3',4'–epoxy–4'–methylcyclohexyl)–6–methylhepta–2,5–diene, the sex pheromone of the southern green stink bug, *Nezara viridula* (L.). J Chem Soc D 6:414–416

Baker TC, Heath JJ (2005) Pheromones – function and use in insect control. In: Gilbert LI, Iatro K, Gill SS (eds) Molecular insect science, vol 6. Elsevier, Academic press, London, pp 407–460

Baker TC (2009) Use of pheromones in IPM (Chapter 21). In: Radcliffe EB, Hutchison WD, Cancelado RE. Integrated Pest Management. Cambridge University Press. Cambridge, UK

Bakke A (1991) Using pheromones in the management of bark beetle outbreaks. In: Baranchikov Y, Mattson WJ, Hain FP, Payne TL (eds) Forest insect guilds: patterns of interaction with host trees; proceedings of a joint IUFRO working party symposium Abakan, Siberia, U.S.S.R. 13–17 Aug 1889

Batista-Fereira LG, Stein K, De Paula AF et al (2006) Isolation, identification, synthesis, and field evaluation of the sex pheromone of the Brazilian population of *Spodoptera frugiperda*. J Chem Ecol 32:1085–1099

Beale MH, Birkett MA, Bruce TJ et al (2006) Aphid alarm pheromone produced by transgenic plants affects aphid and parasitoid behaviour. Proc Natl Acad Sci USA 103:10509–10513

Bergmann J, González A, Zarbin PHG (2009) Insect pheromone research in South America. J Braz Chem Soc 20:1206–1219

Beroza M (2002) More efficient means of detecting insects. US. Patent 0144452, 19 July 2002

Bino RJ, Hall RD, Fiehn O et al (2004) Potential of metabolomics as a functional genomics tool. Trends Plant Sci 9:418–425

Blassioli-Moraes MC, Laumann RA, Oliveira MWM et al (2012) Sex pheromone communication in two sympatric Neotropical stink bug species *Chinavia ubica* and *Chinavia impicticornis*. J Chem Ecol 38:836–845

Blassioli-Moraes MC, Borges M, Laumann RA (2013) Chemical ecology of insect parasitoids. In: Wajnberg E, Colazza S (eds) The application of chemical cues in arthropod pest management for arable crops. Wiley, New York, pp 225–244

Blassioli-Moraes MC, Borges M, Michereff MFF et al (2016) Semiochemicals from plants and insects on the foraging behaviour of Platygastridae egg parasitoids. Pesqui Agropecu Bras 51:454–464

Blassioli-Moraes MC, Khrimian A, Borges M et al (2018) The male produced sex pheromone of *Tibraca limbativentris* revisted: absolute configuration of zingiberenol stereoisomers. 34º International Society Chemical Ecology Meeting. Abstract Book, vol 1, p 68

Bombardi LM (2017) Geografia do Uso de Agrotóxicos no Brasil e Conexões com a União Europeia—São Paulo: FFLCH—USP, 2017.296 p. ISBN:978-85-7506-310-1

Borges M, Jepson PC, Howse PE (1987) Long-range mate location and close range courtship behavior of the green stink bug, *Nezara viridula* and its mediation by sex pheromones. Entomol Exp Appl 44:205–212

Borges M, Aldrich JR (1992) Instar-specific defensive secretions of stink bugs (Heteroptera: Pentatomidae). Experientia 48:893–896

Borges M, Schmidt FVG, Sujii ER et al (1998) Field responses of stink bugs to the natural and synthetic pheromone of the Neotropical brown stink bug, *Euschistus heros*, (Heteroptera: Pentatomidae). Physiol Entomol 23:202–207

Borges M, Costa MLM, Sujii ER et al (1999) Semiochemical and physical stimuli involved in host recognition by Telenomus podisi (Hymenoptera: Scelionidae) toward *Euschistus heros* (Heteroptera: Pentatomidae). Physiol Entomol 24:227–233

Borges M, Moraes MCB, Peixoto MF et al (2011) Monitoring the Neotropical brown stink bug *Euschistus heros* (F.) (Hemiptera: Pentatomidae) with pheromone-baited traps in soybean fields. J Appl Entomol 135:68–80

Borges M, Blassioli-Moraes MC (2017) The semiochemistry of Pentatomidae in stink bugs. In: Čokl A, Borges M (eds) Biorational control based on communication processes, CRC Press, Taylor & Francis Group, Boca Raton, pp 95–124

Borges M, Michereff MFF, Blassioli-Moraes MC et al (2017) Metodologias para o estudo da defesa de memória (Priming) em plantas frente a estresse biótico. Embrapa Recursos Genéticos e Biotecnologia. Circular Técnica, 91

Bortolotto OC, Pomari AF, Bueno RCO et al (2015) The use of soybean integrated pest management in Brazil: a review. Agron Sci Biotechnol 1(25):32

Braasch J, Kaplan I (2012) Over what distance are plant volatiles bioactive? Estimating the spatial dimensions of attraction in an arthropod assemblage. Entomol Exp Appl 145:115–123

Bradbury JW, Vehrencamp SL (2011) Principles of animal communication. Sinauer Associates, Sunderland

Bruce TJA, Pickett JA (2011) Perception of plant volatile blends by herbivorous insects—finding the right mix. Phytochemistry 72:1605–1611

Bruce TJA, Aradottir GI, Smart LE et al (2015) The first crop plant genetically engineered to release an insect pheromone for defence. Sci Rep 5:11183

Butenandt VA, Beckmann R, Stamm D, Hecker E (1959) Über den sexuallockstoff des seidenspinners Bombyx mori. Reindarstellung und konstitution. Z. Naturforsch. B. 14:283–284

Byers JA (2007) Simulation of mating disruption and mass trapping with competitive attraction and camouflage. Environ Entomol 36:1328–1338

Carson R (1962) Silent spring. Mariner Books, New York

Chiozza MV, O'Neal ME, MacIntosh GC (2010) Constitutive and induced differential accumulation of amino acid in leaves of susceptible and resistant soybean plants in response to the Soybean Aphid (Hemiptera: Aphididae). Environ Entomol 39:856–864

Cook SM, Khan ZR, Pickett JA (2007) The use of push-pull strategies in integrated pest management. Annu Rev Entomol 52:375–400

D'Alessandro M, Held M, Triponez Y, Turlings TCJ (2006) The role of indole and other shikimic acid derived maize volatiles in the attraction of two parasitic wasps. J Chem Ecol 32:2733–2748

Dent D (2000) Insect Pest Management. CABI Publishing, Wallingford

Dias AM, Pareja M, Laia M et al (2016) Attraction of *Telenomus podisi* to volatiles induced by *Euschistus heros* in three different plant species. Arthropod Plant Inte 10:419–428

Dickens JC (1989) Green leaf volatiles enhance aggregation pheromone of boll weevil, Anthonomus grandis. Entomologia Experimentalis et Applicata 52(3):191–203

Dowd PF, Vega FE (2003) Autodissemination of *Beauveria bassiana* by sap beetles (Coleoptera: Nitidulidae) to overwintering sites. Biocontrol Sci Technol 13:65–75

Du YJ, Poppy GM, Powell W et al (1998) Identification of semiochemicals released during aphid feeding that attract parasitoid *Aphidius ervi*. J Chem Eco. 24:1355–1368

Dudareva N, Klempien A, Muhlemann JK et al (2013) Biosynthesis, function and metabolic engineering of plant volatile organic compounds. New Phytol 198:16–32

Dunham W (2018) Semiohcemicals may be the fastest growing segment of the biopesticide market. PDF http://dunhamtrimmer.com/wp-content/uploads/2017/07/ProdTrends-2.pdf. Accessed on 01 Nov 2018

Edwards LJ, Siddal JB, Dunam LL et al (1973) *trans*-β-farnesene, alarm pheromone of the green peach aphid, *Myzus persicae* (Sulzer). Nature 214:126–127

El-Sayed AM, Suckling DM, Wearing CH et al (2006) Potential of mass trapping for long-term pest management and eradication of invasive species. J Econ Entomol 99:550–1564

El-Sayed AM, Suckling DM, Byers JA et al (2009) Potential of 'lure and kill' in long-term pest management and eradication of invasive species. J Econ Entomol 102:815–835

Engelberth J, Alborn HT, Schmelz EA et al (2004) Airborne signals prime plants against insect herbivore attack. Proc Natl Acad Sci USA 101:1781–1785

Erb M, Veyrat N, Robert CAM et al (2015) Indole is an essential herbivore-induced volatile priming signal in maize. Nat Commun 6, Article number: 6273. https://doi.org/10.1038/ncomms7273

FAO (2018) AGP—integrated pest management. http://www.fao.org/agriculture/crops/thematic-sitemap/theme/pests/ipm/en/9. Accessed 15 Sept 2018

Francke W, Dettner K (2005) Chemical signalling in beetles. In: Schulz S (ed) Topics in current chemistry 240. Springer, Heildelberg, pp 85–166

Frost CJ, Appel HM, Carlson JE et al (2007) Within-plant signalling via volatiles overcomes vascular constraints on systemic signalling and primes responses against herbivores. Ecol Lett 10:490–498

Geier PW (1966) Management of insect pests. Ann Rev Entomol 11:471–490

Grootaert P, Pollet M, Dekoninck W et al (2010) Sampling insects: general techniques, strategies and remarks. In: Eymann J, Degreef J, Hauser CH et al (eds) Manual on field recording techniques and protocols for all taxa biodiversity inventories and monitoring. Abc Taxa, Belgium, pp 377–399

Groot AT, Marr M, Schöfl G et al (2008) Host strain specific sex pheromone variation in Spodoptera frugiperda. Front Zool 5:20. https://doi.org/10.1186/1742-9994-520

Gordon-Weeks R, Smart L, Ahmad S et al (2010) The role of the benzoxazinone pathway in aphid resistance in wheat. HGCA Project Report 473:1–66

Gurr GM, Wratten SD, Altieri MA (2004) Ecological engineering: a new direction for agricultural pest management. AFBM Journal 1:28–35

Hassemer MJ, Sant'ana J, De Oliveira MW et al (2015) Chemical composition of *Alphitobius diaperinus* (Coleoptera: Tenebrionidae) abdominal glands and the Influence of 1,4-benzoquinones on its behavior. J Econ Entomol 108:2107–2116

Hardie J, Pickett JA, Pow EM et al (1999) In: Hardie J, Minks AK (eds) Pheromones of non-lepidopteran insects associated with agricultural plants. CAB CABI Publishing, Wallingford, pp 227–250

Hellmann C, Greiner A, Wendorff JH (2011) Design of pheromone releasing nanofibers for plant protection. Polym Adv Tecnol 22:407–413

Hilker M, Schwachtje J, Baier M et al (2016) Priming and memory of stress response in organisms lacking a nervous system. Biol Rev 91:1118–1133

Hill AS, Roelofs WL (1981) Sex pheromone of the saltmarsh caterpillar moth, *Estugmene acrea*. J Chem Ecol 7:655–668

Heil M, Bueno JCS (2007) Within-plant signalling by volatiles leads to induction and priming of an indirect plant defense in nature. Proc Natl Acad Sci USA 104:5467–5472

IBAMA Semioquímicos (2017). http://www.ibama.gov.br/agrotoxicos/relatorios-decomercializacao-de-agrotoxicos. Accessed in 28 Jan 2019

Jones VP, Steffan SA, Wiman NG et al (2011) Evaluation of herbivore-induced plant volatiles for monitoring green lacewings in Washington apple orchards. Biol Control 56:98–105

Jones VP, Mills DR, Unruh NJ et al (2016) Evaluating plant volatiles for monitoring natural enemies in apple, pear and walnut orchards. Biol Control 102:53–65

Khan ZR, Midega CAO, Pittchar JO et al (2014) Achieving food security for one million sub-Saharan African poor through push-pull innovation by 2020. Philos Trans R Soc Lond B Biol Sci 369:1–11

Kim J, Felton GW (2013) Priming of anti herbivore defensive responses in plants. Insect Sci 20:273–285

Kreutz J, Zimmermann G, Vaupel O (2004) Horizontal transmission of the entomopathogenic fungus Beauveria bassiana among the spruce bark beetle, *Ips typographus* (Col., Scolytidae) in the laboratory and under field conditions. Biocontrol Sci Technol 14:837–848

Kogan M (1998) Integrated pest management: historical perspectives and contemporary developments. Ann Rev Entomol 43(1):243–270

Kovaleski A, Mumford JD (2007) Pulling out the evil by the root: the codling moth eradication program in Brazil. In: Vreysen MJB, Robinson AS, Hendrichs J (eds) Area wide control of insect pests: from research to field implementation. Springer, Dordrecht, pp 581–590

Kunert G, Otto S, Weisser WW et al (2005) Alarm pheromone mediates production of winged morphs.in aphids. Ecol Lett 8:596–603

Lampson BD, Han YJ, Khalilian A (2013) Automatic detection and identification of brown stink bug, *Euschistus servus*, and southern green stink bug, *Nezara viridula*, (Heteroptera: Pentatomidae) using intraspecific substrate borne vibrational signals. Comp Elect Agr 91:154–159

Landis DA, Wratten SD, Gurr GM (2000) Habitat management to conserve natural enemies of arthropod pests in agriculture. Ann Rev Entomol 45:175–201

Laumann RA, Bottura DM, Čokl A (2017) Use of vibratory signals for stink bug monitoring and control. In: Čokl A, Borges M (eds) Stink bugs: biorational control based on communication. CRC Press, Boca Raton, pp 226–245

Lopes RB, Laumann RA, Moore D, Oliveira WM, Faria M (2014) Combination of the fungus and pheromone in an attract-and-kill strategy against the banana weevil. Entomologia Experimentalis et Applicata 151(1):75–85

McKibben GH, Smith JW, McGovern WL (1990) Design of an attract-and-kill device for the boll weevil (Coleoptera: Curculionidae). J Entomol Sci 25:581–586

Magalhães DM, Borges M, Laumann RA et al (2012) Semiochemicals from herbivory induced cotton plants enhance the foraging behavior of the cotton boll weevil, *Anthonomus grandis*. J Chem Ecol 38:1528–1538

Magalhães DM, Borges M, Laumann RA et al (2016) Influence of two acyclic homoterpenes (tetranorterpenes) on the foraging behavior of *Anthonomus grandis* Boh. J Chem Ecol 42:305–313

Magalhães DM, Borges M, Laumann RA et al (2018) Identification of volatile compounds involved in host location by *Anthonomus grandis* (Coleoptera: Curculionidae). Front Ecol Evol 6:98. https://doi.org/10.3389/fevo.2018.00098

Market & Market (2018) Pheromones Market in Agriculture worth 4.23 Billion USD by 2002. https://www.marketsandmarkets.com/PressReleases/pheromone.asp. Accessed on 05 Nov 2018

Michereff MFF, Laumann RA, Borges M et al (2011) Volatiles mediating plant herbivory-natural enemy interaction in resistant and susceptible soybean cultivars. J Chem Ecol 37:73–285

Michereff MFF, Borges M, Diniz IR et al (2013) Influence of volatile compounds from herbivore-damaged soybean plants on searching behavior of the egg parasitoid. Entomol Exp Appl 147:9–17

Michereff MFF, Borges M, Santos MA et al (2016) The influence of volatile semiochemicals from stink bug eggs and oviposition-damaged plants on the foraging behaviour of the egg parasitoid *Telenomus podisi*. Bull Entomol Res 106:663–671

Michereff MFF, Magalhães DM, Hassemer MJ et al (2018) Variability in herbivore induced defence signalling across different maize genotypes impacts on natural enemy foraging behaviour. J Pest Sci. https://doi.org/10.1007/s10340-018-1033-6

Midega CA, Murage AW, Pittchar JO, Khan ZR (2016) Managing storage pests of maize: Farmers' knowledge, perceptions and practices in western Kenya. Crop Protect 90:142–149

Miller JR, Mcghee PS, Siegert PY et al (2010) General principles of attraction and competitive attraction as revealed by large-cage studies of moths responding to sex pheromone. Proc Natl Acad Sci USA 107:22–27

Moscardi F, Soza-Gomes DR, Corrêa-Ferreira BS (1999) Soybean IPM in Brazil, with emphasis on biological control tactics. In: Proceeding of VI world soybean research conference, Chicago Illinois, USA 1: 331–339

Neves RCS, Torres JB, Barros EM et al (2018) Boll weevil within season and off season activity monitored using a pheromone-and-glue reusable tube trap. Sci Agric 75:313–320

Nordlund DA, Lewis WJ (1976) Terminology of chemical releasing stimuli in intraspecific and interspecific interactions. J Chem Ecol 2:211–220

Oehlschlager AC, Chinchilla C, Castillo G et al (2002) Control of red ring disease by mass trapping of *Rhynchophorus palmarum* (Coleoptera: Curculionidae). Fla Entomol 85:507–513

Oliveira MWM, Borges M, Andrade CKZ et al (2013) Zingiberenol, (1*R*,4*R*,1′*S*)-4 (1′,5′-Dimethylhex-4′-enyl)-1-methylcyclohex-2-en-1-ol, identified as the sex pheromone produced by males of the rice stink bug *Oebalus poecilus* (Heteroptera: Pentatomidae). J Agric Food Chem 61:777–7785

Oliveira CM, Auad AM, Mendes SM et al (2014) Crop losses and economic impact of insect pests on Brazilian agriculture. Crop Prot 56:50–54

Padilha AC, Arioli CJ, Boff MIC, Rosa JM, Botton M (2018) Traps and Baits for Luring Grapholita molesta (Busck) Adults in Mating Disruption-Treated Apple Orchards. Neotropical Entomology 47(1):152–159

Panizzi AR (2013) History and contemporary perspectives of the integrated pest management of soybean in Brazil. Neotrop Entomol 42:119–127

Pavis C, Malosse PH (1986) Mise en evidence d'un attractif sexuel produit par les males de *Nezara viridula* (L.) (Heteroptera: Pentatomidae). C.R Acad Sci Series III 7:272–276

Pareja M, Mohib A, Birkett MA et al (2009) Multivariate statistics coupled to generalized linear models reveal complex use of chemical cues by a parasitoid. Anim Beh 77:901–909

Petschenka G, Agrawal AA (2015) Milkweed butterfly resistance to plant toxins is linked to sequestration, not coping with a toxic diet. Proc R Soc B 282:20151865. https://doi.org/10.1098/rspb.2015.1865

Pherobase (2018) Pherobase database of pheromones and semiochemicals. www.pherobase.com

Pichersky E, Gang DR (2000) Genetics and biochemistry of secondary metabolites in plants: an evolutionary perspective. Trends Plant Sci 5:439–445

Price PW (1997) Insect ecology. Wiley, New York, pp 73–138

Ridgway RL, Inscoe MN, Dickerson WA (1990) Role of the boll weevil pheromone in pest management. In: Ridgway RL, Silverstein RM, Inscoe MN (eds) Behavior modifying chemicals for insect management. Marcel Dekker, New York, pp 437–471

Rodriguez-Saona CR, Rodriguez-Saona LE, Frost CJ (2009) Herbivore-induced volatiles in the perennial shrub, *Vaccinium corymbosum*, and their role in inter-branch signaling. J Chem Ecol 35:163–175

Rostás M (2007) The effects of 2,4-dihydroxy-7-methoxy-1,4-benzoxazin-3-one on two species of *Spodoptera* and the growth of *Setosphaeria turcica* in vitro. J Pest Sci 80(35):41

Sant'Ana J, Bruni R, Abdul-Baki AA et al (1997) Pheromone-induced movement of nymphs of the predator, *Podisus maculiventris* (Heteroptera: Pentatomidae). Biol Control 10:123–128

Schnee C, Köllner TG, Held M et al (2006) The products of a single maize sesquiterpene synthase form a volatile defense signal that attracts natural enemies of maize herbivores. Proc Natl Acad Sci USA 103:1129–1134

Schulz S (2005) The chemistry of pheromones and other semiochemicals I. Part of the topics in current chemistry book series (TOPCURRCHEM, vol 239). Springer, Heidelberg

Simpson M, Gurrr GM, Simmons AT et al (2011a) Insect attraction to synthetic herbivorie-induced plant volatile-treated field crops. Agric For Entomol 13:45–57

Simpson M, Gurrr GM, Simmons AT et al (2011b) Field evaluation of the 'attract and reward' biological control approach in vineyards. Ann Appl Biol 159:69–78

Smith JW, McKibben GH, Villavaso E et al (1995) Management of the cotton boll weevil with attract-and-kill-devices. In: Constable GA, Forrester NW (eds) Challenging the future: proceedings of the world cotton conference, Brisbane, pp 480–484

Smith JW (1998) Boll weevil eradication: area-wide pest management. Ann Entomol Soc Am 91:239–247

Steiner S, Hermann N, Ruther J (2006) Characterization of a female-produced courtship pheromone in the parasitoid *Nasonia vitripennis*. J Chem Ecol 32:1687–1702

Steiner S, Ruther J (2009) Mechanism and behavioral context of male sex pheromone release in *Nasonia vitripennis*. J Chem Ecol 35:416–421

Stenberg JA (2017) A conceptual framework for integrated pest management. Trend Plant Sci 22(9):759–769

Stern VM, Smith RF, van den Bosch R et al (1959) The integrated control concept. Hilgardia 29:81–101

Tewari S, Leskey TC, Nielsen AL, Piñero JC, Rodriguez-Saona CR (2014a) Use of pheromones in insect pest management, with special attention to weevil pheromones. In: Abrol DP (eds) Integrated Pest Management, Academic Press, pp 141–168. ISBN 9780123985293. https://doi.org/10.1016/B978-0-12-398529-3.00010-5

Tewari S, Leskey TC, Nielsen AL, Piñero JC, Rodriguez-Saona CR (2014b) Use of Pheromones in Insect Pest Management, with Special Attention to Weevil Pheromones. In: INTEGRATED PEST MANAGEMENT: Current Concepts and Ecological Perspective. Edited by Dahram P. Abrol. Elsevier, London, UK. 2014

Tilmann PG, Cottrel T (2017) Use of pheromones for monitoring phytophagous stink bugs (Hemiptera: Pentatomidae) In: Colk A, Borges M (eds) Stink bugs: biorational control based on communication. CRC Press, Boca Raton, pp 210–225

Trigo JR (2000) The chemistry of antipredator defense by secondary compounds in Neotropical Lepidoptera: facts, perspectives and Caveats. J Braz Chem Soc 11:551–561

Tognon R, Sant'ana J, Jahnke SM (2014) Influence of original host on chemotaxic behaviour and parasitism in *Telenomus podisi* Ashmead (Hymenoptera: Platygastridae). Bull Entomol Res 104:781–787

Vega FE, Dowd PF, Bartelt RJ (1995) Dissemination of microbial agents using an auto inoculating device and several insect species as vectors. Biol Control 5:545–552

Vieira CR, Moraes MCB, Borges M et al (2013) *cis*-Jasmone indirect action on egg parasitoids (Hymenoptera: Scelionidae) and its application in biological control of soybean stink bugs (Hemiptera: Pentatomidae). Biol Control 64:75–82

Vieira CR, Moraes MCB, Borges M et al (2014) Field evaluation of (*E*)-2-hexenal efficacy for behavioral manipulation of egg parasitoids in soybean. Biocontrol 1:1–13

Yasuda K (1999) Auto-infection system for the sweet potato weevil, *Cylas formicarius* (Fabricius) (Coleoptera: Curculionidae) with entomopathogenic fungi, *Beauveria bassiana*, using a modified sex pheromone trap in the field. App Entomol Zoo 34(501):505

Weber DC, Khrimian A, Blassioli-Moraes MC et al (2018) Semiochemistry of pentatomoidea. In: McPherson JE (ed) Invasive stink bugs and related Species (Pentatomoidea): biology, higher systematics, semiochemistry, and management. CRC Press Boca Raton, pp 677–725

Wermelinger B (2004) Ecology and management of the spruce bark beetle *Ips typographus* a review of recent research. For Ecol Manag 202:67–82

Witzgall P, Kirsch P, Cork A (2010) Sex pheromones and their impact on pest management. J Chem Ecol 36:80–100

Wheeler CA, Cardé RT (2014) Following in their footprints: cuticular hydrocarbons as overwintering aggregation site markers in *Hippodamia convergens*. J Chem Ecol 40:418–428

Wouters FC, Reichelt M, Glauser G et al (2014) Reglucosylation of the benzoxazinoid DIMBOA with inversion of stereochemical configuration is a detoxification strategy in Lepidopteran herbivores. Angew Chem-Ger Edit 126:11502–11506

Chapter 4
Green Nanotechnology for Sustained Release of Eco-Friendly Agrochemicals

Luciano Paulino Silva and Cínthia Caetano Bonatto

Abstract This chapter describes the major concepts related to nanoscience and nanotechnology, role of green nanotechnology as an essential part of a sustainable future of agriculture, and its applicability for the development of innovative solutions to challenging issues. From this, the understanding and use of more efficient and sustainable agrochemicals based on green nanotechnology processes are the major focus of this chapter and one example of a green nanotechnology process to protect postharvested fruits is presented as a case study.

Keywords Nanoagrochemicals · Nanosystems · Nanomaterials · Nanofilm

4.1 Introduction

4.1.1 Definitions of Nanoscience and Nanotechnology

Nanoscience is the area of knowledge that investigates natural or man-made structures and materials reaching few nanometers (billionth part of the meter) in size. Indeed, unique chemical, physical, and/or biological properties emerge from the matter at this ultrasmall scale. This fact is particularly related to features intrinsic to the nanostructures such as the high surface-to-volume ratio and also the unprecedented frameworks allowed by the organization of individual atoms into well-defined positions in space and which are directly related to the emerging features. Last few decades, the rational use of nanoscience principles and fundamentals led to the development of innovative technological applications which ultimately results in the exponentially growing of nanotechnology.

L. P. Silva (✉) · C. C. Bonatto
Brazilian Agricultural Research Corporation, Laboratory of Nanobiotechnology (LNANO), Embrapa Genetic Resources and Biotechnology, PBI, Parque Estação Biológica Final W5 Norte, Asa Norte, Brasilia, DF 70770-917, Brazil
e-mail: luciano.paulino@embrapa.br

C. C. Bonatto
NanoDiversity, Applied Research, TecSinapse, São Paulo, SP, Brazil

© Springer Nature Switzerland AG 2019
S. Vaz Jr. (ed.), *Sustainable Agrochemistry*,
https://doi.org/10.1007/978-3-030-17891-8_4

Nanotechnology is a multidisciplinary and interdisciplinary field that benefits from strategies and techniques of areas like physics, chemistry, biology, medicine, engineering, and information technology (Schummer 2004). Nanostructured systems (nanosystems) present new or improved features when compared to structures of larger dimensions from the same material or even to their atomic–molecular form. Such improvements are related to specific characteristics of nanosystems such as shape, size distribution, aggregation/agglomeration state, composition and/or surface chemistry.

Two distinct approaches are used for nanosystem syntheses which are known in nanotechnology as top-down and bottom-up. The top-down strategy consists of the deconstruction of a certain bulk material, generally by nanolithography techniques or by high-energy grinding, until the final nanostructured product is obtained. This approach is commonly used for scale production, but difficulties are encountered to obtain homogeneity in the characteristics of the final product. The other approach, bottom-up, follows the opposite path in that nucleation for growth occurs from individual atoms and/or molecules for the formation of nanostructures. The bottom-up approach allows controlling and modulating several synthesis parameters, such as the size and shape of nanostructures, being the most used.

Currently, nanoscience and nanotechnology have been applied in several areas of knowledge, such as biology, pharmacy, veterinary medicine, medicine, and agriculture. The systems and devices developed at the nanoscale are often fabricated by bottom-up strategies with inspiration in molecules and structures found in nature, particularly those from living organisms, such as proteins, nucleic acids, membranes, and other biomolecules (German et al. 2006; Sanguansri and Augustin 2006).

4.1.2   Sustainable Nanotechnology

Many chemical and physical routes are used for the bottom-up synthesis of nanosystems from their molecular precursors (Armelao et al. 2002; Maaz et al. 2007; Meffre et al. 2012; Iravani et al. 2014). However, most of these methods include the use of toxic solvents, the generation of residues potentially harmful to health and the environment, or even tend to result in high energy consumption on generally complex and multiple step routes. In this sense, the search for the development of procedures aimed at obtaining nanosystems with broad technological applicability and overcoming some challenges related to traditional synthesis methods is an imminent focus of studies on research, development, and innovation in nanotechnology (Iravani et al. 2014). One promising approach to achieving this goal is to explore the wide range of biological resources available in nature through the so-called green synthesis (Wenbo et al. 2012; Malik et al. 2014; Ahmed et al. 2016).

Green synthesis is the common designation given to synthetic routes which use relatively non-toxic, biodegradable, and low-cost chemicals to synthesize materials, typically having as a primary source or route initiator a biological organism or parts thereof (e.g., organs, tissues, cells, or metabolites) (Silva et al. 2015; Silva et al.

2017). Among living resources, plant and animal products, algae, fungi, bacteria, and the wide range of products, co-products, by-products, and residues derived from agricultural and aquaculture processes involving some of these organisms present potential for use during green synthesis routes, which when applied to the formation of nanosystems gives rise to the so-called green nanotechnology (Joanitti and Silva 2014; Fifovsky and Beilin 2017; Ganachari et al. 2018).

In fact, the rational use of biological resources derived from biodiversity, animal, plant, and microbial production chains is an important step in the application of the concept of bioeconomy in the sustainability of human activities (Silva et al. 2015). Thus, green nanotechnology is an approach perfectly in line with this growing concern with issues related to sustainability as it uses methods and materials aimed at the generation of nanomaterials aligned with the Sustainable Development Goals (SDGs) (Griggs et al. 2013; Silva et al. 2018). This concept offers unique opportunities for the use of primary and secondary metabolites of various living organisms due to the fact that these biomolecules when used to produce nanosystems such as nanoparticles (polymeric, lipidic, or metallic), liposomes, and emulsions present new characteristics that enable a wide range of innovative applications (Silva et al. 2018).

4.1.3   Biological Resources as Inputs for Nanotechnology

Green synthesis of nanosystems (processes based on green chemistry principles, according to Chap. 2) can be performed using prokaryotes or eukaryotes (including microorganisms, plants, and animals) or part of them and may occur in the intracellular or extracellular medium (Iravani 2011; Quester et al. 2013; Srivastava et al. 2013; El-Said et al. 2014). In this case, the biological components present (typically primary and secondary metabolites) act as the main constituents aiming at the formation of nanosystems, which in the case of metallic nanoparticles (MNPs) act as bioreductors and stabilizing agents (Raveendran et al. 2003; Iravani 2011; Narayanan and Sakthivel 2011; Silva et al. 2018); in the case of polymeric nanoparticles (PNPs) act as essential blocks for structuring and coating agents (Medeiros et al. 2014); in the case of liposomes act as essential components for the formation and activity (Bonatto et al. 2015a); and in the case of emulsions act as essential constituents to obtain the distinct properties of the dispersed and continuous phases (Bonatto et al. 2015b). In almost all cases, in particular, when it is desired to obtain a colloidally stable suspension or with specific physicochemical–biological properties, these compounds from biological resources also provide a stabilizing layer (coating) on the surface of the nanostructures and may even allow the emergence of relevant bioactive properties.

In addition, most of the nanosystems obtained through green synthetic routes have as sustainability characteristics the fact that they are eco-friendly (use less toxic solvents and/or renewable resources), biocompatible (they can be used directly to the target organisms), simple (a smaller number of steps for production), biodegradable (can be degraded by biological routes), have low production cost, and still offer high yield (Huang et al. 2008; Sharma et al. 2009; Iravani et al. 2014). Among

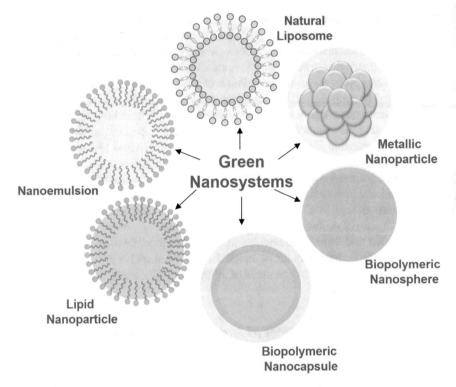

Fig. 4.1 Representative scheme of some common types of nanosystems typically obtained by green synthesis routes

the applications of nanosystems based on green nanotechnology approaches can be highlighted those dedicated to the delivery and release of agricultural inputs (e.g., fertilizers, herbicides, pesticides, and other agrochemical products) (Fig. 4.1).

4.2 Applications of Green Nanomaterials for Agriculture

The molecular systems found in nature have aroused increasing interest as models for nanosystem production. This fact is particularly related to characteristics intrinsic to biological compounds (metabolites) such as the diversity of chemical forms, easy removal of specific regions of the molecule (degradation) through chemical or biological reactions, and the possibility of functionalization (addition of chemical groups) for the development of new properties. This is due to the fact that the nanostructures produced by living organisms carry out chemical reactions and form three-dimensional arrays through the organization of the individual atoms in well-defined positions in space and that is related to the biological functions performed.

Agricultural and forestry products, co-products, by-products, and wastes are recognized as rich sources of primary and secondary metabolites that can be used in green synthesis routes of nanomaterials (Silva and Joanitti 2014). From these biological resources, it is possible to obtain, in many cases at low cost, many grams, kilograms, or even tons of biomolecules that serve as raw materials for the development of products at the nanoscale with high value-added.

Undoubtedly, these nanoproducts can help precision farming and sustainable agriculture development by enhancing production, productivity, and profitability of the processes. Potential applications of green nanotechnology in agriculture include: increasing the productivity using nanoagrochemicals, such as nanopesticides, nanofertilizers, and nanoherbicides; improving the quality of the soil, water, and surrounding environment using nanocomposites and hydrogels; stimulating plant growth and development using advanced nanomaterials; using nanocoatings for improving food quality, safety, and biological/nutritional value; and providing monitoring systems using nanosensors.

Indeed, all the progress made in the last almost three decades of the advent of nanotechnology applied to biology, more recently attributed to the terms bionanotechnology, nanobiotechnology, and green nanotechnology, has been basically achieved by two interlinked approaches: (1) the use of advanced techniques for high-resolution characterization of biological structures at nanoscale and (2) the use of this information about biological structures characterized at nanoscale in order to develop new advanced materials and processes useful in various areas and productive sectors. Among the several practical examples of current nanosystems developed inspired on nanostructural organization models of nature that have attracted interest are MNPs, PNPs, emulsions, and liposomes that can be obtained by routes of green synthesis (Silva et al. 2006), and the use of at least one biological component (Silva et al. 2018).

4.3 Case Study

There are several reports describing the benefits of the use of sustainable nanosystems, including emulsions and PNPs, to transport and efficiently delivery agrochemicals, particularly those useful to control pests and pathogens in the field or those related to the delivery of fertilizers to plants (Vishwakarma et al. 2016; Hazra 2017; Prasad et al. 2017). However, a relatively limited number of approaches report the use of eco-friendly and nano-based agrochemicals applied to postharvest storage of fruit and vegetables. The use of greener approaches at this stage of production is essential to prevent or minimize undesirable contamination that can lead to adverse impacts on human and environmental health. A recent strategy developed by the authors of this chapter is based on the use of compounds originated from the same species desired to be postharvesting conserved that would show potential as eco-friendly raw materials to produce nanosystems useful to protect fruits or vegetables against rot-causing microorganisms and foodborne pathogens. Such nanosystems

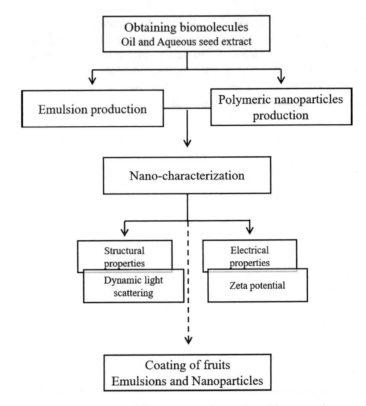

Fig. 4.2 Schematic representation of the experimental setup from this case study

can be applied as edible nanofilms for improving the long-term storage and safety of fruits. Being that, emulsions and PNPs are the most used for the encapsulation of oils and vegetable extracts aiming at the production of nanostructured edible films. Additionally, vegetable oils and extracts from fruits, such as tomatoes, have great antimicrobial potential and can be used to protect the fruit itself. Thus, the goal of the present case study was to produce, characterize, and investigate the in vitro antibacterial activity of tomato oil-based microemulsions and tomato extract-based PNPs associated or not with carboxymethylcellulose polymers to be applied as inputs toward the protection of fruits against microorganisms (Fig. 4.2).

4.3.1 Contextualization

The application of nanotechnology to agriculture and the food sector is relatively recent when compared to its use in the delivery of drugs and pharmaceuticals, focusing mainly on the smart delivery of nutrients, the rapid identification of contaminants, and bioactive packaging, among which are the edible nanocoatings (Sozer and Kokini 2009). Nanofilms or edible nanocoatings are thin colorless and invisible films that completely envelop foods for the maintenance or enhancement of physical, chemical, microbiological, sensory, or nutritional aspects (Mcclements 2011).

Most of the nanosystems used to produce edible nanofilms belong to the group of colloids (emulsions and micelles) and PNPs. Emulsions can be defined as systems composed of two immiscible liquids, one of which is completely dispersed in the other in the form of globules. According to the size of the globule, the emulsions can be classified into macroemulsions, microemulsions, or nanoemulsions. Microemulsions have $1-100$ μm and are said to be thermodynamically stable (Payet and Terentijev 2008). Additionally, some authors report the addition of polysaccharides to the formulations aiming at an increase in the stability and longevity of the emulsion (Gura et al. 2007). Agreed, nanoparticles constituted by biodegradable polymers (PNPs) have also been used in agriculture and the food sector due to their high biocompatibility and biodegradability.

Microemulsions and PNPs based on the application of vegetable and animal oils have been used extensively in pharmacy, medicine, and cosmetics (Hsieh et al. 2006; Gura et al. 2007; Klaypradit and Huang 2008). This is because the lipids, the main constituents of these oils, have presented several biological activities, among them antimicrobial (Silva et al. 2011). In addition, plant extracts have been shown to have antimicrobial activity against various bacterial strains. Tomato is among the most traded fruits in the world, and Brazil is the eighth world producer with more than 1400 million tons of fruits in 2017 (Camargo-Filho and Camargo 2017). However, about 21% of this production is lost at the postharvest due to inadequate transport, which leads to mechanical injuries that become the entry point for fungus and bacterial contamination (Rita et al. 2011).

Commercial processing of tomatoes for the production of juices, extracts, purees, pulps, ketchup, soups, etc., generates large amounts of solid wastes such as skins, seeds, and trimmings. The main part of the solid residue of the tomato seed is an unexplored source of non-traditional oil, containing a percentage of oil above 38% (Kobori and Jorge 2005; Sogi et al. 1999). Tomato oil is mainly composed of lycopene, carotenoid, vitamins A and B complexes, and folic acid, thus possessing a high antioxidant potential (Jones 2007) and thus exhibiting desired features to be applied for green nanotechnology products.

4.3.2   Experimental Methods

4.3.2.1   Obtaining Biomolecules

Tomato oil was purchased commercially from a specialty oil shop, in the city of Brasília-DF, Brazil. The aqueous extract of the seeds was obtained from aqueous extraction. To that end, *Solanum lycopersicum* fruits, popularly known as cherry tomatoes, were purchased in a supermarket. Tomatoes were washed and had their

Table 4.1 The nomenclature used for developed nanosystems (formulations)

Name	Description
CLO	Carboxymethylcellulose + soy lecithin + tomato oil
LO	Soy lecithin + tomato oil
CMC	Carboxymethylcellulose
NPCE	Carboxymethylcellulose + calcium chloride + tomato seeds extract
NPC	Carboxymethylcellulose + calcium chloride

seeds macerated, using a mortar and pestle of porcelain, using liquid nitrogen. Approximately 5 g of the sprayed sample was added to ultrapure water for 3 min. After that time, the aqueous extract was filtered off by gravity with Whatman No. 1 filter paper and protected from environmental light.

4.3.2.2 Emulsion

Carboxymethylcellulose solution was prepared at a concentration of 5 mg mL^{-1} (50 mg dissolved in 10 mL of ultrafiltrated water). The tomato oil was blended to soy lecithin (surfactant) in a ratio of 6:1 v/v (1.5 mL of tomato oil and 250 µL of surfactant). Tomato oil solution containing soy lecithin was added to a carboxymethylcellulose solution in a ratio of 1:5 (v/v) and stirred at 14,000 rpm using an ultraturrax for 5 min at room temperature. The formulation was kept at 20 °C until the moment of use. As controls, the same volumes of carboxymethylcellulose solution and oil were replaced by the identical volumes of ultrapure water (Table 4.1). The pH value of each formulation was determined using test paper strips for pH measurement.

4.3.2.3 Polymeric Nanoparticles

Carboxymethylcellulose (CMC) particles were prepared using the ionic gelation method adapted from Calvo et al. (1997). For the formation of the particles, two solutions were prepared: the first of CMC at a concentration of 5 mg mL^{-1} (50 mg dissolved in 10 mL of ultrapure water under magnetic stirring at room temperature) and the second of calcium chloride at a concentration of 10 mg mL^{-1}. To 5 mL of CMC solution, under magnetic stirring, 5 mL of the aqueous tomato seed extract (50 mg mL^{-1}) was slowly and constantly dropped-wise. One mL of the calcium chloride solution was also slowly and constantly dropped-wise (2500 µL L^{-1}). The solution remained under stirring for 30 min forming the particles. Control particles were developed following the same steps by replacing the aqueous extract with ultrafiltered water (Table 4.1). The pH value of each formulation was determined using test paper strips for pH measurement.

4.3.2.4   Nanocharacterization

Dynamic light scattering (DLS) and Zeta potential: The hydrodynamic diameters and polydispersity indexes (PDIs) of the formed structures were determined by the dynamic light scattering (DLS) technique in Zetasizer Nano ZS (Malvern, UK) using a He–Ne laser (4 mW) operating at 633 nm. Three measurements were performed at room temperature, and light scattering detection was adopted at an angle of 173°. Additionally, the surface Zeta potentials were obtained by means of the same equipment by electrophoretic mobility. For the analyses, each formulation was diluted in ultrapure water 400×.

4.3.2.5   Culture of Microorganisms and Identification by Mass Spectrometry

Sterile swabs were gently wiped on the surface of cherry tomatoes (washed or not with 9 mmol L^{-1} sodium hypochlorite), purchased at local supermarket, and then spun in Luria–Bertani (LB) (10 g L^{-1} peptone, 10 g L^{-1} sodium chloride, 5 g L^{-1} yeast extract, and 15 g L^{-1} agar) in Petri dishes to evaluate the possible growth of microorganisms. The plates were incubated in an oven at 37 °C for 48 h. Isolated colonies were used for the analysis by mass spectrometry directly or by protein extraction. For direct analysis, a small amount of each microorganism's colony was deposited onto a 53 × 41 mm MSP 96-type MALDI target plate, allowed to dry at room temperature. For protein extraction, an abundant amount (pinhead size) of microorganisms obtained from colonies isolated with the help of a toothpick was collected from the culture medium, added to a polypropylene microtube containing 300 μL of ultrapure water, and homogenized with vortex stirring. Then, 900 μL of ethanol (PA) was added, the microtube was centrifuged for 2 min at 11,000 g, and the supernatant was discarded. The pellet was resuspended with the addition of 30 μL of 70% formic acid, followed by the addition of 30 μL of acetonitrile. The sample was then homogenized and centrifuged for 2 min at 11,000 g. Supernatants containing intact proteins and microorganisms were deposited on the MSP 96-type plate surface with 53 × 41 mm allowed to dry, followed by the addition of a saturated matrix solution consisting of alpha-cyano-4-hydroxycinnamic acid (1:2, v/v), and again allowed to dry at room temperature. The components had their molecular masses determined using a MALDI-TOF Microflex mass spectrometer (Bruker Daltonics, Germany) using external calibration under positive linear mode in the range of m/z 2000–22,000. The acquired spectra were compared using a microorganism's protein profile database accompanying the kit, which contained about 5000 microorganisms by May 2014 (Bruker Daltonics, Germany).

4.3.2.6   Coatings of Fruits with Emulsions and Nanoparticles

Cherry tomatoes were subjected to exposure to emulsions and nanoparticles. The fruits were packed in a 6-well polypropylene culture dish. One mL of each emulsion (CLO, LO, and CMC) and nanoparticles (NPCE and NPC) were deposited onto the surface of the tomatoes so that their entire surface was exposed to the formulations. The excess sample was removed using a pipette. The samples were conditioned in a climatic chamber at 30 °C and humidity of 30% for 72 h to evaluate their visual aspects, aiming at the possible inhibition of growth of microorganisms. After each step, the samples were photographed.

4.3.2.7   Statistical Analysis

The obtained results were represented as the mean ± standard deviation (S.D.) or standard error of the mean (S.E.M.) using OriginPro 8.0 program to obtain graphs. The determination of possible statistically significant differences between the groups was determined using the PAST (Hammer Ø et al. 2001) program by analysis of variance (ANOVA) with Tukey's statistical test, with a significance level set at $P <$ 0.05.

4.3.3   Results and Discussion

4.3.3.1   Physicochemical Characterization of the Developed Nanosystems

Dynamic light scattering and electrophoretic mobility showed variable hydrodynamic diameter, polydispersity index, and Zeta potential of the emulsions and nanoparticles (Table 4.2). The hydrodynamic diameters of the emulsions presented significant differences among them, with an increase of the same in the formulations containing tomato oil (Table 4.2). All formulations showed high PDI. Additionally, the CLO presented a significantly higher Zeta potential when compared to the LO and CMC, indicating excellent colloidal stability, as recommended in technical note (ASTM 1985). This fact suggests that the CMC present in the CLO acted as a coating agent surrounding the structures formed by tomato oil and surfactant (Table 4.2). A recent study described the stabilization of emulsions, where the CMC increases the viscosity of the emulsion, causing the migration of oil droplets in the aqueous phase (Malinauskytė et al. 2014). The hydrodynamic diameters of the PNPs indicated the formation of the particles, which showed differences between them, increasing the values in the formulations containing aqueous extract of tomato seeds (Table 4.2). All formulations exhibited high PDI. In addition, the NPCE showed a higher Zeta potential than the NPC and CMC, indicating good colloidal stability, as recommended in the technical note (ASTM 1985).

Table 4.2 Physical–chemical characterization of microemulsions and nanoparticles by DLS and Zeta potential. Polydispersity index (PDI). Values referring to means and standard deviation of three independent readings

System	Z-Average (d. nm)	PDI	Zeta potential (mV)	pH value
CLO	31,553.3 ± 1782.0	0.552 ± 0.134	−75.6 ± 0.8	6.0
LO	2925.0 ± 1820.3	1.000 ± 0.000	−51.2 ± 2.3	6.0
NPCE	3736.7 ± 49.7	1.000 ± 0.000	−43.8 ± 1.2	6.0
NPC	302.3 ± 258.0	0.911 ± 0.066	−34.8 ± 6.7	6.0

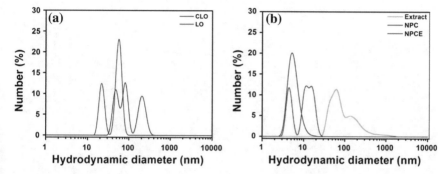

Fig. 4.3 Hydrodynamic diameter of the emulsions (**a**) and nanoparticles (**b**) evaluated by dynamic light scattering and presented in relation to the percentage of structures in number

Although the hydrodynamic diameters represented by Z-Average indicate the formation of microemulsions (CLO and LO) and microparticles (NPCE) when evaluating the distribution curves in relation to the percentage of particles in number, the presence of nanometric particles is evident (Fig. 4.3). The CLO and LO emulsions presented a size distribution in diameter ranging from 22.02 to 56.95 nm and 47.89 to 204.26 nm, respectively (Fig. 4.3a). The NPCE and NPC polymer nanoparticles presented a size distribution varying from 3.00 to 28.54 nm and 2.75 to 20.11 nm, respectively (Fig. 4.3b). The presence of particles of different sizes between the experimental and control groups indicates that the addition of 1% of calcium chloride to the CMC in solution promoted the spontaneous formation of nanoparticles by the establishment of cross-linked structures leading to the potential entrapment of compounds present in the extract. Similar results were obtained by Aswathy et al. (2012), where CMC particles were produced using calcium chloride as polycation and a cross-linking agent (Aswathy et al. 2012).

4.3.3.2 Culture of Microorganisms and Identification by Mass Spectrometry

In order to identify possible microorganisms present in fresh tomato samples, cultures of colonies were carried out in culture Petri dishes, followed by their subsequent iden-

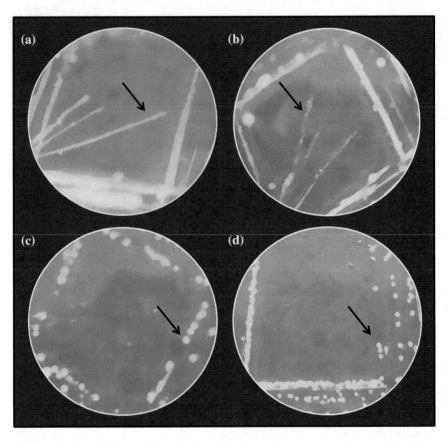

Fig. 4.4 Photographs of Petri dishes containing colonies of microorganisms originated from cherry tomatoes. Smears performed on unwashed (**a**) and washed (**b**) green cherry tomatoes, and unpeeled (**c**) and washed (**d**) red cherry tomatoes. The arrows indicate the colonies which were used for the identification of the microorganism's species

tification by mass spectrometry—MALDI Biotyper. Washed and unwashed cherry tomatoes were used to carry out smears on Petri dishes containing a nutrient medium for the subsequent identification of microorganism present in the fruits. After incubation of the plates for 48 h, it was possible to observe different patterns of growth of microorganisms in all the plaques (Fig. 4.4). Washing fruits and vegetables by immersing them in a disinfectant solution for 10 min is the most commonly used method in the food industry and supermarkets (Keeratipibul et al. 2011). The evaluation of the action of various food sanitizers has been reported, such as chlorine, sodium hypochlorite, chlorine dioxide, acetic acid, ethanol, hydrogen peroxide, and ozone (Keeratipibul et al. 2011). However, some studies claim that sanitizing agents only reduce and do not completely eliminate the pathogens and contaminants present in raw fruits and vegetables (Keeratipibul et al. 2011), corroborating the data presented here.

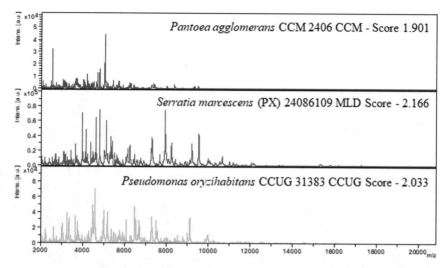

Fig. 4.5 Protein profile mass spectra of the identified bacteria obtained by MALDI-TOF mass spectrometry operated in the positive linear mode in the range of m/z 2000–22000

In this study, it was possible to identify three distinct microorganisms based on their MALDI-TOF MS protein profiles: *Pseudomonas oryzihabitans* (score 2.033), *Pantoea agglomerans* (score 1.901), and *Serratia marcescens* (score 2.166), which are not considered as frequent nor relevant regarding public health (Fig. 4.5). Previous studies have reported that several organisms have already been isolated from tomatoes, among which are *Escherichia coli*, *Pseudomonas aeruginosa*, Salmonella, Serratia, Citrobacter, Proteus, *Erwinia herbicola*, *Hafnia alvei*, Klebsiella, Shigella, and *Listeria monocytogenes* (Keeratipibul et al. 2011; Aswathy et al. 2012). However, while the main agent occurring in tomatoes is Salmonella, *E. coli* is among the most worrying pathogenic bacteria in terms of public health. This is because several strains are responsible for numerous foodborne diseases caused by the consumption of raw contaminated vegetable salads (Payet and Terentijev 2008).

4.3.3.3   Coatings of Fruits with Emulsions and Nanoparticles

Cherry tomatoes were exposed to emulsions and nanoparticles, aiming at the visual inspection of the fruits for the growth of microorganisms. During the 72 h, no visible changes were observed to the naked eye for all treated groups, except for the control group with no treatment that the development of some possible fungus occurred (Fig. 4.6). It is worth mentioning that the monitoring time may not have been sufficient for a more effective evaluation of nanosystems. Additional studies are required to evaluate the fruit surface after the period of exposure to the samples, re-cultivation of the microorganisms to identify potential microorganisms present on the surface, and internal morphological investigations of the fruits.

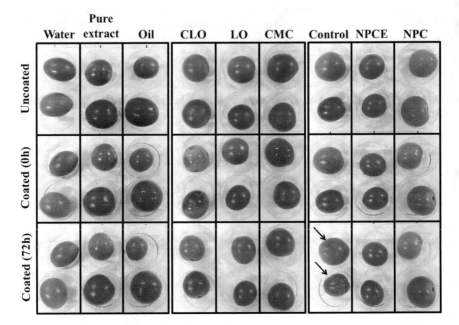

Fig. 4.6 Photographs of cherry tomatoes before and after exposure to emulsions and polymeric nanoparticles and their respective controls. The arrows point damaged tomato fruits with holes caused by microorganisms associated with deterioration

4.3.4 Conclusion of the Case Study

The results obtained from this case study show that the microemulsions and PNPs composed of tomato oil and tomato extract represent a new approach against microorganisms associated with postharvested fruits and vegetables.

References

Ahmed S, Ahmad M, La B et al (2016) A review on plants extract mediated synthesis of silver nanoparticles for antimicrobial applications: a green expertise. J Adv Res 7(1):17–28. https://doi.org/10.1016/j.jare.2015.02.007

Armelao L, Bertoncello R, Cattaruzza E, Gialenella S (2002) Chemical and physical routes for composite materials synthesis: Ag and Ag2S nanoparticles in silica glass by sol–gel and ion implantation techniques. J Mater Chem 12(8):2401–2407. https://doi.org/10.1039/b203539c

ASTM D4187 (1985) American society for testing and materials. Zeta potential of colloids in water and waste quarter. ASTM Standard D 4187–4182

Aswathy RG, Sivakumar B, Brahatheeswaran D et al (2012) Multifunctional biocompatible fluorescent Carboxymethylcellulose nanoparticles. J Biomater Nanobiotechnol 3:254–261. https://doi.org/10.4236/jbnb.2012.322031

Bonatto CC, Joanitti GA, Silva LP (2015a) Method for obtaining bioactive molecules in microstructured and nanostructured carrier systems. Patent, App/Pub Number: WO2016119030A1

Bonatto CC, Joanitti GA, Silva LP (2015b) In vitro cytotoxic activity of chitosan-bullfrog oil microemulsion against melanoma cells. IET Nanobiotechnol 9(4):172–177. https://doi.org/10.1049/iet-nbt.2014.0010

Calvo P, Remuñán-López C, Vila-Jato JL, Alonso MJ (1997) Novel hydrophilic chitosan-polyethylene oxide nanoparticles as protein carriers. J Appl Polym Sci 63:125–132

Camargo-Filho WP, Camargo FP (2017) A quick review of the production and commercialization of the main vegetables in Brazil and the world from 1970 to 2015. Hortic Bras 35(2):160–166. https://doi.org/10.1590/s0102-053620170202

El-Said WA et al (2014) Synthesis of metal nanoparticles inside living human cells based on the intracellular formation process. Adv Mater 26(6):910–918. https://doi.org/10.1002/adma.201303699

Figovsky O, Beilin D (2017) Green nanotechnology. Pan Stanford, United States, p 538

Ganachari SV et al (2018) Green nanotechnology for biomedical, Food, and Agricultural Applications. In: Martínez L, Kharissova O, Kharisov B (eds) Handbook of Ecomaterials. Springer, Cham

German JB, Smilowitz JT, Zivkovic (2006) Lipoproteins: when size really matters. Curr Opin Colloid Interface Sci 11:171–183. https://doi.org/10.1016/j.cocis.2005.11.006

Griggs D, Stafford-Smith M, Gaffney O, Rockström J et al (2013) Sustainable development goals for people and planet. Nature 495(7441):305–307. https://doi.org/10.1038/495305a

Gura KM, Lee S, Valim C et al (2007) Safety and efficacy of a fish-oil-based fat emulsion in the treatment of parenteral nutrition-associated liver disease. Pediatrics 121:678–686. https://doi.org/10.1542/peds.2007-2248

Hammer Ø et al (2001) Paleontological statistics software package for education and data analysis. Palaeontol Electron 4:1–9

Hazra DK (2017) Nano-formulations: High definition liquid engineering of pesticides for advanced crop protection in agriculture. Adv Plants Agric Res 6(3):00211. https://doi.org/10.15406/apar.2017.06.00211

Huang L, Zhai ML, Long DW, Peng J et al (2008) UV-induced synthesis, characterization and formation mechanism of silver nanoparticles in alkalic carboxymethylated chitosan solution. J Nanopart Res 10:1193–1202

Hsieh WC, Chang CP, Gao YL (2006) Controlled release properties of chitosan encapsulated volatile Citronella Oil microcapsules by thermal treatments. Colloids Surf B Biointerfaces 53(2):209–214. https://doi.org/10.1016/j.colsurfb.2006.09.008

Iravani S (2011) Green synthesis of metal nanoparticles using plants. Green Chem 13:2638–2650

Iravani S, Korbekand H, Mirmohammadi SV, Zolfaghari D (2014) Synthesis of silver nanoparticles: chemical, physical and biological methods. Res Pharm Sci 9(6):385–406

Joanitti GA, Silva LP (2014) By-products as scaffolds for drug delivery systems. Curr Drug Targets 15:5:477–477. https://doi.org/10.2174/1389450115051404041502013

Jones JB (2007) Tomato plant culture in the field, greenhouse and home garden, 2nd edn. CRC Press LLC., Florida

Keeratipibul S, Phewpan A, Lursinsap C (2011) Prediction of coliforms and *Escherichia coli* on tomato fruits and lettuce leaves after sanitizing by using Artificial neural networks. LWT—Food Sci Technol 44:130–138. https://doi.org/10.1016/j.lwt.2010.05.015

Klaypradit W, Huang YW (2008) Fish oil encapsulation with chitosan using ultrasonic atomizer. LWT 41:1133–1139. https://doi.org/10.1016/j.lwt.2007.06.014

Kobori CN, Jorge N (2005) Characterization of some seed oils of fruits for utilization of industrial residues. Ciênc Tecnol Aliment 29(5):1008–10014

Maaz K, Mumtaz A, Hasanain SK, Ceylan A (2007) Synthesis and magnetic properties of cobalt ferrite ($CoFe_2O_4$) nanoparticles prepared by wet chemical route. J Magn Magn Mater 38(2):289–295. https://doi.org/10.1016/j.jmmm.2006.06.003

Malik P, Shankar R, Malik K et al (2014) Green chemistry based benign routes for nanoparticle synthesis. J Nanopart 2014(302429):1–14. https://doi.org/10.1155/2014/302429

Malinauskytė E, Ramanauskaitė J, Leskauskaitė D et al (2014) Effect of human and simulated gastric juices on the digestion of whey proteins and carboxymethylcellulose-stabilised O/W emulsions. Food Chem 15(165):104–112. https://doi.org/10.1016/j.foodchem.2014.05.078

Mcclements DJ (2011) Edible nanoemulsions: fabrication, properties, and functional performance. Soft Matter 7(6):2297–2316

Medeiros KA, Joanitti GA, Silva LP (2014) Chitosan nanoparticles for dermaseptin peptide delivery toward tumor cells *in vitro*. Anticancer Drugs 25(3):323–331. https://doi.org/10.1097/CAD.0000000000000052

Meffre A, Mehdaoui B, Kelsen V, Fazzini PF et al (2012) A Simple chemical route toward monodisperse iron carbide nanoparticles displaying tunable magnetic and unprecedented hyperthermia properties. Nano Lett 12(9):4722–4728

Narayanan KB, Sakthivel N (2011) Green synthesis of biogenic metal nanoparticles by terrestrial and aquatic phototrophic and heterotrophic eukaryotes and biocompatible agents. Adv Colloid Interface Sci 169(2):59–79

Payet L, Terentijev EM (2008) Emulsification and stabilization mechanisms of o/w emulsions in the presence of chitosan. Langmuir 24:12247–12252. https://doi.org/10.1021/la8019217

Prasad R, Bhattacharyya A, Nguyen QD (2017) Nanotechnology in sustainable agriculture: recent developments, challenges, and perspectives. Front Microbiol 8:1014

Quester K et al (2013) SERS properties of different sized and shaped gold nanoparticles biosynthesized under different environmental conditions by neurospora crassa extract. PLoS ONE 8(10):1–8

Raveendran P, Fu J, Wallen (2003) Completely "Green" synthesis and stabilization of metal nanoparticles. J Am Chem Soc 125(46):13940–13941

Rita FLB, Salles LB, Barboza RA et al (2011) Atividade antimicrobiana de biofilme com óleos essenciais para conservação pós-colheita de tomate cv rasteiro. Rev Bras Eng Agríc Ambient 5:466–474. https://doi.org/10.3895/S1981-36862011000100010S1

Sanguansri P, Augustin MA (2006) Nanoscale materials development-a food industry perspective. Trends Food Sci Technol 17:547–556. https://doi.org/10.1016/j.tifs.2006.04.010

Schummer J (2004) Multidisciplinarity, interdisciplinarity, and patterns of research collaboration in nanoscience and nanotechnology. Scientometrics 59:425–465

Sharma VK, Yngard RA, Lin Y (2009) Silver nanoparticles: green synthesis and their antimicrobial activities. Adv Colloid Interface Sci 145(1–2):83–96. https://doi.org/10.1016/j.cis.2008.09.002

Silva LP et al (2006) Effects of fish oil treatment on bleomycin-induced pulmonary fibrosis in mice. Cell Biochem Funct 24:387–396. https://doi.org/10.1002/cbf.1237

Silva LP et al (2017) Nanotecnologia verde para síntese de nanopartículas metálicas. In: Resende RR (ed) Biotecnologia aplicada à agro&indústria, vol 4, issue no. 26. Blucher, São Paulo, pp 967–1012. https://doi.org/10.5151/9788521211150-26

Silva LP, Joanitti GA, Leite JRSA, Azevedo RB (2011) Comparative study of the antimicrobial activities and mammalian cytotoxicity of 10 fatty acid-rich oils and fats from animal and vegetable. Nat Prod Res 1(1):40–46

Silva LP, Pereira TM, Bonatto CC (2018) Frontiers and perspectives in the green synthesis of silver nanoparticles. In: Shukla A, Iravani S (ed) (Org). Green synthesis, characterization and applications of nanoparticles. Elsevier, pp 137–164

Silva LP, Reis IG, Bonatto CC (2015) Green synthesis of metal nanoparticles by plants: current trends and challenges. In: Basiuk VA, Basiuk EV (eds) Green processes for nanotechnology. Springer, Cham, pp 259–275. https://doi.org/10.1007/978-3-319-15461-9_9

Sogi DS, Kiran J, Bawa AS (1999) Characterization and utilization of tomato seed oil from tomato processing waste. J Food Sci Technol 36(3):248–249

Sozer N, Kokini JL (2009) Nanotechnology and its applications in the food sector. Trends Biotechnol 27(2):1–8. https://doi.org/10.1016/j.tibtech.2008.10.010

Srivastava P et al (2013) Synthesis of silver nanoparticles using haloarchaea isolate *Halococcus salifodinae* BK3. Extremophiles 17(5):821–831

Vishwakarma GS, Gautam N, Badu JN et al (2016) Polymeric encapsulates of essential oils and their constituents: a review of preparation techniques, characterization, and sustainable release mechanisms. Polym Rev 56(4):668–701. https://doi.org/10.1080/15583724.2015.1123725

Wenbo L, Qin X, Liu S et al (2012) Economical, green synthesis of fluorescent carbon nanoparticles and their use as probes for sensitive and selective selection of mercury(II) ions. Anal Chem 84(12):5351–5357. https://doi.org/10.1021/ac3007939

Chapter 5
Magnetic Resonance Spectroscopy Techniques to Improve Agricultural Systems

Sílvio Vaz Jr., Etelvino Henrique Novotny and Luiz Alberto Colnago

Abstract Nuclear magnetic resonance (NMR) and electron paramagnetic resonance (EPR) are two types of magnetic resonance (MR) spectroscopy that has been used to study physical and chemical of agriculture inputs and products; biomass; and environmental organic matter, such as soil, sedimentary and aquatic organic matter. Both techniques are very useful in agricultural sciences—highlighting NMR applications—to understand the constitution, properties, functionality and quality of food and non-food crops and soil organic matter, contributing to improve the productive systems related to the agroindustrial chains. This chapter deals with the physical phenomena involved in each one, their uses in agriculture and some examples of practical uses.

Keywords Spectroscopy · Chemical characterization · Biomass products · Organic matter

5.1 Introduction

Nuclear magnetic resonance (NMR) and electron paramagnetic resonance (EPR) are two types of magnetic resonance (MR) spectroscopy that has been used to study physical and chemical of agriculture inputs and products; biomass; and environmental organic matter, such as soil, sedimentary and aquatic organic matter. NMR and EPR

S. Vaz Jr. (✉)
Brazilian Agricultural Research Corporation, National Research Center for Agroenergy (Embrapa Agroenergy), Embrapa Agroenergy, Parque Estação Biológica, s/n, Av. W3 Norte (final), Brasilia, DF 70770-901, Brazil
e-mail: silvio.vaz@embrapa.br

E. H. Novotny
Brazilian Agricultural Research Corporation, Embrapa Soils, Rua Jardim Botânico, 1024, Rio de Janeiro, RJ 22460-000, Brazil

L. A. Colnago
Brazilian Agricultural Research Corporation, Embrapa Instrumentation, Rua XV de Novembro, n° 1.452, São Carlos, SP 13560-970, Brazil

© Springer Nature Switzerland AG 2019   131
S. Vaz Jr. (ed.), *Sustainable Agrochemistry*,
https://doi.org/10.1007/978-3-030-17891-8_5

are based on the magnetic properties of these atomic particles. Conversely to others spectroscopies, NMR and EPR signal is detected only in the presence of an external magnetic field (**B**) (Zeeman effect), since in the absence of **B** the energy levels are degenerated. In the presence of **B**, the spin energy levels are separated, and the MR phenomena are observed when an electromagnetic radiation match the energy difference between the energy levels causing the spin transition and this energy absorption can be monitored and the spectrum recorded.

The NMR phenomenon can also be explained using classical mechanics. In this model, the magnetic properties of electron and nucleus come from the fact that electrically charged particles, rotating around the axis itself (spin), generate a magnetic field. Therefore, these particles behave like small magnets (dipoles). In the absence of a magnetic field, the electron or nucleus magnetic dipoles are randomly aligned and no net magnetization is observed. However, in the presence of an external magnetic field (**B**) the particles are polarized generating a net magnetization. In the presence of the magnetic field, the magnetic moment precesses with a frequency that depends on magnetic field strength and particle properties. The MR phenomena are observed when an oscillating magnetic field matches the precession frequency and energy can be transferred to the system and energy absorption is detected. In EPR, the detected magnetic probe is the unpaired electrons while in NMR the probes are nucleus with unpaired nucleons (protons or neutrons). Both magnetic probes provide information about the chemical environment they are in, i.e., the chemical structure.

5.2 Electron Paramagnetic Resonance

Electron paramagnetic resonance (EPR) is the branch of spectroscopy in which electromagnetic radiation, usually at the microwave frequency, is absorbed by molecules, ions or atoms having electrons with unpaired spins, which are called paramagnetic centers (Drago 1992), when they are subjected to a magnetic field. The EPR technique is based on the existence of a liquid magnetic moment, or resultant, associated to the electron spin (Parish 1990). This technique concerns the detection of unpaired electrons and the characterization of their chemical environment. Diamagnetic substances, that is, those that do not have unpaired electrons, cannot be detected by EPR and, therefore, do not interfere in the experiments involving paramagnetic substances.

With the EPR, it is possible to analyze non-destructively solid, liquid and gaseous samples. It is an extremely sensitive technique and, under favorable conditions, the limit of detection for paramagnetic centers is in the range of 10^{11} to 10^{12} spins g^{-1}, which is equivalent to ng g^{-1}. The resolution, however, is lost when the paramagnetic centers are close enough that significant dipole interaction occurs between them. Therefore, this technique is mainly applied to the characterization of magnetically diluted species (Goodman and Hall 1994); however, unlike infrared spectroscopy, EPR does not obtain the dilution of solid samples by simply mixing with a "silent" powder (diamagnetic) but there is a need to dilute the paramagnetic centers at the molecular level. In the case of solutions, solvents with a high dielectric constant (e.g., water) are not recommended because there is a loss of power of the energy applied, in the microwave range of the electromagnetic spectrum, by the interaction of the solvent with the electrical component of it, as well as heating and other problems (Parish 1990). However, it is possible to acquire spectra under these conditions using flat cells or special capillary tubes.

Most of the EPR experiments are performed at a frequency around 9 GHz, which is known as X-band or 35 GHz (Q-band). At these frequencies, the magnetic field ($\mathbf{B_r}$—resonance magnetic field) of the free electron will be 321.1 and 1,248.9 mT, respectively. These frequencies were chosen because they were already used in marine (9 GHz) and airports (35 GHz) radar equipments (Parish 1990). There are, however, experiments performed on other microwave frequencies such as 1–2 GHz (L-band, $\mathbf{B_r}$ = 35.7 and 71.4 mT, respectively); 3–4 GHz (S-band, $\mathbf{B_r}$ = 107 and 142 mT); 24 GHz (K-band, $\mathbf{B_r}$ = 856.4 mT); 50 GHz (V-band, $\mathbf{B_r}$ = 1.7841 T); and 95 GHz (W-band, $\mathbf{B_r}$ = 3.3898 T). Normally, the EPR experiments are performed at room temperature, but under certain circumstances, it is interesting to perform them at low temperature using liquid N_2 or even liquid He (77 and 4.2 K, respectively) for sample cooling.

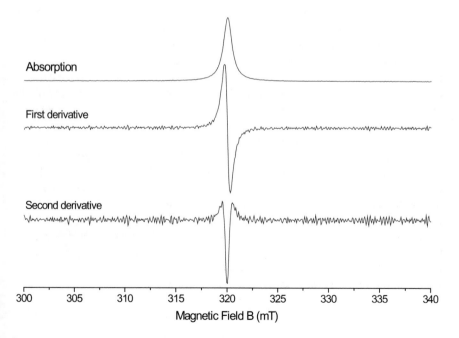

Fig. 5.1 Simulated free electron EPR spectra with noise

The frequently observed paramagnetic species include organic free radicals, paramagnetic metal complexes and excited triplet states of diamagnetic molecules (Bunce 1987). For the acquisition of an EPR spectrum, the sample is placed inside a resonant cavity whose dimensions are adjusted to the frequency of used microwave source and therefore, it is impossible to vary that frequency without varying the dimensions of the cavity. Thus, for practical reasons, it is more convenient to subject the sample to microwave radiation with fixed and known frequency and to vary magnetic field (Parish 1990). This causes the variation of energetic level difference between the spin states until the resonance condition is reached and the spin transition occurs with energy absorption that is detected by the equipment.

Generally, the EPR spectra are acquired as the first derivative of the absorption spectrum (Fig. 5.1). The resolution can be improved by the acquisition of the second derivative spectrum (Parish 1990); however, the signal-to-noise ratio in spectroscopy decreases geometrically with successive derivations (Starsinic et al. 1984).

5.2.1 EPR Probes Frequently Detected in Bioresources

Example of paramagnetic species frequently detected by EPR in natural samples of agricultural interest are Fe^{3+}; Mn^{2+}; Cu^{2+}; VO^{2+}; Cr^{3+}; defect centers in clay minerals; organic free radicals, etc. (Fig. 5.2). When comparing the obtained parameters of the EPR spectra with those of the literature, it is possible to infer about the oxidation state, the symmetry and the ligands of these paramagnetic centers among other information. This information can be related to the stability of these complexes and their consequent plant availability and environmental release potential, as well as to evaluate the redox potential of natural organic substances. However, it is also possible, in more elaborate works, to obtain more information about the bonds between the atoms, on the distribution of the unpaired electron in the molecules, and on the ordering of energy levels in the paramagnetic compounds (Mangrich and Vugman 1988).

Other experimental techniques using EPR involve the use of spin-trap and spin-labels. The former "capture" and stabilize unpaired electrons formed in reactions where the free radicals formed are so ephemeral that they would not be detected conventionally. The second ones are EPR probes that are added to substances that do not have unpaired electrons or whose own signal does not provide the desired information. Such EPR probes are readily detectable free radicals normally with hyperfine or super-hyperfine structure sensitive to the environment in which they are exposed. The coupling is called hyperfine for the case where the unpaired electron interacts with the nuclear spin of its own nucleus and super-hyperfine when the interaction occurs with the nuclear spin of adjacent nuclei.

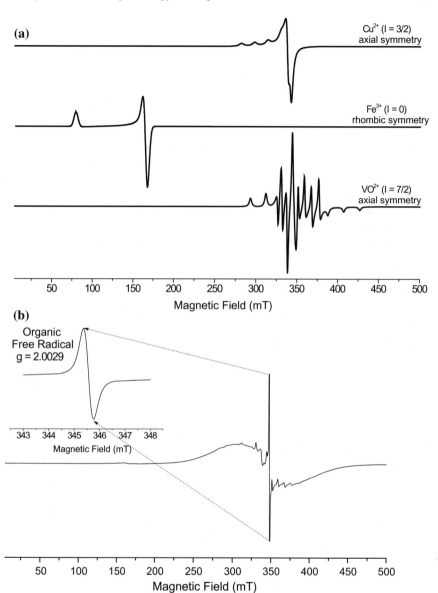

Fig. 5.2 **a** Simulated EPR spectra of some paramagnetic ions commonly found in natural samples; **b** Experimental spectrum with the same paramagnetic ions in (**a**), the inset is the experimental organic free radical spectrum

5.3   Nuclear Magnetic Resonance

Nuclear magnetic resonance spectroscopy (NMR) has been one of the most powerful analytical techniques to analyze environmental materials such as biomass and their transformation products; natural products; soil and water organic matter; live plant and animals; besides organic and inorganic materials. ^{1}H; ^{13}C; ^{15}N; and ^{31}P NMR have been used to analyze carbohydrates; proteins; lipids; and secondary metabolites from plants; animals; fungi; and algae. NMR has been used in qualitative and quantitative analyses of almost all materials, determination of the chemical composition and structure and dynamics of monomer; oligomers; and polymeric materials. One of the major NMR advantages is its non-destructive nature that maintains sample integrity and the analyzed samples can be analyzed by other methods.

When available, the NMR spectroscopy is one of most important analytical tool to analyze agriculture products and inputs, biomass and its transformation products and environmental organic matter, especially in solid state. It requires minimal sample preparation and the low solubility of several materials of agricultural interest, not preclude the analysis.

The basic NMR experiment consists of exposing a sample with magnetic nucleus ($I \neq 0$) to a static magnetic field ($\mathbf{B_0}$) and an oscillating magnetic field ($\mathbf{B_1}$), with resonant frequency (v_L) given by: $v_L = \gamma \, \mathbf{B_0} \, (2\pi)$ where γ is the magnetogyric ratio. Thus, in a 9.4 T magnet apparatus the ^{1}H ($\gamma = 2{,}675 \times 10^8$ T^{-1} s^{-1}) will resonate at about 400 MHz, which is v_L in this magnetic field and the usual way of describing the NMR equipment (Silverstein et al. 1991), while the ^{13}C in the same equipment will resonate at about 100 MHz since their γ is approximately four times smaller than those of the ^{1}H.

Carbon, which is obviously of great importance for organic chemistry, and consequently in studies of environmental organic matter, agriculture products and some inputs, has two stable isotopes, ^{12}C and ^{13}C, whose natural abundances are 98.9 and 1.1%, respectively. Of these, only ^{13}C has a magnetic moment, so carbon NMR spectroscopy is limited to this isotope, since the resulting spin of ^{12}C is zero.

As the magnetic moment (μ_n) of ^{13}C is about four times smaller than that of ^{1}H, its sensitivity in NMR experiments is lower. In addition to this, its low natural abundance makes detection of ^{13}C much more difficult, which is expressed by its less receptivity. These relationships are given by:

$$\text{Sensitivity} = (I + 1)I^{-2} \, \mu_n^3 \, \mathbf{B_0^2} \tag{5.1}$$

$$\text{Receptivity} = N\gamma^3 I(I + 1) \tag{5.2}$$

where N is the natural abundance of the nuclide in question. As the ^{1}H has the highest sensitivity and consequently the greater receptivity in the same $\mathbf{B_0}$, it is frequently used as a reference for the other nuclides. The relative receptivity and sensitivity values of ^{13}C and other nuclides of main interest for analysis of agriculture materials by NMR are given in Table 5.1.

Table 5.1 NMR properties of some nuclides utilized as probe for NMR analysis of materials of agricultural interest

Isotope	I	Natural abundance (%)	Sensitivity (same number of nucleus)	Relative receptivity
^{1}H	½	99.9844	1	1
^{13}C	½	1.108	1.59×10^{-2}	1.76×10^{-4}
^{15}N	½	0.365	1.04×10^{-3}	3.80×10^{-6}
^{31}P	½	100	6.63×10^{-2}	6.63×10^{-2}
^{19}F	½	100	8.32×10^{-1}	8.33×10^{-1}
^{29}Si	½	4.7	7.86×10^{-3}	3.68×10^{-4}
^{27}Al	5/2	100	2.07×10^{-1}	2.08×10^{-1}
^{39}K	3/2	93.1	5.10×10^{-4}	4.75×10^{-4}

The possibility or NMR analysis in Agriculture Science is virtually infinite, the search (December 14, 2018) for papers in the Web of Science platform with the terms "*NMR* or *Nuclear Magnetic Resonance* and *Soil Science* or *Agricul** or *Animal* or *Veteri** or *Food* or *Biomass*" returned more than 17,000 results, and "*NMR* or *Nuclear Magnetic Resonance*" more than 500,000 results. And so, it is impossible a detailed review about all the possibilities, instead, we will give below few examples of application obtained by the authors of this chapter.

5.3.1 Low-Field NMR in Bioresources

The NMR signal is obtained in low field pulsed NMR spectrometer usually is analyzed directly in time domain without Fourier transform, because it is a single and broad line in frequency domain (Colnago et al. 2011, 2014). Therefore, this type of NMR spectroscopy is known as time domain NMR (TD-NMR); low-field NMR (LF-NMR that will be used in this text); low-resolution NMR (LR-NMR); or relaxometry.

The intensity of the NMR signal after a radiofrequency pulse ($\mathbf{B_1}$) has been used in quantitative analyses of oils and fats in oilseeds, fruits and algae (van Duynhoven et al. 2010). The analysis is based on the linear correlation between the NMR signal intensity at a short time after the pulse (in order of 50 μs) and lipid content (van Duynhoven et al. 2010; Colnago et al. 2011, 2014). This method has been used for more than four decades and has been recognized by several agencies as standard method (Colnago et al. 2011) and is widely used in selection and breeding program to increase the oil content in fruits, seeds and algae. Comparing to conventional wet chemical methods, LF-NMR is fast; non-destructive and non-invasive; simple calibration; and does not use solvents (van Duynhoven et al. 2010). Comparing to

Fig. 5.3 LF NMR signals of linseed (LIN), soybean (SOY) and macadamia (MAC) nut obtained with the Carr-Purcell-Meiboom-Gill (CPMG) sequence

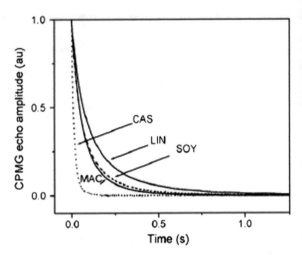

high-field NMR, it is a cheap bench top, cryogenic free spectrometer and does not need specialized operator (van Duynhoven et al. 2010; Colnago et al. 2014).

Figure 5.3 shows the LF-NMR signal of oilseed with different fatty acid composition, obtained with the Carr-Purcell-Meiboom-Gill (CPMG) sequence. The exponential decays are governed by the transverse relaxation time or T_2 and are related to oil viscosity that consequently depends on fatty acid composition. The linseed (LIN) decay shows the longest T_2 since it is composed mainly by linolenic acid that has three double bonds and is the fatty acid with the lowest viscosity. The soybean (SOY) has the intermediate decay and is composed mainly by linoleic acid with two double bonds and intermediate viscosity. The macadamia (MAC) decay shows the shortest T_2 as is composed principally by oleic acid with a single double bond and is the most viscous unsaturated fatty acid. These NMR signals have been used also for measuring sugar content in fresh fruits, meat quality, and fat content in mayonnaises among many other chemical and physical properties.

5.3.2 High-Resolution Liquid–State NMR

Analyses of samples in liquid state or solution are the major application of NMR. These analyses are known as high-resolution NMR (HR-NMR) and usually are performed in high magnetic field (>9 T) with high-field homogeneity (<0.01 ppm) and have been used to study the structure and dynamics of molecules, in qualitative and quantitative analysis; monitor chemical reactions; and among many others applications (Levitt 2008). Because they are the most common application of NMR they will not be discussed in details. In biomass, HR-NMR has been used in the determination of the structure and dynamics of fatty acids, fatty acid methyl esters (biodiesel) (Berman et al. 2015), triacylglycerides (oil and fat) as well as the structure

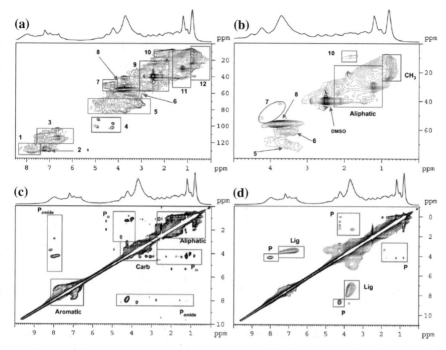

Fig. 5.4 Examples of 2D NMR spectra of hydrophobic acid fractions of soil drainage water. **a** Heteronuclear Multiple Quantum Coherence (HMQC) spectrum; **b** is an expanded region of the HMQC; **c** is the Total Correlation Spectroscopy (TOCSY); and **d** is the Nuclear Overhauser Effect Spectroscopy (NOESY). Detailed interpretation of these spectra can be found in Byrne et al. (2010)

and dynamics of mono and oligosaccharides (Vliegenthart and Woods 2006). These analyses have been performed using ^{1}H and ^{13}C NMR and are based on chemical shift, spin-spin coupling, signals area, relaxation times using uni- and multidimensional (Fig. 5.4) pulse sequences (Levitt 2008).

5.3.3 High-Resolution Solid-State NMR in Bioresources

In organic matter, such as biomasses; agriculture products and inputs; and environmental organic matter, solid-state NMR (SSNMR—Fig. 5.5) is mostly restricted to spin ½ nuclei, mainly ^{13}C and ^{1}H. The main advantages of SSNMR compare to liquid state are: No dilution of the samples, and so insoluble materials, can be analyzed, besides the higher signal intensity than in liquid state, saving equipment time for the spectra acquisition due to no sample dilution and technical concerning, such as ^{1}H–^{13}C cross-polarization, that increase the signal intensity of the low sensitivity and rare ^{13}C nuclei (Table 5.1) and decrease of the experimental time due to the shortness of the ^{13}C relaxation time (theoretical and technical details about these are

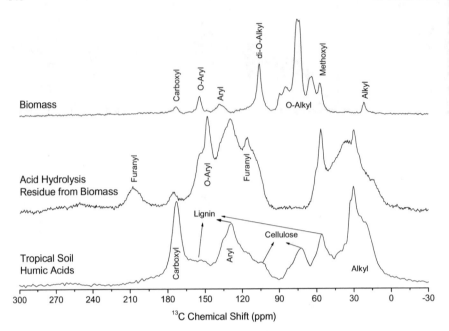

Fig. 5.5 Examples of ^{13}C SSNMR spectra of sugar cane bagasse (Biomass); its acid hydrolysis residue; and soil humic acids, a fraction of the soil organic matter, extracted from Amazonian soil

not the scopes of this review and can be found in the excellent available textbooks about NMR); and the possibility of experimental observation of the space mediated phenomena that can provide unique conformational and structural information about the sample. And as disadvantages, we can mention the loss of resolution (broader resonance lines than in liquid state) and less possibility of multidimensional experiments.

5.4 Case Study

EPR and NMR in the study of components of the soil organic matter from tropical soils (adapted from Vaz Jr 2010).

The soil organic matter (SOM) originates from the decomposition of residues of plant and animal biomass through chemical, physical and biological processes, undergoing by structural modification giving rise to a series of organic compounds, whose main representatives are humic substances (HS), which contribute fundamentally to the soil properties, such as the functionality for agricultural crops and capacity of sequestration, mobilization and redox effect of organic molecules foreign to the environment (xenobiotics) (Stevenson 1994).

HS are, undoubtedly, the most abundant sources of organic components in nature and are present in all soils and natural waters containing organic matter (OM)

(Stevenson 1994). It is now accepted that peat and soil OM are the main active carbon sources in terms of the reversibility of entry and exit of this element in the biogeochemical cycle of nature (Jenkinson et al. 1991).

The constitution of HS, based on operationally on acid or alkaline solubility, is as follows (Stevenson 1994):

- Fulvic acids (FA): fraction soluble in all acid and alkaline mediums;
- Humic acids (HA): fraction soluble in alkaline medium;
- Humin: fraction insoluble in any acid or alkaline medium.

Of these three fractions, HA is more susceptible to structural changes due to soil management practices and MO degradation processes (Tatzber et al. 2008). From the structural point of view, the definition of a chemical structure of HA is a controversial topic, since there are two proposals under discussion: macromolecular (Swift 1999) and supramolecular (Piccolo 2001; Burdon 2001).

Structural information is very significate to understand the nature, properties and functions of HA and OM in order to develop, establish and improve productive systems of food and non-food crops, as can be seen in the Chap. 7 dedicated to this theme. To reach these information, NMR and EPR, are analytical tools of large applicability.

The samples originated from two different management situations: absence of cultivation (sample A) and no-tillage for 20 years (sample B). These samples were originated from red podzolic soils from Brazilian Southeast region (São Paulo State) and were prepared by mixing the subsamples of their individual points of lower depth, followed by drying at room temperature for 15 days and homogenization in a 2 mm granulometric sieve. The HAs were extracted according to the procedure suggested by the International Humic Substances Society (2019).

EPR

For this spectroscopic technique, the sample of interest is subjected to a magnetic field applied so that the unpaired electrons lose the spin degeneration and the electronic levels of Zeeman are established, being used for the observation of paramagnetic species, such as organic free radicals in HA (Weil et al. 1994). In the case of AH, the chemical species of interest is the semiquinone-type radical (Fig. 5.6), which is detected at a magnetic field value around 3,400 G of the X-band of electromagnetic radiation at approximately 9 GHz (Senesi 1990). Following these conditions, the quantitative parameters can be determined: (a) isotropic factor-g value; (b) line width (ΔG); and (c) concentration of sample radicals (spins g^{-1}), which can provide information on the chemical ambient to which the semiquinone-type radical is bound, as well as on the presence of the electron transfer interaction mechanism due to the increase in the radical concentration (Christoforidis et al. 2007).

An EPR system EMX (Brüker) was used for the characterization of the semiquinone organic radical, where approximately 15 mg of solid sample, in duplicate, were transferred to quartz tubes in order to have both the same volume filled, and soon thereafter inserted individually into the rectangular cavity of the spectrometer, with each measurement performed at room temperature and at 16 scans, with a

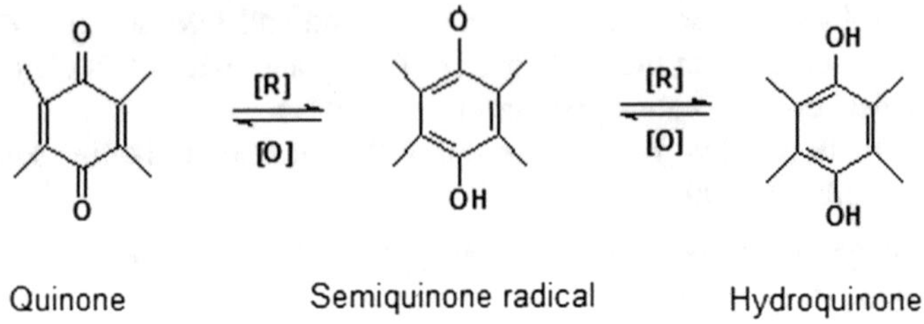

Fig. 5.6 The semiquinone radical present in humic acids and its stabilization

Table 5.2 Values for the EPR parameters determined for samples of HA extracted from tropical soils	HA sample	Spin g^{-1} (s.d.)	Factor-g (s.d.)	Line width (ΔG average)
	A	2.43×10^{17} (052)	2.0033 (0.0000)	4.6
	B	6.27×10^{16} (0.57)	2.0033 (0.0005)	4.3

spectral gain of 1.00×10^4, in the X-band; were used as standards: KCl/Cr^{3+} (strong pitch), synthetic ruby (secondary pattern for correction of undesirable resonance cavity effects) and synthetic ruby/Cr^{3+} (internal standard for determination of Factor-g). The values of the parameters, determined in duplicate, are presented in Table 5.2.

The values for absorption line width and for the g-factor are in agreement with those characteristic of HAs, that is, from 4 to 6 ΔG line width (Christoforidis et al. 2007) and from 2.003 to 2.005 for the G-factor (Steelink and Tollin 1962). Regarding to the concentration of semiquinone radicals, expressed in spin g^{-1}, the HA *A* presented a value close to that found for lignin in soil (5.0×10^{16} spin g^{-1}) (Steelink and Tollin 1962), with a lower degree of humification (Martin-Neto et al. 1998). The sample *B* presented a characteristic value for HA extracted from soil (Saab and Martin-Neto, 2003).

NMR

NMR technique is one of the instrumental analytical techniques conventionally used for the characterization of HA (Stevenson 1994), in the mode of cross-polarization with variable amplitude ^{13}C in the solid state. It can be mentioned its use for the characterization of natural and synthetic, modified and unmodified HA (Sachs et al. 2002), and in the evaluation of the effect of the humification process under the molecular structure of the HA, confronting HAs extracted from soil with historical cultivation with HAs originated from composting of food and vegetable residues (Adani et al. 2006).

An Unity Inova 400 MHz spectrometer (Varian) was used to characterize the carbons—^{13}C nuclei—as a function of their chemical ambient, operating at a resonance frequency of 100.58 MHz, a spectral band of cross-polarization of 50 kHz, contact time of 1 ms, 500 ms repetition time, 0.0128 ms acquisition time, 0–230 ppm scan, with the samples analyzed in 5 mm zirconia rotors with magic angle of 6.4 kHz

Table 5.3 Percentage of relative area for the main functional groups to which the carbons have been associated in the HA molecule and the derived molecular parameters

C-type assigned	Chemical shift range (ppm)	Relative area (%)	
		Sample A	Sample B
C-alkyl	0–45	28.2	21.4
C- methoxyl C-N-alkyl	45–65	12.1	11.9
C-O-alkyl	60–110	28.2	27.8
C-aromatic	110–140	11.8	16.0
C-phenolic	140–160	5.9	6.9
C-carboxylic	160–185	10.2	10.5
C-carbonyl	185–230	5.0	2.2
Degree of aromaticity (%)[a]		*32.4*	*36.8*
C aliphatic content[b]		*68.5*	*61.1*
Polarity index[c]		*1.2*	*1.3*

[a]Degree of aromaticity (%) = C-aromatic total area (110–230 ppm)/all peaks total area (0–230 ppm)
[b]C aliphatic content = non-substituted C-aliphatic area (0–45 ppm) + C associated to O and N area (45–65 ppm) + sugar C-derived area (60–110 ppm)
[c]Polarity index = Σ C linked to O area (60–110 ppm, 140–160 ppm, 160–185 ppm and 185–230 ppm)/C-aromatic area (0–45 ppm) + C-alkyl area (110–140 ppm)

and channel ramp for ^{1}H from 110 to 60% (in kHz), according to Novotny (2002). Obtained results are presented in Table 5.3.

These values present the molecular characteristics and properties of the HA molecule, which can vary according to MO source, whether conditions and tillage or non-tillage historic of the soil. Furthermore, these values can be directly applied in soil science, e.g., in strategies to improve the soil quality.

5.5 Conclusions

EPR and NMR can be seen as high-advanced analytical techniques to understand the constitution, properties, functionality and quality of food and non-food crops and soil organic matter, contributing to improve the productive systems related to the agroindustrial chains.

Although they are typically laboratory-manipulated techniques—except for some NMR spectrometers—their advantages for the advancement of knowledge and the technological applications of biomass derivatives—as renewable and sustainable raw materials—make them of considerable relevance to agricultural chemistry.

References

Adani F, Genevini P, Tambone F, Montoneri E (2006) Compost effect on soil humic acid: a NMR study. Chemosphere 65:1414–1418

Berman P, Meiri N, Colnago LA, Moraes TB, Linder C, Levi O, Parmet Y, Saunders M, Wiesman Z (2015) Study of liquid-phase molecular packing interactions and morphology of fatty acid methyl esters (biodiesel). Biotechnol Biofuels 8:12

Bunce NJ (1987) Introduction to the interpretation of electron spin resonance spectra of organic radicals. J Chem Educ 64:907–914

Burdon J (2001) Are the traditional concepts of the structures of humic substances realistic? Soil Sci 166:752–769

Byrne CMP, Hayes MHB, Kumar R, Novotny EH, Lanigan G, Richards KG, Fay D, Simpson AJ (2010) Compositional changes in the hydrophobic acids fraction of drainage water from different land management practices. Water Res 44:4379–4390

Christoforidis KC, Un S, Deligiannakis Y (2007) High-field 285 GHz electron paramagnetic resonance study of indigenous radicals of humic acids. J Phys Chem A 111:11860–11866

Colnago LA, Azeredo RB, Marchi Netto A, Andrade FD, Venancio T (2011) Rapid analyses of oil and fat content in agri-food products using continuous wave free precession time domain NMR. Magn Reson Chem 49:113–120

Colnago LA, Andrade FD, Souza AA, Azeredo RB, Lima AA, Cerioni LM, Osán DJ, Pusiol DJ (2014) Why is inline NMR rarely used as industrial sensor? Challenges and opportunities. Chem Eng Technol 37:191–203

Drago RS (1992) Physical methods for chemists, 2nd edn. Saunders, Orlando

Goodman BA, Hall PL (1994) Clay mineralogy: spectroscopic and chemical determinative methods, Chapter 5. In: Wilson MJ (ed). Chapman & Hall, London

International Humic Substances Society (2019) Isolation of IHSS samples. http://humic-substances.org/isolation-of-ihss-samples/

Jenkinson EJ, Adamns DE, Wild A (1991) Model estimates of CO_2 emissions from soil in response to global warming. Nature 351:304–306

Levitt MH (2008) Spin dynamics: basics of nuclear magnetic resonance, 2nd edn. Wiley, Chichester

Mangrich AS, Vugman N (1988) Bonding parameters of vanadyl ion in humic acid from the Jucu river estuarine region, Brazil. Sci Total Environ 75:235–241

Martin-Neto L, Rossel R, Sposito G (1998) Correlation of spectroscopic indicators of humification with mean annual rainfall along a temperate grassland climosequence. Geoderma 81:305–311

Novotny EH (2002) Estudos espectroscópicos e cromatográficos de substâncias húmicas de solos sob diferentes sistemas de preparo [Spectroscopic and chromatographic studies of humic substances from soils under different preparation systems]. Doctoral thesis, Universidade of São Paulo, São Carlos. https://doi.org/10.11606/t.75.2002.tde-29032004-182153

Parish RV (1990) NMR, NQR, EPR and Mössbauer spectroscopy in inorganic chemistry. Elis Horwood, London

Piccolo A (2001) The supramolecular structure of humic substances. Soil Sci 166:810–832

Saab SC, Martin-Neto L (2003) Use of the EPR technique to determine thermal stability of some humified organic substances found in soil organic-mineral fractions. Quím Nova 26:497–498

Sachs S, Bubner M, Schmeide K, Choppin GR, Heise KH, Bernhard G (2002) Carbon-13 NMR spectroscopic studies on chemically modified and unmodified synthetic and natural humic acids. Talanta 57:999–1009

Senesi N (1990) Applications of ESR spectroscopy in soil chemistry. In: Stewart BA (ed) Advances in soil science, vol 14. Springer, New York, pp 77–130

Silverstein RM, Bassler GC, Morrill TC (1991) Spectrometric identification of organic compounds, 5th edn. Wiley, New York

Starsinic M, Otake Y, Walker PL Jr, Painter PC (1984) Application of FT-ir spectroscopy to the determination of COOH groups in coal. Fuel 63:1002–1007

Stevenson FJ (1994) Humus chemistry: genesis, composition, reaction, 2nd edn. Willey, New York

Steelink C, Tollin G (1962) Stable free radicals in soil humic acid. Biochim Biophys Acta 59:25–34

Swift RS (1999) Macromolecular properties of soil humic substances: fact, fiction, and opinion. Soil Sci 164:760–802

Tatzber M, Stemmer M, Spiegel H, Katzlberger C, Hanernhauer G, Gerzabek MH (2008) Impact of different tillage practices on molecular characteristics of humic acids in a long-term field experiment—an application of three different spectroscopic methods. Sci Total Environ 406:256–268

van Duynhoven J, Voda A, Witek M, Van As H (2010) Time-domain NMR applied to food products. Annu Rep NMR Spectrosc 69:145–197

Vaz Jr S (2010) Estudo da sorção do antibiótico oxitetraciclina a solos e ácidos húmicos e avaliação dos mecanismos de interação envolvidos [Study of the antibiotic oxytetracycline sorption to soils and humic acids and evaluation of the interaction mechanisms involved]. Doctoral thesis, Universidade of São Paulo, São Carlos. https://doi.org/10.11606/t.75.2010.tde-30062010-155624

Vliegenthart JFG, Woods RJ (2006) American chemical society. Meeting: NMR spectroscopy and computer modeling of carbohydrates: recent advances. American Chemical Society, Washington, DC

Weil JA, Bolton JR, Wertz JE (1994) Electron paramagnetic resonance: elementary theory and practical applications. Wiley, New York, p 568

Chapter 6
Chemical Analyses for Agriculture

Sílvio Vaz Jr.

Abstract Chemical analyses play an important role in agriculture, as supporting technologies at all stages of agro-industrial chains as grains, forests, pulp and paper, waste and agricultural residues, among other sources of agricultural products. Furthermore, chemical analyses give the knowledge of chemical composition and the presence or absence of contaminants in food and animal feed, and the soil and water quality and their pollution levels. A set of relevant analytical techniques are discussed in accordance with their application in the agriculture.

Keywords Analytical techniques · Analytical chemistry · Agricultural matrices · Green analytical chemistry

6.1 Introduction

The technological development of modern society has stimulated the need for control of products and processes, both to ensure that final products are consumed according to quality standards, and to prevent negative impacts on the environment. The concern of the society demanding a sustainable supply became a point of strong commercial appeal to the productive sectors such as agribusiness, since the latter has been proposed in recent years a reduction in the greenhouse gases, increased productivity combined with lower tillage per area, decrease in agrochemicals and application of sustainable practices.

Chemical analyses play an important role in agriculture, as supporting technologies at all stages of agro-industrial chains as grains, forests, pulp and paper, waste and agricultural residues, among other sources of agricultural products. Furthermore, chemical analyses give the knowledge of chemical composition and the presence or absence of contaminants in food and animal feed, and the soil quality and its pol-

S. Vaz Jr. (✉)

Brazilian Agricultural Research Corporation, National Research Center for Agroenergy (Embrapa Agroenergy), Embrapa Agroenergia, Parque Estação Biológica, s/n, Av. W3 Norte (Final), Brasilia, DF 70770-901, Brazil
e-mail: silvio.vaz@embrapa.br

© Springer Nature Switzerland AG 2019
S. Vaz Jr. (ed.), *Sustainable Agrochemistry*,
https://doi.org/10.1007/978-3-030-17891-8_6

lution. Then, it ensures the quality and the reliability of agricultural products and processes from the producer to the consumer.

Initially, it should be taken into account that chemical analysis can be applied in three different or complementary situations:

- Characterization: observation of some physical property attributed to the *analyte*—the species of interest in the analytical process. For example, the absorption of visible radiation in the wavelength range of 400–450 nm or the behavior of the molecule against the incidence of radiation of other wavelengths—this is the typical application of certain spectroscopic techniques (infrared, nuclear magnetic resonance) and microscopic techniques.
- Identification: qualitative information on the presence or absence of the analyte—a good example is mass spectrometry, which identifies the compounds from the fractionation of their molecular structure.
- Determination: quantitative information on the analyte concentration in the sample—an example is the elemental analysis of the composition and the chromatographic analyzes coupled to the detection techniques.

This chapter comprises the use of several analytical techniques to be applied in large agricultural necessities, according to these three purposes, together or separated.

6.2 Most Used Analytical Techniques

6.2.1 Analytical Techniques Based on Mass, Volume, Mol, and Charge

The analytical techniques applied in the quantification of the analytes can be divided into two classes: the *classical techniques* based on the measurement of mass, moles, and charge, which provide absolute values, and the *instrumental techniques*, which work with values expressed as mg L^{-1}, mg kg^{-1}, μg m^{-3}, and so on.

The following equations express the fundamentals of these two sets of techniques.

$$A_S = kn_A \tag{6.1}$$

$$A_S = kC_A \tag{6.2}$$

Equation 6.1 applies to classical techniques, where A_S is the measured signal—or response—of the analyte, k is the proportionality constant to be standardized, and n_A is the number of moles, charge, or grams obtained for the measurement. Equation 6.2

Table 6.1 Physical properties employed in the most commonly used analytical techniques in chemical analysis

Properties	Instrumental techniques
Absorption of radiation	Spectrophotometry and photometry (ultraviolet and visible) Atomic spectrometry Infrared spectroscopy (near, medium, and far)
Electric current	Voltammetries (cyclic, square wave, anodic, cathodic, polarography)
Emission of radiation	Emission spectroscopy (X-ray, ultraviolet, and visible) Optical emission spectrometry Fluorescence (X-ray, ultraviolet, and visible)
Mass	Gravimetry
Electric potential	Potentiometry
Mass/charge ratio	Mass spectrometry
Refraction of radiation	Refractometry and interferometry
Electrical resistance	Conductometry

Modified from Skoog et al. (2014)

applies to instrumental techniques, where A_S is also the measured signal (or response) of the analyte, k is again the proportionality constant to be standardized, and C_A is the relative concentration of the measured analyte. However, some spectroscopic techniques treated in this section do not obey these two concepts, since they provide information about the structural characteristics of the sample.

In Table 6.1 are listed some physical properties explored by the analytical techniques in order to provide the response of the measurement.

Instrumental techniques measure a physical phenomenon resulting from a molecular or atomic property that will be qualitatively or quantitatively related to the analyte; that is, the physical phenomenon will produce a signal that will be directly correlated to the presence or concentration of the analyte in the sample.

Those techniques from Table 6.1—commonly considered detection techniques—can be hyphenated with separation techniques, as chromatography. It will:

- Minimize or eliminate matrix effects on the final result;
- Improve the analyte signal;
- Decrease the amount of sample and residues generated during the analytical process;
- Decrease time.

This chapter will describe detection, separation, and hyphenated techniques. Furthermore, other techniques and technologies of highlighted importance for the analysis of agricultural matrices will be treated as sensors and probes.

6.2.2   Spectroscopic, Spectrophotometric, and Spectrometric Techniques

This diverse set of analysis techniques is functionally based on the extent of absorption or emission of electromagnetic radiation by the analyte. The techniques are classified according to the wavelength or wave number of the spectral region. Table 6.2 shows the main regions of the electromagnetic spectrum as a function of wavelength (λ), indicating which types of energy transitions are produced when the radiation reaches the sample. According to the spectral region, we have different interactions of incident radiation with matter, and from there, different techniques can be developed and applied.

Electromagnetic radiation exhibits wave and particle properties. While wave has characteristics as speed, wave number, and frequency, it is quite common to use the wave number in cm^{-1} to describe the radiation. The wave number of the electromagnetic radiation (k) is directly proportional to its energy and, consequently, to its frequency (v), as can be evidenced by Eqs. 6.3 and 6.4:

Table 6.2 Approximate regions of the electromagnetic radiation spectrum and the types of energy transitions produced

Wavelength range (λ)	Region of the electromagnetic radiation	Associated transitions	Derived techniques of analysis
10^{-12}–10^{-8} m	X-rays	Electronic, in the inner layer of the atom	X-ray fluorescence
200–380 nm	Ultraviolet	Electronic, in the valence layer	Spectrophotometry of absorption or emission (fluorescence)
380 nm (violet) 480 nm (blue) 530 nm (green) 580 nm (yellow) 630 nm (orange) 730 nm (red)	Visible	Electronic, in the valence layer	Spectrophotometry of absorption or emission (fluorescence)
730–1,000 nm	Infrared	Molecular vibration	Fourier-transformed infrared spectroscopy
10^{-3}–10^{1} m	Radiofrequency and microwaves	Nuclear and electron spin	Nuclear magnetic resonance and electron paramagnetic resonance spectroscopies

Adapted from Basset et al. (1989) and Harvey (2000)

$$E = h\nu \qquad (6.3)$$

$$E = hc/\lambda = hck \qquad (6.4)$$

where

E energy (J);
h Planck's constant (6.626×10^{-34} J s);
ν frequency (Hz);
c speed of light ($2{,}998 \times 10$ m s^{-1});
λ wavelength (nm);
k wave number (cm^{-1}).

It should be remembered that the frequency ν is directly proportional to c/λ; while the wave number k is proportional to $1/\lambda$.

The following are the main techniques used for electromagnetic radiation, which can be applied to organic and inorganic analytes in agricultural matrices.

Absorption of UV–vis radiation, or molecular spectrophotometry

This technique is widely used for the identification and determination of organic, inorganic, and biological species. Usually, molecular absorption spectra are more complex than atomic absorption spectra due to the higher number of energy states of the molecule compared to the isolated atoms (see ahead in the atomic spectrometry item).

As can be seen in Table 6.1, the UV region of the electromagnetic spectrum is approximately comprised between 200 and 400 nm, and the region of the visible is comprised between 400 and 750 nm. The absorption of radiation by molecules in these regions results from the interactions between photons and electrons that participate in a chemical bond, or between electrons that are not bound in atoms like oxygen, sulfur, nitrogen, and halogens. The wavelength where absorption occurs depends on the type of bond that these electrons participate. Electrons shared in single carbon–carbon or hydrogen–hydrogen bonds are so tightly bound that they require high energy at wavelengths below 180 nm and are not observed by the most common methods of analysis. Due to experimental difficulties in working in this region, single-bond spectra are poorly explored. The electrons involved in double and triple bonds are not so strongly trapped, and consequently, they are excited more easily and produce more useful absorption peaks.

Absorption spectroscopy in UV–vis is mainly used in quantitative analysis of several organic compounds containing mainly C=O and C=C bonds, as the intensity of the absorption peaks can be directly correlated to the concentration of the analyte, now called spectrophotometry is widely used as a detector after separation by liquid chromatography, to be seen later.

The Lambert–Beer law (Eq. 6.5) correlates the signal intensity at a given wavelength value directly with the analyte concentration, which allows quantitative data to be obtained—it is worth noting that it is necessary to have the respective curve with linear behavior.

Table 6.3 Examples of chemical groups present in agricultural samples, which absorb UV radiation and their associated electronic transitions

Chemical group	Structure	Electronic transitions	λ_{max} (nm), nearly
Carbonyl (ketone)	RR'C=O	$\pi \rightarrow \pi^*$ $n \rightarrow \pi^*$	180 271
Carbonyl (aldehyde)	RHC=O	$\pi \rightarrow \pi^*$ $n \rightarrow \pi^*$	190 293
Carboxyl	RCOOH	$n \rightarrow \pi^*$	204
Amide	RC=ONH$_2$	$\pi \rightarrow \pi^*$ $n \rightarrow \pi^*$	208 210
Conjugated diene	RCH–CH=CH–CHR	$\pi \rightarrow \pi^*$	250
Aromatic	C$_6$H$_6$	$\pi \rightarrow \pi^*$	256

Adapted from Settle (1997) and from Pavia et al. (2001)

Fig. 6.1 Absorption spectra in the UV–vis region of organic matter in basic aqueous medium, with absorption bands for conjugated diene (–C=C–C=C–) and carbonyl (C=O). Adapted from Vaz (2010)

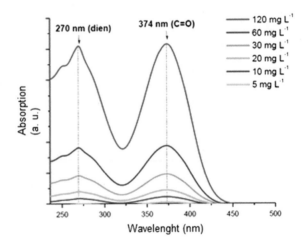

$$A = \varepsilon bc \qquad (6.5)$$

where

A absorbed radiation (arbitrary units);
ε molar absorptivity of the medium (cm^{-1} L mol^{-1});
b cell length (cm);
c concentration (mol^{-1} L).

The electron transitions give rise to the absorption spectrum as the electron passes from a lower energy state to a higher energy state. In Table 6.3 are described information on electronic transitions and wavelengths of UV absorption of some chemical groups present in several agricultural samples.

Figure 6.1 shows absorption spectra for organic matter presents in the soil.

Advantages of UV–vis spectrophotometry are (EAG Laboratories 2018):

- Fast sample analysis;
- Suitable for a wide variety of analytes;
- Much simpler than chromatographic techniques;
- User-friendly interface;
- Little maintenance required.

Limitations are:

- Subject to fluctuations from scattered light and temperature changes;
- Relatively low sensitivity;
- Other sample components may cause interferences;
- Not as specific as chromatography, for instance;
- Requires a relatively large sample volume, >0.2 mL.

Emission of UV–vis radiation, or fluorescence

The process of UV–vis emission, commonly known as fluorescence, occurs when molecules are excited by the absorption of electromagnetic radiation, upon returning to the ground state and releasing the excess of energy as photons.

The sample is excited at a given wavelength, called the excitation wavelength, and its emission is measured at a higher wavelength, called the fluorescence wavelength. This phenomenon is usually associated with systems with electrons π, that is, systems commonly with a double bond. Fluorescence usually has a sensitivity and range of work greater than those of UV–vis spectrophotometry. However, it has limited application due to the limited number of systems that fluoresce. It is a technique widely used in the analysis of molecules with aromatic rings, and it is used as a detector after a separation by liquid chromatography. Figure 6.2 presents the emission spectra of antibiotic and humic acid from soil.

Infrared molecular spectroscopy

Vibrational spectroscopy refers to a type of interaction of the radiation with vibrational states of the chemical bonds. Therefore, there is no electronic transition. Here, we can highlight infrared (IR) absorption spectroscopy in its three wavelength ranges: near, medium, and far. Polarity has a direct influence on the IR spectrum, modifying its form.

The electromagnetic region of the IR is located between the visible region and the microwaves, that is, from 12,800 to 10 cm^{-1}, remembering that the unit cm^{-1} refers to the wavenumber. As discussed above, the IR spectrum is subdivided into three regions: near-infrared (NIR), medium (MIR, mid-infrared), and far (FIR, far-infrared). The mid-infrared (MIR), which is the most used technique in organic analysis, is divided into two regions: frequency groups, from 4,000 up to 1,300 cm^{-1}, and absorption of functional groups of two atoms, or vibration, of 1,300 to approximately 700 cm^{-1}, also called *fingerprint*. In the near-infrared (NIR), the radiation is comprised between 12,800 and 4,000 cm^{-1}. The absorption bands in this region are harmonic or combinations of fundamental stretching bands, often associated with

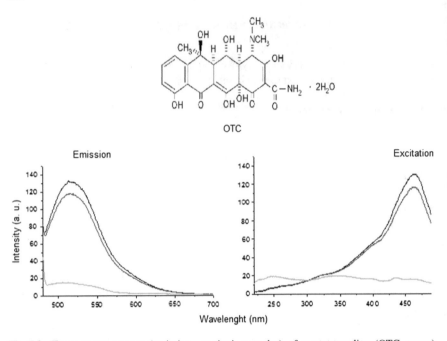

Fig. 6.2 Fluorescence spectra (emission—excitation modes) of oxytetracycline (OTC, green), humic acid from soil (black), and the product of molecular interaction between both molecules (red). Adapted from Vaz (2010)

hydrogen atoms—this has been a technique in growing in agriculture, due to the ease of handling of the sample.

IR spectra are typically employed to identify pure organic compounds or impurities, interactions, and binding formation.

Table 6.4 describes the main possible correlations for the assignment of the absorption bands in the MIR as a function of the type of bound. Figure 6.3 shows an absorption spectrum in this region.

Advantages of FTIR are (EAG Laboratories 2018):

- Capable of identifying organic functional groups and often specific organic compounds;
- Extensive spectral libraries for compound and mixture identifications;
- Ambient conditions (vacuum is not necessary, applicable for semi-volatile compounds);
- Minimum (limit of detection) analysis area: ~15 μm. *Rule-of-thumb: if you can see the sample by eye, it most likely can be analyzed*;
- Can be quantitative with appropriate standards and uniform sample thicknesses.

Table 6.4 Characteristic bands of deformations and vibrational stretches, which may be present in pesticide molecules

Band position (cm^{-1})	Assignment	Intensity
3,500–3,000	Intramolecular stretching of O–H and N–H	Medium absorption
2,940–2,900	Asymmetric stretching of aliphatic C–H	Strong absorption
1,725–1,720	Stretching of C=O in COOH and ketones	Strong absorption
1,660–1,630	Stretching of amide groups (amide band I) and quinone; C=O stretching of hydrogen bonded to conjugated ketones; stretching of COO$^-$	Strong absorption
1,620–1,600	Stretching of aromatic C=C; stretching of COO$^-$	Medium-to-weak absorption
1,460–1,450	Stretching of aromatic C–H	Medium absorption
1,400–1,390	Deformation of O–H and stretching of C–O and OH phenolic; deformation of C–H in CH$_2$ and CH$_3$; asymmetric stretching of COO$^-$	Medium absorption
1,170-950	Stretching of C–O in polysaccharides or polysaccharides-like compounds	Strong absorption

Adapted from Stevenson (1994) and Engel et al. (2011)

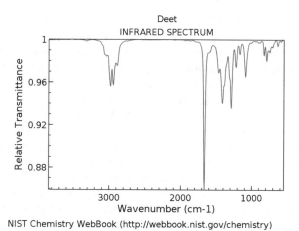

NIST Chemistry WebBook (http://webbook.nist.gov/chemistry)

Fig. 6.3 FTIR absorption spectra of *N*, *N*-diethyl-*m*-toluamide (DEET) molecule, with the sample prepared in KBr pellet; resolution of 4 cm^{-1}. Band assignments: 3,000 cm^{-1} = intramolecular stretching of N–H; 1,600 cm^{-1} = stretching of –NH$_3{}^+$; 1,540 cm^{-1} = stretching of aromatic C–H; 1,300 cm^{-1} = deformation of C–H in CH$_2$ and CH$_3$; 1,100 cm^{-1} = stretching of C=C–H. Adapted from National Institute of Standard and Technology (2018)

Limitations are:

- Limited surface sensitivity (typical limit of detection is a film thickness of 25 nm);
- Only specific inorganic species exhibit an FTIR spectrum (e.g., silicates, carbonates, nitrates and sulfates);
- Sample quantitation requires the use of standards;
- Glass absorbs infrared light and is not an appropriate substrate for FTIR analysis;
- Water also strongly absorbs infrared light and may interfere with the analysis of dissolved, suspended, or wet samples;
- Simple cations and anions, e.g., Na^+ and Cl^-, do not absorb FTIR light and hence cannot be detected by FTIR; identification of mixtures/multiple sample components may require additional laboratory preparations and analyses;
- Metals reflect light and cannot be analyzed by FTIR.

Xing et al. (2019) used FTIR coupled to photoacoustic spectroscopy to carry out a depth profiling of three types of agricultural soils (fluvo-aquic soil, red soil, and black soil) at micrometer scales associated with chemometrics. The compositional and structural differences in soil organic matters were explored from the surface to a depth of about 4 μm. Furthermore, Bekiaris et al. (2016) using also FTIR-photoacoustic spectroscopy determined the phosphorous speciation presented in biochar samples obtained from pyrolysis of biomass; hydroxylapatite and calcium phosphates were the abundant species.

Atomic absorption spectrometry

When electromagnetic radiation is applied to atoms in the gaseous state, some of these atoms can be brought to a level of energy that allows the emission of the characteristic radiation of that atom. However, most can remain in the ground state and absorb energy, which in general would correspond to the energy in the gaseous state at the wavelength they would emit if they were excited from the ground state. Thus, when atoms absorb energy, an attenuation of the intensity of the radiation beam occurs. Thus, atomic absorption spectrometry (AAS) is based on the absorption of the electromagnetic radiation by gaseous atoms in the ground state.

AAS is widely used in the analysis of metals, semi-metals, and non-metals in agricultural matrices, as fertilizers. There are three different types of atomizer: combustion flame of different gases (hydrogen, acetylene, or natural gas), graphite furnace (or electrothermal), and cold mercury vapor (for determination of the mercury present by reduction to elemental mercury), with the application of each of them depending mainly on the analyte to be determined and the limit of detection (LOD) required by the method—the flame AAS is the most common technique. The radiation absorbed has a direct relation with the analyte concentration what turns this technique very useful in quantitative analyses of metals, especially for soil analysis.

In general, the spectra obtained by AAS are simpler than those obtained by atomic or optical emission (see ahead). A particular chemical element absorbs energy at certain wavelengths. Typically, for analysis of an element, the highest absorption wavelength is chosen if there is no interference due to the absorption of the radiation

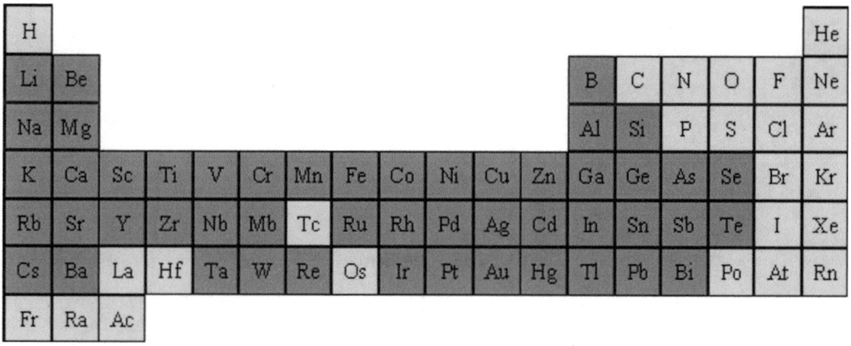

H																	He
Li	Be											B	C	N	O	F	Ne
Na	Mg											Al	Si	P	S	Cl	Ar
K	Ca	Sc	Ti	V	Cr	Mn	Fe	Co	Ni	Cu	Zn	Ga	Ge	As	Se	Br	Kr
Rb	Sr	Y	Zr	Nb	Mb	Tc	Ru	Rh	Pd	Ag	Cd	In	Sn	Sb	Te	I	Xe
Cs	Ba	La	Hf	Ta	W	Re	Os	Ir	Pt	Au	Hg	Tl	Pb	Bi	Po	At	Rn
Fr	Ra	Ac															

Fig. 6.4 Elements detectable by AAS are highlighted in pink in this periodic table. Adapted from New Mexico State University (2018)

by another element at that wavelength. Due to its simplicity and cost, AAS is the most widely used atomic method.

In Fig. 6.4 are highlighted those elements that are detectable by AAS.

Atomic emission spectrometry or optical emission spectrometry

Atomic emission spectrometry (AES), or optical emission spectrometry (OES), is based on the measurement of the emission of the electromagnetic radiation in the UV–visible region by neutral and ionized atoms, not in excited state, being widely used in elemental analysis. The most common OES system uses an argon plasma torch that can reach up to 9000 K (inductively coupled plasma, ICP) for the electrons excitation in gaseous state. ICP can also be coupled to a quadrupole mass analyzer (ICP-MS); it offers extremely high sensitivity to a wide range of elements.

The technique has high stability, sensitivity, low noise, and low background emission intensity. However, because it involves relatively expensive methods that require extensive operator training, it is not as applied as AAS. All metals or non-metals of agricultural interest, determined by AAS, can be determined by OES—the latter can favor, for some elements, the achievement of lower values of LOD and LOQ (limit of quantification).

Advantages of OES are (EAG Laboratories 2018):

- Bulk chemical analysis technique that can determine simultaneously up to 70 elements in a single sample analysis.
- The linear dynamic range is over several orders of magnitude.
- Instrumentation is suitable for automation, thus enhancing accuracy, precision, and throughput.

Limitations are:

- The emission spectra are complex, and inter-element interferences are possible if the wavelength of the element of interest is very close to that of another element.

Fig. 6.5 X-ray emission
spectra of calcium in a
sample

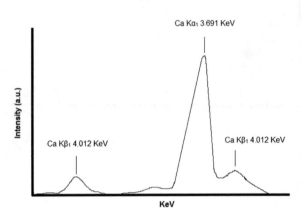

- In mass spectrometry, determination and quantification of certain elements can be affected by interference from polyatomic species, matrix elements, and atmospheric elements.
- The sample to be analyzed must be completely digested or dissolved prior to analysis in order to determine the element(s) of interest.

X-ray emission spectrometry

This technique allows a rapid and non-destructive multielement analysis for solid and liquid samples (identification and quantification). When an atom is excited by the removal of an electron from its inner layer, it emits X-rays when returns to its ground state; such radiation has a typical signal intensity for each element, which is used in the analysis (Fig. 6.5).

There are two XRF systems available: the wavelength dispersive spectrometer (WDXRF) and the energy dispersive spectrometer (EDXRF); the latter has higher signal throughput, which enables small area analysis or mapping.

Advantages of XRF are (EAG Laboratories 2018):

- Non-destructive technique;
- Can analyze areas as small as ~150 μm;
- Can analyze any solid material;
- Sampling depth ranging from a few micrometers to several millimeters depending on the material.

Limitations are:

- Cannot detect elements lighter than Al using small spot EDXRF;
- Highest accuracy measurements require reference standards similar in composition and/or thickness to the test sample.

Miller et al. (2005) used micro X-ray fluorescence to study heterogeneous soil particle samples. As observed, both single particle as well as bulk analyses must be performed on the sample to insure full elemental characterization due to the heterogeneous chemical constitution of soil particles.

6.2.3 Mass Spectrometry

Mass spectrometry (MS) is essentially a technique for detecting molecular components having the mass/charge ratio (m/z) as the unit of measurement. Depending on the ionization technique used, analytes may present with one or multiple charges. In single charge components, the m/z ratio corresponds to the total mass of the ion in Daltons. In cases where ions with two or more charges are more frequent, the calculation of the original ion mass will depend on deconvolutions of the original signal.

The direct analysis of the sample in the mass spectrometer seldom generates results that can be considered quantitatively, even if the sample is pure. This is a consequence of the high sensitivity of the technique and the efficiency of the ionization process, besides the intrinsic characteristics of each sample that allow greater or less easiness of ionization.

MS is often associated with a separation technique, usually gas chromatography or liquid chromatography—are the hyphenated techniques, where a separation technique coupled to detection and quantification technique is used. In this case, the mass spectrometer functions as a detector. Such hyphenated techniques make it possible to separate complex mixtures, identify the components, and quantify them in a single operation. Almost all measurements of MS are done under high vacuum, as this allows the conversion of most of the molecules into ions, with a lifetime enough to allow their measurement. The mass spectrometer consists essentially of three components: ionization source, mass analyzer, and ion detector.

There are several commercially available ionization systems: electron impact ionization (EI), chemical ionization (CI), fast atom bombardment (FAB), particle beam bombardment (PBB), matrix-assisted laser desorption ionization (MALDI), electrospray ionization (ESI), atmospheric pressure photoionization (API), and atmospheric pressure chemical ionization (APCI). For high molecular weight, non-volatile and heat sensitive materials, such as some pesticides and food components, MALDI, APCI, and FAB techniques are used. The most common analyzers are: quadrupole, quadrupole ion trap, and time-of-flight tube. The detection is done by electron multiplier tube.

Figure 6.6 presents a fragmentogram for the insect repellent molecule DEET.

Advantages of high-resolution MS are (EAG Laboratories 2018):

- Provides comprehensive accurate mass information in a single analysis by MS^n technology (tandem mode);
- Detects more low-level components in complex samples;
- Designed and well suited for large molecule analysis.

Main limitation is that a large volume of data to process requires an experienced analyst to operate.

Fig. 6.6 Fragmentogram of the *N*, *N*-diethyl-*m*-toluamide (DEET) molecule obtained by electron ionization (EI) technique. National Institute of Standard and Technology (2018)

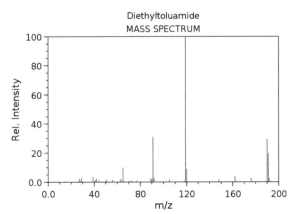

NIST Chemistry WebBook (http://webbook.nist.gov/chemistry)

6.2.4 Chromatographic Techniques

Chromatography is, conceptually, a technique of separating components from a sample according to their *retention time*, for further identification and determination/quantification; an instrumental approach is used. The following equations conceptually define the chromatography and its application.

Firstly, resolution is the quantitative measure of the degree of separation between two peaks *A* and *B*, referring to two different molecules and is defined as:

$$R = t_{R(B)} - t_{R(A)}/0.5(W_B + W_A) \tag{6.6}$$

where

$t_{R(A)}$ retention time for the peak *A*;
$t_{R(B)}$ retention time for the peak *B*;
t_M dead time;
W_A baseline width for the peak *A*;
W_B baseline width for the peak *B*.

In Fig. 6.7, we have a conceptual chromatogram, with the representations of the terms used in Eq. 6.6.

Separation between peaks *A* and *B* is governed by the *partition coefficient*, K_D, which measures the solute distribution, or analyte, from its concentration in the *mobile phase* (S_m) and in the *stationary phase* (S_s), in a condition of equilibrium:

$$K_D = [S_m]/[S_e] \tag{6.7}$$

Thus, the higher the K_D value, the lower the t_R of the analyte, and vice versa.

Fig. 6.7 Description of the components of a chromatogram, with the presence of two peaks corresponding to two different molecules (A and B). The dead time (t_M) is the time required for the solute not retained in the column to move from the injection point to the detector, not being considered in the quantitative interpretation of the chromatographic analysis

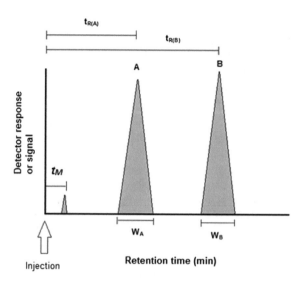

The efficiency of a separation column for the chromatographic analysis can be verified by the number of *theoretical plates* of the column:

$$N = 16(t_R/W)^2 \tag{6.8}$$

The column efficiency increases with the increase in the value of N, which also leads to an increase in the peak resolution.

In the vast majority of cases, chromatographic techniques are coupled with *detection techniques*—what is known as *hyphenated techniques*—which require an instrumental approach to their understanding and application. As forms of hyphenation, we can mention:

- Coupling of solid-phase extraction systems, known as SPE, and solid-phase microextraction (SPME)—these systems allow increased extraction performance from equilibrium phenomena, or thermal sorption-desorption or with organic solvents, which may help to reduce LOD and LOQ values.
- Liquid chromatography (LC) coupling with gas chromatography (GC), or vice versa, promoting the so-called multidimensional separation techniques that allow to work with complex mixtures, such as: LC-GC, LC-GC × GC, and LC × LC; however, the use of chemometrics for the treatment of the generated data is required for this type of hyphenation.

One way to classify the chromatographic techniques is by the physical form of the mobile and stationary phases. Thus, the first classification would be planar or column—from planar originates the *thin layer chromatography* and from column *liquid* and *gas phase chromatographies*. Table 6.5 provides a description of the functional division categories for GC and LC.

Table 6.5 Description of categories of chromatographic techniques according to the stationary phase, considering only the case where the separation takes place in chromatographic columns, which is the type of separation most applied in agricultural matrices

General classification	Category	Stationary phase	Equilibrium type
Gas chromatography	Gas-liquid	Liquid bound to solid	Gas-liquid partition
	Gas-solid	Solid	Adsorption
Liquid chromatography	Liquid-liquid partition	Liquid bound or adsorbed to solid	Liquid-liquid partition (immiscible)
	Liquid-solid or adsorption	Solid	Adsorption
	Ion exchange	Resin for ion exchange	Ion exchange
	Size exclusion	Liquid in the interstices of polymeric solid	Partition or penetration
	Affinity	Liquid bound to solid surface	Liquid-liquid partition

Adapted from Skoog et al. (2014)

The division presented above is due to physicochemical equilibrium phenomena, which are those that govern the transfer of analyte mass between the mobile and stationary phases. Partitioning is emphasized here, through the chemisorption (involving covalent bonds) and physisorption (involving intra- or intermolecular interactions, usually Van der Waals forces).

6.2.4.1 Gas Chromatography

In GC, the components of a sample are separated as a function of their partition between a gaseous mobile phase, usually the helium gas, and a liquid or solid phase contained within the column. One limitation of GC is when the analyte to be analyzed is not volatile (i.e., it is thermally stable); an alternative is the *derivatization*, when the formation of another molecule is from the analyte with lower boiling values. The elution of the components is done by an inert mobile phase (carrier gas) flow; that is, the mobile phase does not interact with the molecule of analyte.

The modernization of the equipments, through the development of new stationary phases and data processing software, also led to an investment in systems that provide higher speed during the chromatographic analysis. The shortest analysis time has the direct consequence of reducing the cost of the analytical process and increasing the analytical capacity of the laboratory. The increase in the speed of the chromatographic analysis can be related to the reduction of the size of the column, and reduction of its internal diameter, which compensates the loss of resolution in the determinations.

Table 6.6 Most common GC detectors

Detector	LOD
Flame ionization detector (FID)	0.2 pg
Thermal conductivity detector (TCD)	500 pg
Electron capture detector (ECD)	5 fg
Thermal-ionic detector (TID) or nitrogen-phosphorus detector (NPD)	0.1 pg
Mass spectrometer (MS)	<100 pg

Adapted from Harvey (2000) and Rouessac and Rouessac (2007)
$1 \text{ pg} = 10^{-12} \text{ g}; 1 \text{ fg} = 10^{-15} \text{ g}$

Fig. 6.8 Chromatogram of a mixture of organic compounds from a GC analysis with a FID detector. Courtesy of Shimadzu

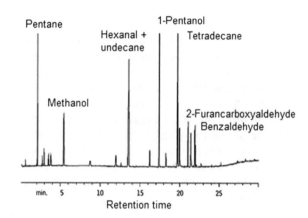

Regarding the choice of the most suitable detector to be used, the nature of the sample (matrix + analyte) should be taken into account. Several detectors are commercially available for use in GC, with thermal conductivity, flame ionization, electron capture, and mass spectrometer detectors being most commonly used. An ideal detector should meet the following characteristics:

- Adequate sensitivity;
- Good stability and reproducibility;
- Linear response to analytes, extending to several orders of magnitude;
- Temperature range from ambient to at least 400 °C;
- Ease of use;
- Similarity of response to all analytes in the sample.

In practice, the detectors do not group all of the features described above. Table 6.6 shows the most common detectors used in GC and their LOD.

Figure 6.8 presents an example of a chromatogram.

Liquid Chromatography

LC can be applied in a variety of operating modes, with the best mode depending on the structural characteristics of the analyte to be separated by the chosen analytical method. As shown in Table 6.5, the most commonly used modes or categories are: partition chromatography—or ion chromatography, adsorption chromatography, ion exchange chromatography, size exclusion chromatography, and affinity chromatography.

High-performance liquid chromatography (HPLC)

The use of low-pressure and high-pressure columns, called high-performance liquid chromatography (HPLC), outperforms GC in the analysis of semi-volatile and non-volatile organic compounds. In its many variants, it allows the analysis of complex mixtures, difficult to separate by other techniques, especially mixtures of biomolecules.

Typically, the HPLC equipment is equipped with two or more solvent reservoirs. Elution with a single solvent or a mixture of solvents of constant composition is called *isocratic elution,* while the use of a mixture of solvents at different polarity, with composition varying in a programmed manner, is a *gradient elution.* Generally, gradient elution improves the efficiency of the separation process. The pumping system is an important component whose function is to ensure a constant and reproducible flow from the mobile phase to the column. They have a pressure of 0.1–350 bar. The columns are generally stainless steel with lengths ranging from 10 to 30 cm and internal diameters between 2 and 5 mm. The column fillings (or stationary phase) typically have particles with diameters between 3 and 10 μm. Systems with particles smaller than 2 μm and pressures in the range of 1,000 bar are called ultra-high-performance liquid chromatography (UHPLC) or ultra-performance liquid chromatography (UPLC)—this mode of liquid chromatography can provide a higher resolution in a shorter retention time. Stationary phases for most chromatography modes consist of a silica material, or a polymer such as a polysaccharide or polystyrene, with functional groups of interest attached to the surface of this substrate—they may be either *normal phase* (polar stationary phase) type or *reverse phase* (non-polar stationary phase) type.

Selection of the mobile phase is critical for partitioning, adsorption and ion exchange chromatography, and less critical for the other modes. For the solvents used to form this phase, properties such as the UV–vis cut-off wavelength and the refractive index are important parameters when working with UV–vis and/or refractive index detectors. The polarity index (P') and the eluent force (ε^0) are polarity parameters that aid in choosing the phase for partitioning and adsorption chromatography, respectively.

As for GC, there are several types of detectors available commercially, and the choice usually depends on the type of analyte and the number of analyses required. Detectors may be concentration sensitive, when the analytical signal produced is proportional to the analyte concentration in the effluent or eluted; or mass sensitive,

Table 6.7 Characteristics of the main HPLC detectors

Detector	LOD
UV–vis absorption or diode array detector (DAD)	10 pg
Mass spectrometer (MS)	1 pg
Fluorescence detector (FD)	1 ng

$1 \text{ pg} = 10^{-12} \text{ g}; 1 \text{ ng} = 10^{-9} \text{ g}$

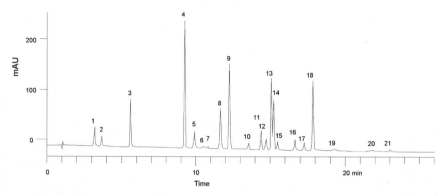

Fig. 6.9 Chromatogram of carbamate and urea pesticides obtained with an UV–vis detector at 250 nm. Reverse phase; gradient elution: water (solvent A) and acetonitrile (solvent B). Peaks: (1) aminocarb, (2) barban, (3) carbaryl, (4) carbofuran, (5) clhoropropham, (6) diuran, (7) fenuron, (8) fenuron-TCA, (9) fluometuron, (10) linuron, (11) methiocarb, (12) methomyl, (13) mexacarbate, (14) monuron, (15) monuron-TCA, (16) neburon, (17) oxamyl, (18) propham, (19) propoxur, (20) siduron, (21) swep. Courtesy of Phenomenex

when the signal produced is proportional to the mass flow rate. Table 6.7 lists the main detection systems for HPLC.

Figure 6.9 shows an example of a chromatogram.

Ion chromatography

The principle of separation of the ion chromatography (IC) is based on the ion exchange, from the electrostatic attraction between opposing charges—positive and negative. The stationary phase is a polymer resin, such as divinylbenzene bound to polystyrene by crosslinking, with ionic functional groups covalently attached to the resin. The counter ion of this ionic grouping, which may be positive or negative, should be displaced by competition with other ions of the same charge, which have greater affinity for the binding site of the functional group attached to the resin. Thus, the exchange resin can be both cationic and anionic. Retention time values will depend on the intensity of the electrostatic attraction of the analyte charge by the binding site present on the resin—higher attraction intensity leads to a longer retention time. IC and HPLC have common operational characteristics.

The types of resins available are strongly acidic cation exchange (sulfonic acid as functional group), weakly acidic cation exchange (carboxylic acid), strongly basic

Fig. 6.10 Chromatogram of an ion mixture. Detector of electrical conductivity suppression; gradient elution: NaOH aqueous solution at 10 and 35 mmol L^{-1}. Courtesy of Thermo Fisher Scientific

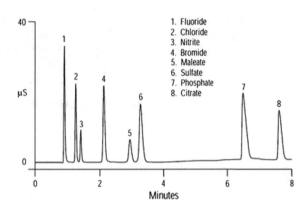

1. Fluoride
2. Chloride
3. Nitrite
4. Bromide
5. Maleate
6. Sulfate
7. Phosphate
8. Citrate

anion exchange (quaternary amine), and weakly basic anion exchange (amine). The choice will depend on the analyte superficial charge and the pH value of the medium.

IC is widely used in the analysis of anions and inorganic cations in aqueous medium, such as Cl^-, Br^-, NO^{3-}, SO_4^{2-}, CO_3^{2-}, K^+, Na^+, as well as organic compounds such as carboxylic acids, amines, and amides. Detector is usually electrochemical (electric conductivity detector, ECD). An example of a chromatogram is shown in Fig. 6.10.

6.2.5 Electrochemical Techniques

Electrochemistry studies the conversion of electrical energy into chemical energy, and vice versa, considering the transport of charges of ionic species. Some electrochemical techniques are based on oxidation-reduction reactions, such as potentiometry, coulometry, electrogravimetry, and voltametries and others, in Faradaic processes, as in the case of conductimetry.

Potentiometry

Potentiometry is a technique based on measuring the potential of electrochemical cells without appreciable current consumption. The potentiometric measurements are perhaps the most accomplished in the instrumental chemical analysis, being that of the hydrogenation potential is the best known and applied.

The basic structure of a potentiometer is composed of reference electrode, indicating electrode (or work electrode) and a potential measuring device. Ideally, the reference electrode is a half cell that has a known and constant electrode potential at a given temperature, independent of the composition of the analyte solution. Potentiometric methods were initially developed to determine the end point of a titration; later, they were used to determine the concentration of ionic species through the so-called *direct potentiometry*. The technique requires only the comparison of the potential developed in the cell, after immersion of the indicator electrode in the analyte solution, with its potential when immersed in standard solutions of known concentrations of the analyte.

One of the applications of direct potentiometry is the determination of the hydrogen ionic potential (pH) of aqueous media using, for this purpose, a glass electrode and a pH meter. This potentiometric method is possibly the most common analytical method ever created. In infinitely diluted solutions, the *activity* of an ionic species is approximately equal to its concentration. Thus, the concentration of the species to be determined is related to the potential of the electrode, according to the Nernst equation.

Voltammetry

Voltammetry is a technique that involves the determination of substances in solution that can be oxidized or reduced on the surface of an electrode. For these determinations, the relationships between current, voltage, and time during electrolysis in a cell are studied. The equipment for voltammetry employs three electrodes immersed in the solution containing the analyte and an excess of non-reactive electrolyte, called support electrolyte.

The current of analytical interest is the faradaic current, which arises due to the oxidation or reduction of the analyte in the working electrode. The current due to the migration of ions under the influence of an electric field is called the *capacitive current*. The voltage at the working electrode varies systematically as the current response is measured (Fig. 6.11). Various voltage-time functions called excitation signals can be applied to the working electrode; in function of these signals of excitation is that one has the type of voltammetry: *square wave*, *linear sweep in anodic or cathodic direction*, *cyclic*, and *polarography*. The simplest type is linear sweep voltammetry, where the potential in the electrode of work increases or decreases linearly while the current is recorded. With the development of differential pulse and square wave voltammetries, it became possible analyte determinations of the order of 10^{-7}–10^{-8} mol L^{-1}—measurements of lower concentrations are affected by the residual current.

Analytical pre-concentration processes have been used for trace analysis in order to increase the faradaic current. One of the techniques used is the anodic dissolution voltammetry, which can be used in the determination of toxic metals in soil.

6.2.6   Sensors and Miniaturized Probes

In the broadest definition, a sensor is an electronic component, module, or subsystem whose purpose is to detect events or changes in its environment and send the information to other electronics, frequently a computer processor. A sensor is always used with other electronics, whether as simple as a light or as complex as a computer.

A chemical sensor is a self-contained analytical device that can provide information about the chemical composition of its environment, that is, a liquid or a gas

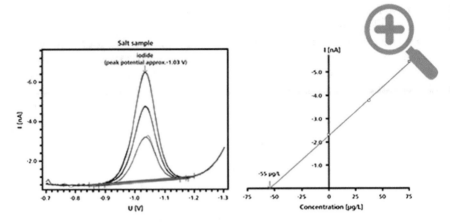

Fig. 6.11 A voltammogram for a trace-level analysis. Courtesy of Metrohm

phase (mainly) (Banica 2012). The information is provided in the form of a measurable physical signal that is correlated with the concentration of a certain chemical species (analyte). Two main steps are involved in the functioning of a chemical sensor, namely *recognition* and *transduction*. In the recognition step, analyte molecules interact selectively with receptor molecules or sites included in the structure of the recognition element of the sensor. Consequently, a characteristic physical parameter varies and this variation is reported by means of an integrated transducer that generates the output signal (the transduction step). A chemical sensor based on recognition material of biological nature is a *biosensor*. Nowadays, the development of new materials for sensors, as molecularly imprinted polymers and aptamers, has eliminated the differentiation of chemical/biochemical sensors.

Electrochemical sensors are well-recognized as easy to hand and fast devices for chemical analyses. As traditional techniques (e.g., spectroscopies and chromatographies) require either lengthy sample preparation events or complicate instrumentation and hence are time-consuming techniques, electrochemical sensors present as a good alternative, since they are rapid and stable response, with a high sensitivity and selectivity, and ease of miniaturization.

Miniaturized probes are frequently based on spectroscopic and electrochemical technologies. For instance, colorimetry can be applied for luminescent probe for field screening of Pb^{2+} in water, with a simple, fast, cost-effective, and highly selective, without requirements for additional instrumentation (Zeng et al. 2017).

NIR spectroscopic probes highlight according to their fast and easy-to-handle characteristics to analyze agricultural samples such as fruit, grains, fish, and meat. According to Bingemann (2017), radiation at NIR wavelengths penetrates samples with less scattering than other techniques, allowing internal composition to be analyzed nondestructively. Although NIR spectra are often broad, overlapping, and complex, statistical modeling can be used to unlock their secrets—chemometrics, seen ahead. Spectral data for NIR light reflected from an agricultural sample such as grain

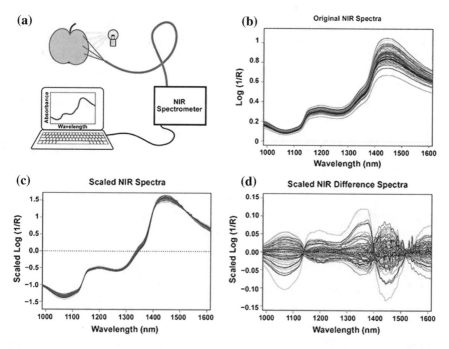

Fig. 6.12 A NIR spectrometer setup (**a**) becomes a sophisticated analytical tool for determining fruit sweetness by recording diffuse reflectance spectra (**b**) and applying a standard normal variate step that accounts for differences in sampling geometry (**c**). Differences between the scaled spectra and the overall average spectrum relate to sweetness (**d**). Courtesy of Ocean Optics

or produce are acquired and compared with a calibration model generated from spectral data of samples with known levels of the constituents of interest (Fig. 6.12).

For fruit or produce covered by a peel, the longer wavelengths used for NIR analyses are weakly absorbed; they pass through the peel, enabling sampling of the fruit pulp beneath. Little or no sample preparation is required. The measurement of NIR reflection is rapid and nondestructive. With the assistance of carefully constructed calibration models, it is also quantitative. With fruit and produce, starches and sugars (primarily fructose, glucose, and sucrose) are commonly measured to determine fruit maturity and quality. Starch, for example, has some specific wavelengths that enable construction of a multi-parametric model that can be used to determine fruit quality. For grains, NIR spectroscopy techniques enable quick characterization of samples for moisture, protein, fat, and starch content, with quantitative results obtained using statistical modeling and the appropriate calibration protocols for the parameters of interest.

The techniques are used during grain processing and in storage, and can determine nutritional content data for product labeling. NIR spectroscopy also plays an important role in quality control and the classification of finished goods, including sorting gluten-free from standard grains. Moisture content analysis of grains is important for determining proper product handling during processing to avoid spoilage by mold or fungi. Moisture analysis also provides information used to determine the types of grains best suited for a particular use or type of grinding. In addition, moisture content is used to determine if processes such as freeze-drying should be employed to preserve the grains.

In a typical quantitative chemometric analysis, a number of spectra are recorded for a variety of samples spanning a wide range of values for the property of interest, each of which is tagged to the measured value as determined in a subsequent laboratory analysis. The spectra are preprocessed to eliminate setup variations and correlated to the laboratory results, yielding a model that allows the user to predict the property of interest from the spectrum alone. This prediction is tested with known samples before it is used on unknown samples. To account for long-term changes in the instrument or the seasonality of natural samples, the model needs to undergo continued tests and maintenance with additional validation samples.

6.2.7 Chemometrics

In some applications, an analytical methodology alone is not sufficient to provide qualitative or quantitative information of the sample, using only data such as the intensity of absorption or emission, and/or the region of absorption of the electromagnetic spectrum—called *univariate analysis*. Often, the analysis is associated with chemometric tools to provide the best information.

Chemometrics can be understood as an area of knowledge of chemistry that uses mathematic models, along with formal logic to interpret and predict data, thus extracting the maximum of relevant information. It is largely used for spectroscopic and chromatographic data—in the case of spectroscopic data, each wavelength is a variable. Spectra or complete chromatograms, parts of them or selection of variables can be used. Since several variables are treated at the same time, the data analysis is called *multivariate analysis*. In order to carry out the multivariate analysis, the data are first organized in matrix form, called matrix X of original data, where the columns correspond to the predictor variables (such as absorbance) and the lines correspond, for example, to the concentration of an analyte (Martens and Naes 1989).

After organizing the data in the matrix, sometimes it is necessary to preprocess, eliminating irrelevant information or standardizing the data. The objective of the multivariate analysis can be from an exploratory analysis to the quantification of an analyte (Brereton 2003). The exploratory analysis is performed with the objective of obtaining initial information from a set of samples, such as the formation of clusters according to a certain chemical property. The main chemometric tool used in the exploratory analysis is the principal component analysis (PCA). When it is desired to verify similarities between samples of a certain class, samples are classified, with the most common methods being *k*-nearest neighbor (KNN), linear discriminant analysis (LDA), hierarchical cluster analysis (HCA), soft independent modeling of class analogy (SIMCA). When it is intended to predict analyte concentration, cali-

Table 6.8 Relative
intensities of fluorescence
emission at four different
wavelengths (300, 350, 400
and 450 nm) for 12
compounds, A–L

Compound	Wavelength (nm)			
	300	350	400	450
A	16	62	67	27
B	15	60	69	31
C	14	59	68	31
D	15	61	71	31
E	14	60	70	30
F	14	59	69	30
G	17	63	68	29
H	16	62	69	28
I	15	60	72	30
J	17	63	69	27
K	18	62	68	28
L	18	64	67	29
Mean	15.75	61.25	68.92	29.25
Standard deviation	1.485	1.658	1.505	1.485

Adapted from Miller and Miller (2005)

Table 6.9 Covariance matrix
for the data from Table 6.8

λ (nm)	λ (nm)			
	300	350	400	450
300	2.20455			
350	2.25000	2.75000		
400	−1.11364	−1.15909	2.26515	
450	−1.47727	−1.70455	1.02273	2.20455

Adapted from Miller and Miller (2005)

bration models are constructed, with patterns of known concentration and working range that contemplate the analyte concentration. The most widely used method for this purpose is partial least squares (PLS). For instance, Tables 6.7 and 6.8 illustrate the application of PCA on analytical data. Initially, Table 6.8 presents data for a fluorescence hypothetical analysis.

From the data of Table 6.8, we obtain the *covariance matrix*—a joint variance[1] of two variables, in Table 6.9.

This shows that, for example, the covariance for the fluorescence intensities at 350 and 400 nm is −1.15909. The table also gives the variances of the fluorescence intensities at each wavelength along the leading diagonal of the matrix. For example, for the fluorescence intensities at 350 nm the variance is 2.75. We can consider this

---

[1] Variance is the square of the standard deviation (s); covariance is the sum of variance for a certain measurement.

kind of information in a practical way to understand the propagation of errors, and consequent reliability, for a certain analysis.

According to Szymánka et al. (2015), chemometrics application for quantitative and qualitative purposes can be generated:

- For qualitative results: compound identification, compound classification, and sample classification.
- For quantitative results: sample calibration.

Chemometrics does not only apply to measurements, but also to extractions. Because it is based on multiparametric analyzes, it allows to evaluate the effect of the variation of the operational parameters on the recovery percentage values of the extraction method. It is possible, for example, to verify among several extraction methods the most suitable for a group of analytes, or the effect of the environmental matrices on the analyte group against more than one extraction method.

6.3 Main Agricultural Matrices and Their Analytical Necessities

The main agricultural matrices for chemical analyses are:

- Animal feed
- Biofuels
- Fertilizers
- Food
- Plant-based materials
- Soil

Table 6.10 describes the analytical necessities for each one.

Detailed information about the applied methodology for each analysis can be accessed inside the respective reference. In the most of cases, free of charge.

6.4 Green Analytical Chemistry

The 12 fundamental principles of green chemistry (GC) and their relation with agriculture were deeply treated in Chap. 2 as follows. Here, we will apply some principles on the chemical analysis.

Armenta et al. (2008) have discussed the term *green analytical chemistry*, its milestones, and examples of application, namely:

- Sample treatment;
- Oriented scanning methodologies;
- Alternatives to toxic reagents;

Table 6.10 Required analytical information and the corresponding analytical technique for the main analytical agricultural matrices

Agricultural matrix	Required analytical information and technique	Purpose	Reference
Animal feed	• Acid detergent fiber (ADF) and lignin (ADL) by filtration • Aflatoxins by HPLC-fluorescence • Ash insoluble in hydrochloric acid • Calcium by spectrophotometry • Chlorine by titration • Crude ash by thermogravimetry • Crude fat by ether extraction • Crude fiber by filtration • Deoxynivalenol (DON) by HPLC-UV • Dry matter by thermogravity • Dry matter digestibility—in vitro using rumen liquor • Elements by flame AAS • Fumonisins by HPLC-fluorescence • Gross energy by oxygen bomb calorimetry • Lactic acid in silages by enzymatic catalysis and spectrophotometry at 340 nm • NIR analysis • Nitrogen and calculation of crude protein by Kjeldahl method based on titration • Nitrogen and calculation of crude protein by thermogravimetry • Neutral detergent fiber (NDF) by filtration • Phosphorus by spectrophotometry • Reducing sugar by Luff-Schoorl reagent and titration • Starch by enzymatic catalysis and spectrophotometry • Urea by spectrophotometry • Volatile fatty acids (VFA) in silage by GC-FID • Zearalenone (ZON) by HPLC-fluorescence	Quality control of raw materials and products	Food and Agriculture Organization of the United Nations (2011)

(continued)

Table 6.10 (continued)

Agricultural matrix	Required analytical information and technique	Purpose	Reference
Biofuels (ethanol)	• Acidity (max.) by volumetry • Chloride content (max.) by IC • Ethanol content (min.) by GC-FID • Iron content (max.) by flame AAS • pH by electrochemistry (direct potentiometry) • Sulfate content (max.) by IC • Residues (max.) by gravimetry	Quality control of the product	Vaz (2014)
Fertilizers	• 2-amino-4-chloro-6-methylpyrimidine (AM) by HPLC-UV (295 nm) • 1-amidino-2-thiourea (ASU) by HPLC-UV (262 nm) • Activity coefficient of nitrogen by titration • Alkalinity by chelatometric titration with EDTA • Ammoniac nitrogen by distillation and (neutralization) titration • Ammonium thiocyanate (sulfurized cyanide) by IC-ECD or HPLC-ECD • Arsenic by hydride generation AAS or ICP-OES • Ash content by ignition and gravimetry • Biuret nitrogen by HPLC-UV (190 nm) • Cadmium by flame AAS or ICP-OES • Carbon dioxide by thermogravimetry • Carbon-nitrogen ratio • Citrate-soluble boron by azomethine-H and spectrophotometry • Citrate-soluble magnesium by flame AAS • Citrate-soluble manganese by flame AAS	Quality control of products	Japan Food and Agricultural Materials Inspection Center (2016)

(continued)

Table 6.10 (continued)

Agricultural matrix	Required analytical information and technique	Purpose	Reference
	• Citrate-soluble phosphoric acid by ammonium vanadomolybdate solution and spectrophotometry (420 nm) analysis		
	• Chlorine by IC-ECD		
	• Chromium by flame AAS		
	• Clopyralid and its degradation products by HPLC-MS tandem		
	• Cold buffer solution soluble nitrogen (water-soluble nitrogen) by cold buffer solution and (neutralization) titration		
	• Dicyandiamide nitrogen by HPLC-UV (215 nm)		
	• Electrical conductivity by electrical conductivity meter		
	• Granularity by dry-type sieving testing		
	• Guanidine nitrogen by HPLC-UV (190 nm)		
	• Guanylurea nitrogen by HPLC-UV (190 nm)		
	• Heat buffer solution soluble nitrogen (hot water-soluble nitrogen) by heat buffer solution and (neutralization) titration		
	• Humic acid (acid insoluble—alkali soluble component) by gravimetry		
	• Initial elution rate by standing-in-water and flow rate		
	• Lead by flame AAS or ICP-OES		
	• Melamine and its degradation products by GC-MS		
	• Mercury by cold vapor AAS		
	• Moisture or moisture content by loss heating		

(continued)

Table 6.10 (continued)

Agricultural matrix	Required analytical information and technique	Purpose	Reference
	• Nickel by flame AAS or ICP-OES • Nitrate nitrogen by distillation and titration • Nitrous acid by HPLC-UV (210 nm) • Oil content by diethyl ether extraction (using a Soxhlet extractor) and gravimetry • Organic carbon by dichromate oxidation and titration • pH by electrochemistry (direct potentiometry) • Residue of agrochemicals by multi-component analysis by HPLC-MS • Sodium by flame AAS • Soluble lime by flame AAS • Soluble magnesium by flame AAS • Soluble manganese by flame AAS • Soluble phosphoric acid by ammonium vanadomolybdate and spectrophotometry (420 nm) • Soluble silicic acid by precipitation titration • Sulfamic acid (amidosulfuric acid) by IC-ECD or HPLC-ECD • Total copper by flame AAS or ICP-OES • Total lime by flame AAS • Total nitrogen by Kjeldahl (neutralization) titration or combustion and gravimetry with a total nitrogen analyzer • Total phosphoric acid by spectrophotometry (420 nm) • Total potassium by flame AAS method or flame photometry • Total sulfur content by gravimetry		

(continued)

Table 6.10 (continued)

Agricultural matrix	Required analytical information and technique	Purpose	Reference
	• Total zinc by flame AAS or ICP-OES • Urea nitrogen by urease catalysis and titration • Water-soluble boron by ICP-OES • Water-soluble calcium by flame AAS or ICP-OES • Water-soluble cobalt by flame AAS or ICP-OES • Water-soluble copper by flame AAS • Water-soluble iron by flame AAS or ICP-OES • Water-soluble magnesium by flame AAS or ICP-OES • Water-soluble manganese by flame AAS or ICP-OES • Water-soluble molybdenum by spectrophotometry or ICP-OES • Water-soluble phosphoric acid by spectrophotometry or ICP-OES • Water-soluble potassium by flame AAS, ICP-OES or flame photometry • Water-soluble silicic acid by titration • Water-soluble zinc by flame AAS or ICP-OES		
Food	• Carbohydrates by gravimetry • Component amino acids (for proteins) by IC-ECD • Fats by gravimetry • Pesticides residues by HPLC-UV • Protein content as total nitrogen content by titration	Food and nutritional security	Food and Agriculture Organization of the United Nations (2003) Food and Agriculture Organization of the United Nations (1992)

(continued)

Table 6.10 (continued)

Agricultural matrix	Required analytical information and technique	Purpose	Reference
Plant-based materials	• Chemical speciation and quantification of metals and non-metals (e.g., catalysts for glycerin use) or verification of glucose or starchy oxidation processes by voltammetry (e.g., cyclic and square wave) • Determination of combustion properties of biomass (exothermic or endothermic) by DSC • Determination of crystallinity and chemical composition of cellulose by XRD • Determination of structural carbohydrates and lignin by HPLC-RID and spectrophotometry • Determination of sugars, byproducts, and degradation products in liquid fraction process samples by HPLC-RID • Multi-elemental quantification in solid and liquid samples from biomass residues by XRF • Structural identification and quantification of several organic compounds based on m/z ratio by MS • Structural identification of organic compounds and lignocellulosic components by IR spectroscopy (near and medium) • Surface and structural analysis of materials by SEM • Structural identification of organic compounds from biomass processing (e.g., lignocellulosic and oleaginous) by NMR (e.g., ^{13}C in solid state) • Total solids by automatic IR moisture analyzer	Raw material analysis	Vaz (2016) National Renewable Energy Laboratory (2018)

(continued)

Table 6.10 (continued)

Agricultural matrix	Required analytical information and technique	Purpose	Reference
Soil	• Calcium, copper, magnesium, iron, potassium, free and total SO_2, urea, ammonia by automated discrete photometry • Carbon content by organic elemental analysis • Effective cation exchange capacity by titration • Hydrogen content by organic elemental analysis • Nitrogen content by organic elemental analysis • Organic matter by combustion or loss on ignition • pH in $CaCl_2$ by electrochemistry (direct potentiometry) • Phosphorous (resin extractable) by spectrophotometry (625 nm) • Sulfur content by organic elemental analysis • Multi-elemental analysis (elemental nutrients, contaminants/pollutants) by ICP-OES • Total exchangeable bases by distillation and titration • Trace elemental analysis (elemental nutrients, contaminants/pollutants) by flame AAS or ICP-MS	Nutrient analysis, metal and organic contaminants/pollutants	Thermo Fischer Scientific (2018) Faithfull (2002)

AAS = atomic absorption spectrometry; FID = flame ionization detector; DSC = differential scanning calorimetry; ECD = electric conductivity detector; EDTA = ethylenediaminetetraacetic acid; GC = gas chromatography; HPLC = high-performance liquid chromatography; IC = ion chromatographic; ICP-OES = inductively coupled plasma-optical emission spectrometry; IR = infrared; MS = mass spectrometry; NMR = nuclear magnetic resonance; RID = refractive index detector; SEM = scanning electron microscopy; UV = ultraviolet; XRD = X-ray diffractometry; XRF = X-ray fluorescence

- Waste minimization;
- Recovery of reagents;
- The online decontamination of wastes;
- Reagent-free methodologies.

Chemical analyses should be based on the 12 principles of green chemistry (Anastas and Werner 1998) as the context of their application is reflected in the sustainability of the raw materials and processes. For instance, the application of seven of the most representative principles for analytical chemistry will contribute to achieving a more sustainable analytical methodology.

Waste prevention, safe solvents and auxiliaries, energy efficiency, and inherently safer chemistry for accident prevention are obvious requirements for all chemical operations. Safer chemicals, the reduction of derivatives, and the use of catalysts should be taken into account for each analysis because each analytical process has its own technical particularities. For example, the use of real-time analysis for pollution control is a good opportunity for technological development in analytical chemistry in the use of an in situ system for effluent analyses (gaseous and liquids). A large number of cases is not possible to apply all of these principles due to the particularities of either the sample or the matrix, but is very important to consider these individually in an analytical process. This exercise will ensure the "greening" of the analysis.

As a practical guidance, De la Guardia and Garrigues (2011) established the main objectives to be considered in green analytical chemistry:

- Simplification;
- The selection of reagents to be avoided based on toxicity, renewability, or degradability data;
- The maximization of information;
- The minimization of consumables, taking into consideration the number of samples, the volumes or masses of reagents, and energy consumption;
- The detoxication of wastes.

These objectives will define the best strategy to be applied as a result of the principles of green chemistry.

6.5 Conclusions

As seen in this chapter, chemical analyses play an important role in agriculture, as supporting technologies at all stages of agro-industrial chains as grains, forests, pulp and paper, waste and agricultural residues, among other sources of agricultural products. Furthermore, chemical analyses give the knowledge of chemical composition and presence or absence of contaminants in food and animal feed, and the soil and water quality and their pollution levels.

Nowadays, there are a large number of analytical techniques available to the laboratories according to their necessities based on matrices and analytes (samples). For agricultural purposes, we can highlight spectroscopic and spectrometric techniques (e.g., UV–vis, FTIR, NIR, AAS, and OES); chromatographic techniques (liquid and gaseous phases) hyphenated to a large variety of detectors; electrochemical techniques (e.g., voltammetries); and sensors and probes.

References

Anastas P, Werner JC (1998) Green chemistry: theory and practice. Oxford University Press, New York

Armenta S, Garrigues S, de la Guardia M (2008) Green analytical chemistry. TrAC Trend Anal Chem 27:497–511

Banica F-G (2012) Chemical sensors and biosensors: fundamentals and applications. Wiley, Chichester

Basset J, Denney RC, Jeffery GH, Mendham J (1989) Vogel's textbook of quantitative chemical analysis, 5th edn. Wiley, New York

Bekiaris G, Peltre C, Jensen LS, Bruun S (2016) Using FTIR-photoacoustic spectroscopy for phosphorus speciation analysis of biochars. Spectrochim Acta Part A 168:29–36

Bingemann D (2017) Near-infrared spectroscopy probes food freshness. https://www.photonics.com/Articles/Near-Infrared_Spectroscopy_Probes_Food_Freshness/a61511. Accessed Sept 2018

Brereton RG (2003) Chemometrics: data analysis for the laboratory and chemical plant. Wiley, Chichester

De la Guardia M, Garrigues M (eds) (2011) Challenges in green chemistry. RSC Publishing, Cambridge

EAG Laboratories (2018) Techniques. https://www.eag.com/. Accessed Sept 2018

Engel RG, Kriz GS, Lanpman GM, Pavia DL (2011) Introduction to organic laboratory techniques—a small scale approach, 3rd edn. Cengage Learning, New York

Faithfull NT (2002) Methods in agricultural chemical analysis: a practical handbook. CABI Publishing, Wallingdorf

Food and Agriculture Organization of the United Nations (1992) Manual of food control 13. Pesticide residues analysis in food control laboratory. FAO Food and Nutrition Paper. FAO, Rome

Food and Agriculture Organization of the United Nations (2003) FAO food and nutrition paper 77. http://www.fao.org/docrep/006/Y5022E/y5022e00.htm#Contents. Accessed Sept 2018

Food and Agriculture Organization of the United Nations (2011) Quality assurance for animal feed analysis laboratories. FAO Animal Production and Health Manual N°. 14. FAO, Rome

Harvey D (2000) Modern analytical chemistry, 1st ed, 798 pp. McGraw Hill, Boston, ISBN 0-07-237547-7

Japan Food and Agricultural Materials Inspection Center (2016) Testing methods for fertilizers. http://www.famic.go.jp/ffis/fert/obj/TestingMethodsForFertilizers2016.pdf. Accessed Sept 2018

Martens H, Naes T (1989) Multivariate calibration. Wiley, Chichester

Miller TC, DeWit HL, Havrilla GJ (2005) Characterization of small particles by micro X-ray fluorescence. Spectrochim Acta, Part B 60:1458–1467

Miller JN, Miller JC (2005) Statistics and chemometrics for analytical chemistry, 5th edn. Pearson, Harlow

National Renewable Energy Laboratory (2018) Biomass compositional analysis laboratory procedures. https://www.nrel.gov/bioenergy/biomass-compositional-analysis.html. Accessed Sept 2018

New Mexico State University (2018) Atomic absorption spectroscopy. https://web.nmsu.edu/~esevosti/report.htm. Accessed Sept 2018

National Institute of Standard and Technology. NIST (2018) Standard reference data program collection. https://www.nist.gov/srd. Sept 2018

Pavia DL, Lampman GM, Kriz GS (2001) Introduction to spectroscopy, 3rd edn. Thomson Learning, New York

Rouessac F, Rouessac A (2007) Chemical analysis—modern instrumentation methods and techniques, 2nd edn. Wiley, Chichester

Settle F (ed) (1997) Handbook of instrumental techniques for analytical chemistry. Prentice Hall, New Jersey

Skoog DA, West DM, Holler FJ, Crouch SR (2014) Fundamentals of analytical chemistry, 9th edn. Cengage Learning, Belmont

Stevenson FJ (1994) Humus chemistry: genesis, composition, reaction, 2nd edn. Willey, New York

Szymánka E, Gerretzen J, Engel J, Geurts J, Blanchet L, Buydens LMC (2015) Chemometrics and qualitative analysis have a vibrant relationship. TrAC Trends Anal Chem 69:34–51

Thermo Fischer Scientific (2018) Soil analysis. https://www.thermofisher.com/br/en/home/industrial/environmental/soil-analysis.html. Accessed Sept 2018

Vaz Jr S (2010) Study of the antibiotic oxytetracycline sorption on soil and humic acids and evaluation of the interaction mechanisms involved. Doctoral thesis, University of São Paulo. https://doi.org/10.11606/t.75.2010.tde-30062010-155624

Vaz Jr S (2014) Analytical techniques for the chemical analysis of plant biomass and biomass products. Anal Methods 6:8094–8105. https://doi.org/10.1039/C4AY00388H

Vaz Jr S (2016) (ed) Analytical techniques and methods for biomass. Springer, Cham

Xing Z, Tian K, Lu C, Li C, Zhou J, Chen Z (2019) Agricultural soil characterization by FTIR spectroscopy at micrometer scales: depth profiling by photoacoustic spectroscopy. Geoderma 335:94–103

Zeng X, Zhang Y, Zhang J, Hu H, Wu X, Long Z, Hou X (2017) Facile colorimetric sensing of Pb^{2+} using bimetallic lanthanide metal-organic frameworks as luminescent probe for field screen analysis of lead-polluted environmental water. Microchem J 134:140–145

Chapter 7
The Soil Humeome: Chemical Structure, Functions and Technological Perspectives

Alessandro Piccolo, Riccardo Spaccini, Davide Savy, Marios Drosos and Vincenza Cozzolino

Abstract Humus or humic substances (HS) are of pivotal importance in the global ecosystem dynamics, since fluctuation in their amount affects not only the growth of both plants and soil microorganisms, but also the main biogeochemical cycles. The development of technologies aimed at controlling HS in the agroecosystem processes is hindered by the limited knowledge of their chemical structure and dynamics. The recent acknowledgement of the supramolecular nature of soil HS allowed to devise a fractionation procedure, called Humeomics, that enables a detailed characterization of the structure of humic molecules in soil. Humeomics produces homogeneous fractions by progressively breaking esters and ether C–O bonds but not carbon—carbon bonds. The molecules in fractions are then identified by means of advanced spectroscopic and mass spectrometric techniques, thereby providing a body of structures that may well represent the soil Humeome. Humeomics enabled to unravel the effects of different soil management practices on soil carbon dynamics and to explain the recalcitrance of HS in soil. Moreover, the application of Humeomics allowed to corroborate the novel concept of humification, that is unambiguously described as the progressive accumulation of hydrophobic molecular components, which are no longer biotically accessible, due to their rapid thermodynamically driven partitioning from liquid to the solid soil phases. Conceiving HS as supramolecular associations of relatively small compounds also helped to unravel the reactivity of HS with respect to plant and microbial development, as well as towards xenobiotics. Finally, the supramolecular understanding of HS encouraged the proposal of an innovative technology for the control of organic matter stabilization in soil. This is based on the in situ photo-polymerization of humic molecules catalysed by metal porphyrin biomimetic

A. Piccolo (✉) · R. Spaccini · V. Cozzolino
Interdepartmental Research Centre on Nuclear Magnetic Resonance for the Environment, Agro-Food and New Materials (CERMANU), University of Napoli Federico II, Portici, Italy
e-mail: alessandro.piccolo@unina.it

Department of Agricultural Sciences, University of Napoli Federico II, Portici, Italy

D. Savy
Plant Biology Laboratory, University of Liège, Gembloux Agro-Bio Tech, Gembloux, Belgium

M. Drosos
Faculty of Biology and Environment, Institute of Resource, Ecosystem and Environment of Agriculture (IREEA), Nanjing Agricultural University, Nanjing, China

© Springer Nature Switzerland AG 2019
S. Vaz Jr. (ed.), *Sustainable Agrochemistry*,
https://doi.org/10.1007/978-3-030-17891-8_7

catalysts. The resulting increase in the molecular mass of humic molecules was found not only to increase soil aggregate stability but also to sequester in soil significant yearly amounts of organic carbon. It is expected that the research findings presented here will prompt novel studies on the man-driven control of the soil Humeome in order to increase its content in soil, and contribute to positively affect both crop yields and soil microbial activity.

Keywords Humus supramolecular structure · Soil Humeome · Humeomics · Humification · Recalcitrance of soil carbon · Carbon sequestration

7.1 Introduction

Soil organic matter (SOM) can vary greatly with soil properties. Though SOM content is generally low in agricultural soils, it is nonetheless a key factor in determining the fertility of soil and the yield and quality of crops (Wood and Baudron 2018). SOM represents the largest carbon pool on Earth, approaching 1,300 Pg C in the upper 1 m and 3,000 Pg C when including deeper soil carbon (Köchy et al. 2015). SOM is composed of up to 80% by humus or humic substances (HS), which are organic materials ubiquitously found in soils, whose reactivity and dynamics are directly linked to soil quality and microbial activity, as well as to the soil C sequestration capacity and plant productivity (Canellas and Olivares 2014; Woo et al. 2014). HS have been traditionally and operationally defined by their solubility in water at different pH values: fulvic acids (FA), soluble under all pH values; humic acids (HA), insoluble at pH value < 2; and humin, insoluble at any pH value (Stevenson 1994). It is important to note that intensive agricultural practices have determined an alarming acceleration in HS mineralization, thus leading to both an increase in greenhouse gases emission from soils and a degradation of soil quality (Smith et al. 2014). Hence, preserving or even increasing the amount of soil HS in soil is essential for keeping high crop yields (Canellas and Olivares 2014) and sequestering carbon in soil (Piccolo 2012). This task can be adequately reached only once, and a thorough understanding of the molecular nature of HS is attained, in order to then wittingly and profitably intervene on their dynamics and relationships with both soil microbiota and plant roots.

7.2 Humic Substances, Chemical Composition and Physicochemical Properties

The scientific community of humic scientists has tried to unravel HS structure since the eighteenth century, when Achard (1786) for the first time isolated and fractionated peat HS. The difficulties met in precisely describing the structures and reactivity of HS, due to the HS chemical complexity and heterogeneity continued even when more sensitive and powerful analytical techniques have become available. In fact, since HS arise from the decay of plant tissues and microbial metabolites, their molecular heterogeneity is difficult to unravel. Only at the end of last century, the real nature of the HS began to be clarified (Piccolo 2001, 2002). Before then, HS were described as macropolymers, with dispersive molecular masses from ten to hundred thousand daltons. However, all efforts to isolate and characterize such macropolymers failed together with the provision of unequivocal evidence for the occurrence of any polymerization process in natural soil systems (Hayes 2009 and references therein).

Only in the mid-1990s, the supramolecular nature of HS has been called upon to account for the conformational behaviour of HS, based on the results of, first, low-pressure size-exclusion chromatography (Piccolo et al. 1996) and, then, by high-performance size-exclusion chromatography (HPSEC). Aqueous HS solution that was injected in the HPSEC system before and after the addition of very small amount of acetic acid (AcOH) showed significant changes in the elution profile (Conte and Piccolo 1999; Cozzolino et al. 2001). In particular, lowering the solution pH value from 7.0 to 3.5 produced a hypochromic effect (a dramatic reduction in peak absorbance) for some elution peaks, which were concomitantly shifted towards larger elution volumes. This phenomenon could only be explained with a new understanding of the humic molecular structure: a supramolecular association of heterogeneous and relatively small molecules (not more than 400–1000 Da) self-assembled by weak dispersive forces (van der Waals, $\pi-\pi$, $\pi-CH$) into apparently large molecular sizes. In fact, the pH value change due to the AcOH addition to the loosely bound humic suprastructures causes the formation of new intra- and intermolecular hydrogen bonds and disruption of the weak dispersive forces which metastabilize the humic clusters at pH value of 7.0 (Cozzolino et al. 2001; Piccolo 2002). The result was the separation of the suprastructures into smaller but more stable associations showing lesser absorbance of incident electromagnetic radiation. Further evidence was obtained by repeating the experiment with dioic acids, namely oxalic, malonic, succinic and glutaric acids, which should have mimicked the effects of commonly root-exuded organic acids on HS conformations (Piccolo et al. 2003a). The authors reported modifications in the HPSEC profiles comparable to those observed when adding AcOH. The employed dioic acids had progressively longer carbon chains corresponding to different pK_a values and increasing degrees of protonation at pH 3.5. Hence, concomitant to the formation of new strong hydrogen bonds, a better contact with the hydrophobic moieties of the loosely bound humic superstructures was reached by the longer chain acids, thus causing a more extensive HS conformational rearrangement. The HPSEC thus proved to be an useful tool to assess whether a complex mixture displays a supramolecular or a polymeric nature (Piccolo et al. 2001), and it is still currently employed to confirm the supramolecular nature of humic-like materials from composts or lignin-like extracts from lignocellulosic biomasses (Maia et al. 2008; Savy and Piccolo 2014; Savy et al. 2016a, 2017). The novel understanding of the supramolecular nature of HS shed new light on their possible interactions with plant exudates and microbially derived soil molecules, and provided plant physiologist better chemical bases to unravel the mode of action of HS towards plant nutrition (Canellas et al. 2008).

The conformational changes observed by HPSEC were also confirmed by diffusion-ordered nuclear magnetic resonance spectroscopy (DOSY-NMR). This technique was applied to several FA and HA in order to measure their diffusion coefficients (D) (Šmejkalová and Piccolo 2008a, b). An inverse correlation between FA and HA concentration and the corresponding D values were noted and explained with the capacity of humic molecules to self-assemble with increasing concentration, thus concomitantly decreasing the diffusion of larger associations. On the other hand, when the pH value of the humic suspensions was lowered down to 3.6, the DOSY-NMR spectra showed a significant increase in diffusivity due to the disruption of large assemblies in more rapidly diffusing smaller aggregates, thus proving also by this technique the supramolecular structure of both FA and HA (Šmejkalová and Piccolo 2008a, b).

7.2.1 Literature Evidence of the Supramolecular Structure of HS

Once the traditional paradigm of the polymeric nature of HS has been challenged by the supramolecular view, a series of experiments have been carried out, in order to verify the validity of the hypothesis postulated by Piccolo and co-workers. Varga et al. (2000), for example, suggested that the dramatic modifications of pH value in the HPSEC columns, as obtained by adding organic acids, may provide misleading results. Hence, in response, Piccolo et al. (2001) compared the HPSEC profiles of HS with those of real polymers of well-known molecular mass (namely, polysaccharides and polystyrene sulphonates) and showed that the real polymers added with AcOH did not show any change in HPSEC behaviour as it did in the case of HS, thereby showing that only metastable humic conformations can be disrupted by organic acids additions and not those of covalently stabilized polymers. Simpson (2002) analysed the DOSY diffusivity of HA at two different concentrations and also found a negative correlation between humic concentration and diffusivity. He explained the observed results by invoking the ability of organic acids of causing a modification of the supramolecular humic conformation (Simpson 2002), thus supporting the evidence brought about by Piccolo (2001, 2002).

The supramolecular concept of HS served to explain the behaviour of HS in various experiments. For example, Peuravuori and Pihlaja (2004) and Peuravuori (2005) studied HS from freshwaters derived from spodosol watersheds by both HPSEC and NMR spectroscopy. They showed that the humic substrates were actually composed of nano-scaled supramolecular associations of structurally similar compounds of various ranges of molecular sizes.

Furthermore, two-dimensional NMR spectroscopy and transmission electron microscopy were applied to study the structure of commercial and river water HS in solution as a function of cation type and concentration, pH and solution salinity (Baalousha et al. 2005, 2006). Results were explained by considering HS as associations of relatively small-sized compounds, interconnected in more complex assemblies by dispersive forces and through cation binding.

A plausible explanation for the interaction of metals with humic compounds was reached when the HS supramolecular structure was introduced. Buurman et al. (2002) analysed the thermal stability of soil HS saturated with H, Na, Ca, or Al and after a treatment with either methanol or formic and acetic acids. Interestingly, the thermal behaviour of protonated humic molecules (i.e. the control material) did not show any difference upon addition of polar organic molecules, while the thermal oxidation of Na-humates was shifted to larger temperatures than control. Moreover, a significantly lower effect of the polar compound addition was recorded for Ca-humates, whereas *non*-significant modifications of the thermal characteristics of Al-humates were noted. According to the authors, the humic components were strongly bound to each other by hydrogen bonds in protonated HS and by electrostatic bridges in Ca- and Al-humates. Such binding forces could not be overcome by adding polar organic molecules, thereby maintaining the stability of such substrates which could be only slightly altered. Conversely, in Na-humates, humic molecules were held together only by *non*-specific weaker hydrophobic interactions, which could be altered by adding polar organic molecules. It is noteworthy that such a different thermal behaviour of the metal-complexed humates can be hardly explained by considering HS as macromolecular polymers, while it is fully understandable by conceiving HS as supramolecular assemblies (Buurman et al. 2002).

Other workers exploited a wide array of analytical techniques for the characterization of the aggregates formed by supramolecular humics and hydrolysed Fe species under several pH values and mixing conditions (Siéliéchi et al. 2008). At low Fe concentration, the humic molecules self-organized into compact structures, while at increasing Fe amount the rearrangement of humic colloids was significantly reduced. The authors concluded that the interaction between HS and Fe ions led to a competition between the rearrangement of the humic network and the collision rate of metastable colloids. Again, these results were only explained by assuming HS as small self-associated molecules, rather than real polymers (Siéliéchi et al. 2008).

These conclusions were corroborated by the work of Nuzzo et al. (2013), who reported a decreasing molecular weight (MW) distribution of HA when they were progressively complexed with increasing Fe amount. The authors explained their findings by the disruption of the weak hydrophobic bonds stabilizing the humic conformational structures by the strong iron complexation that forced the humic superstructures into smaller-sized aggregates but of greater conformational stability.

Furthermore, Nebbioso and Piccolo (2009) studied the mobility and the diffusivity of Al- and Ca-humates by measuring the spin–lattice relaxation time in the rotating frame $(T1\rho(H))$ by cross polarization magic angle spinning (CPMAS NMR) spectra, and the diffusion constants (D) by DOSY-NMR spectroscopy, respectively. Results suggested that the molecular rigidity of HA significantly increased with metal addition throughout the full carbon spectral region, and that it was more pronounced for triple-charged Al ions than for double-charged Ca ions. The $T1\rho(H)$ values of the various spectral ranges indicated that the increase in molecular rigidity followed the order: aliphatic C > aromatic/double bonds C > carboxyl C. DOSY spectra also showed that complexing HA with Al^{3+} or Ca^{2+} resulted in lowering D values for alkyl components, while concomitantly increasing those for aromatic and hydroxyl-alkyl moieties. The findings suggested that metal complexation prompted a molecular size increase in the former and a decrease in the latter, thus pointing out the preferential role of saturated and unsaturated long-chain alkanoic acids to be involved in complexation with Al and Ca, with a consequent increase in conformational rigidity and molecular size of humic hydrophobic domains. On the contrary, the more hydrophilic or mobile humic components seemed to be relatively less affected by the rearrangements of humic molecules induced by metal complexation (Nebbioso and Piccolo 2009).

7.2.2 The Question of the HS Molecular Mass

Since HS were no longer regarded as polymers, questions were raised about their real molecular size. In fact, before being aware of their supramolecular nature, scientists had proposed that HS molecular mass could be up to several hundred thousand Da (Stevenson 1994), whereas the real size of humic molecules may be much less. A useful technique to investigate this point appeared the modern liquid chromatography hyphenated with mass spectrometry. The electrospray ionization interphase to mass spectrometry (ESI-MS), being a soft ionization process, has the advantage of providing unfragmented ions, thus helping to obtain not only an absolute MW value but also an unambiguous information on the chemical structure of the analysed molecules when the MS/MS mode is applied to humic samples (Nebbioso et al. 2010).

By using a low-resolution ESI-MS instrumentation, Piccolo and Spiteller (2003) showed that the average molecular size of a soil humic acid was much lower than 1,000 Da. Furthermore, sample heating by changing the ESI cone voltage was found not to produce any molecular fragmentation, thereby implying that ESI-MS enabled the assessment of the real mass of humic molecules (Peuravuori et al. 2007; Piccolo et al. 2010). However, the size fractions separated by a preparative HPSEC from the same soil humic acid used by Piccolo and Spiteller (2003) showed similar molecular sizes as the unfractionated HA, thus suggesting that the heterogeneous supramolecular associations of humic molecules responded differently to the ESI ionization. The ESI-MS technique may then have some limitation in adequately ionizing all molecules present in complex hydrophobic and hydrophilic systems, and such constraints can explain the similarities in the average molecular mass noted for both fractionated and unfractionated HA. In fact, Nebbioso et al. (2010) found that a mixture of both polar and apolar molecules had a heterogeneous electrospray ionization, depending on their most probable positioning at the surface of the evaporating droplet during the electrospray ionization process. Since hydrophobic compounds are most likely positioned at the aqueous-gas interphase, the ionized apolar components of the droplet are preferentially transferred in the gas phase, as compared to the more polar molecules, which are instead preferably retained in the inner part of the droplet and never reach the droplet surface before it is discarded. Hence, the ESI-MS detection of hydrophobic compounds is comparatively larger than for hydrophilic molecules.

The role of hydrophobic molecules in stabilizing humic biosuprastructures was pointed out by several studies. Chilom et al. (2009) reported that the apolar-containing moieties play a key role in both initiating the aggregation process and facilitating the formation of micelle-like structures. Instead, Conte et al. (2006, 2007) separated by preparative HPSEC several HA fractions of decreasing molecular size which were then characterized by advanced physical–chemical techniques. By applying the variable contact-time pulse sequence of CPMAS NMR spectroscopy, the authors were able to calculate both the cross polarization (T_{CH}) and the proton spin–lattice relaxation (T1ρH) times for the separated size fraction, and related such parameters to structural differences among the various fractions. They found that larger size fractions were characterized by a greater number of hydrophobic domains with slower molecular motion, whereas the smaller size fractions contained more polar moieties, which showed faster local molecular motion. These results were in line with the supramolecular understanding of humic matter and confirm that the apparently large sizes are stabilized by dispersive bonds among hydrophobic components.

7.2.3 Humeomics and Humeomes

Conceiving the HS as associations of relatively small molecules (<1,000 Da) held together in only apparently large molecular sizes by weak dispersive forces, such as hydrogen bonds, hydrophobic intermolecular interactions and metal-bridged inter-molecular electrostatic linkages, led to develop novel procedures to separate humic superstructures into less complex and more homogeneous sub-fractions, in order to facilitate their structural identification (Fiorentino et al. 2006; Spaccini and Piccolo 2007; Nebbioso and Piccolo 2011, 2012; Drosos et al. 2018a). The new chemi-cal fractionation of humic matter coupled to the advanced spectroscopic and chro-matographic techniques for the characterization of humic molecules was called Humeomics, similarly to other -omics approaches, and had the aim to separate the single humic components from the bulk matrix without breaking any C–C bond. Hence, the overall humic molecules separated and identified by Humeomics have been grouped under the term Humeome (Monda et al. 2018). Humeomics has been described as "a stepwise separation of molecules from the complex bulk suprastruc-ture by progressively cleaving esters and ether bonds and characterizing the separated molecules by advanced analytical instrumentation" (Fig. 7.1, Nebbioso and Piccolo 2011).

The Humeomics fractionation (Fig. 7.2) entails first of all a treatment of the bulk HS (also called RES0) with a mixture of organic solvents (dichloromethane and methanol), in order to isolate a fraction called ORG1 composed of free and unbound humic compounds, which are associated with each other only by weak dispersive forces. Then, the readily accessible ester linkages of the Humeome are transesteri-fied by means of a BF_3/CH_3OH solution, and the resulting supernatant is separated

Fig. 7.1 Conceptual
approach for the Humeomics
sequential fractionation
(Nebbioso and Piccolo 2011)

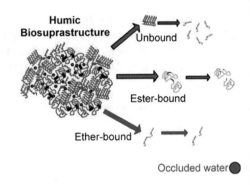

into dichloromethane- and water-soluble fractions (called ORG2 and AQU2, respectively), obtained by liquid–liquid separation of the transesterified product. A further cleavage of ester linkages with a KOH/CH₃OH mixture allows the extraction of those ester-bound compounds that were not physically accessible during the first transesterification. Again, the hydrolysed components are separated into organosoluble (ORG3) and hydrosoluble (AQU3) fractions. In the last step, the remaining humic matrix is treated with a 47% HI aqueous solution, in order to cleave ether bonds, yielding again both organosoluble (ORG4) and hydrosoluble (AQU4) components. The remaining solid residue (RES4) can then be finally treated with a traditional alkaline solution to solubilize the *non*-extractable humic compounds.

A complete Humeomics fractionation was first performed to reach the molecular composition of HA extracted from a volcanic soil (Nebbioso and Piccolo 2011). The bulk RES0 was studied by high-resolution Orbitrap electrospray (ESI) mass spectrometry after elution through a hyphenated HPSEC column. The preliminary separation of this Humeome by HPSEC simplified the Orbitrap mass spectra and allowed to ascertain that RES0 was mainly composed by alkanedioic and hydroxydioic acids. For example, alkanoic acids, di- and tri-hydroxylated C18 acids, monounsaturated C14, C16 and C18 acids, hydroxy-unsaturated C6–C20 acids, C4–C24 dioic, C16–C24 hydroxydioic, cyclic acids were found in the RES0 fraction with large elution time. In the same elution interval, a cyclic acid with empirical formula $C_7H_6O_8$ was also identified, and, due to its large unsaturation, it was suggested that it had an aromatic character, such as a hydroxylated and carboxylated furan ring. Conversely, the still great molecular aggregation in the HPSEC fraction eluted at smaller elution time prevented the correct assignment to any definite empirical formula or class structure (Nebbioso and Piccolo 2011).

The ORG1 fraction, comprising the unbound organosoluble humic components, was analysed by both gas chromatography-mass spectrometry (GC-MS) and NMR spectroscopy. The two techniques showed that this fraction contained a predominance of alkyl and saturated components, particularly mono- and dicarboxylic acids and some iso- and anteiso-branched alkanoic acids and hydroxyacids, such as ω-C16-18, β-C14 (Nebbioso and Piccolo 2011).

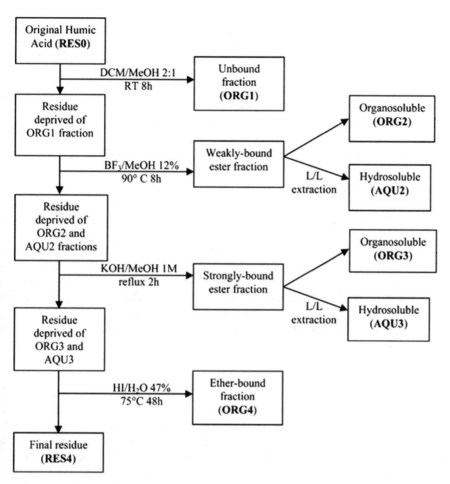

Fig. 7.2 Scheme of Humeomics fractionation (Nebbioso and Piccolo 2011). M = mol L^{-1}

The transesterification with BF$_3$/CH$_3$OH solution and the subsequent separation of the supernatant in its dichloromethane (DCM)- and water-soluble components produced the ORG2 and AQU2 fractions. The NMR spectrum of ORG2 showed that its chemical structure was similar to that of ORG1, while its GC-MS characterization revealed significant amounts of unsaturated C9-29 linear and branched alkanoic acids, α/β C9-26 mono- and di-hydroxyacids, C18 unsaturated acid, C8-30 α,ω-diacids, sterols, C12-28 n-alkanols, and substituted benzoic and cinnamic acids. It is important to note that such an abundance of both hydroxyacids and unsaturated compounds in ORG2 could have been unveiled only after the removal of the unbound alkanoic and saturated acids that were identified in ORG1 (Nebbioso and Piccolo 2011). The AQU2 contained less unsaturated compounds and larger oxygen substitution than in RES0 and the previous ORG fractions. Also, a large content of N-containing compounds was reported and ascribed to the larger affinity of nitrogen-rich components for the aqueous phase. However, such fraction still contained a large amount of odd- and even-numbered C10-18 saturated acids, C15-18 unsaturated acids, C15-18 hydroxy-unsaturated acids and C5-9 diacids. Moreover, the ESI technique allowed identifying C7-18 negatively charged compounds containing one to five N atoms, as well as cyclic acids and $C_xH_{2y}O_y$ carbohydrate structures. Several other unidentified N-containing compounds were mostly represented by the $C_8H_7O_3N_3$ empirical formula (Nebbioso and Piccolo 2011).

The resulting residue was then subjected to another transesterification reaction, carried out with a KOH/CH$_3$OH mixture, in order to cleave the remaining ester bonds. A subsequent liquid–liquid extraction allowed a further separation of molecules according to their solubility in DCM or water. The NMR spectra of ORG3 showed several peaks attributed to either aliphatic or CH-X compounds, such as hydroxylated acids, and these findings were confirmed by GC-MS. The latter technique revealed the occurrence of C9-32 saturated and unsaturated n-alkanoic acids, C6-8 α,ω-diacids, C16,22,24 ω-monohydroxyacids, C22-26 β-monohydroxyacids, C18 di- and tri-hydroxyacids, C12-28 n-alkanols, phenolic acids and steroids. Also, two-dimensional NMR techniques (COSY, TOCSY and HSQC) were applied and suggested that ORG3 contained several aromatic and CH-X molecules. The small amount of the AQU3 fraction prevented any characterization of its composition (Nebbioso and Piccolo 2011).

Finally, the ether linkages in the humic residue were cleaved by means of a 47% HI aqueous solution, thus yielding ORG4 and a remaining solid residue, named RES4. While the former was not obtained in sufficient amount, the latter was enough to be studied by liquid- and solid-state NMR spectroscopy, as well as by high-resolution ESI-MS (Nebbioso and Piccolo 2011). Results showed that RES4 contained low amount of alkyl structures and was completely depleted of alkandioic acids, as also revealed by measuring the proton relaxation time (T1ρH), that is an indicator of changes in molecular rigidity. Specifically, a tighter molecular packing was observed for RES4 fraction than for RES0. Furthermore, the solid-state dipolar-dephasing NMR technique indicated that RES4 contained a large amount of quaternary carbons, thus suggesting that humic matter remaining after Humeomics was largely formed by totally substituted or condensed aromatic moieties. Finally, Nebbioso and Piccolo (2011) solubilized RES4 in an alkaline solution and studied it by HPSEC-ESI-MS. Results revealed the presence of odd- and even-numbered C10-18 saturated acids, C14-18 unsaturated acids, C6,8 hydroxy-unsaturated acids and cyclic acidic structures. Overall, the proposed chemical fractionation allowed to identify up to 60% of the Humeome, while the lack of determination of the remaining 40% was attributed to loss of occluded hydration water, small volatile organic compounds, and/or decarboxylation reactions (Nebbioso and Piccolo 2011).

In order to further simplify the complex Humeome, Nebbioso and Piccolo (2012) size-fractionated by preparative HPSEC the same HA formerly studied by Humeomics, and the three separated fractions were again subjected to the Humeomics procedure. They identified a far larger number of compounds in the Humeome of the three size fractions than in that of the unfractionated HA. These results were explained by the ability of the HPSEC preliminary separation to weaken the conformation of the humic superstructures with the consequent separation into less complex humic fractions, which can be then studied in further details by advanced analytical tools. The hydrophobic components were mainly in the largest size fraction, while the hydrophilic compounds were eluted in the smallest size fraction. Such results were in line with previous studies (Conte et al. 2006, 2007) and corroborated the supramolecular view of humic nature, according to which the intermolecular hydrophobic interactions are of pivotal importance in stabilizing the humic associations of apparently large sizes. The N-containing molecules were mostly present in the middle size fraction, possibly owing to their amphiphilic nature. Altogether, these results indicated that the conformation of supramolecular clusters was driven by the structural characteristics of the prevalent humic molecules. Interestingly, the lack of residual RES4 fractions in any of the size fractions corroborated the idea that the preliminary HPSEC separation enabled a reduction in the strength of the humic intermolecular association, thereby allowing Humeomics to fully dissolve the molecules composing the Humeome (Nebbioso and Piccolo 2012).

A size fractionation by preparative HPSEC was also adopted to further simplify the Humeome of the RES4 fraction, after its solubilization in an alkaline solution (Nebbioso et al. 2014a). Ten different size fractions were separated and then injected into an analytical HPSEC system hyphenated with a high-resolution Orbitrap ESI-MS. The resulting total ion chromatograms of the fractions showed two eluting peaks, whose empirical formulae could be mostly associated with linear alkanoic, unsaturated, hydroxylated, hydroxy-unsaturated acids and cyclic acids. Quantitative measurements highlighted the importance of long, saturated and unsubstituted linear acids in the stabilization of large supramolecular assemblies, their interaction leading to the formation of a more favoured packing among such compounds, compared with that arising by the interaction of irregularly shaped cyclic, unsaturated or hydroxylated compounds (Nebbioso et al. 2014a). Several unknown compounds had been also eluted by HPSEC, and they were studied by running tandem MS. The double fragmentation allowed to infer the nature of compounds whose m/z was 129, 141, 155 and 217, and whose empirical formulae were $C_6H_{10}O_3$, $C_6H_6O_4$, $C_6H_4O_5$ and $C_7H_6O_8$, respectively. The analysis of daughter ions helped to assign the ESI-derived peaks to hydroxy-unsaturated hexanoic acid, two furane rings and a norbornane-like ring. Importantly, the latter three compounds had never been reported before for terrestrial HA and recall the carboxyl-rich molecules already proposed earlier for dissolved organic matter (DOM) (Hertkorn et al. 2006). The size fractions obtained by subjecting RES4 to preparative HPSEC were further examined by several two-dimensional NMR techniques, which indicated that the chromatographic separation caused profound conformational modifications on the RES4 structure and provided simpler humic material. Consequently, less NMR signal overlapping in size fractions was noted, and more homo- and hetero-correlations were obtained for the size fractions than for the bulk RES4, showing mainly hydroxy-alkyl and alkyl protons (Nebbioso et al. 2014b). DOSY-NMR spectra confirmed the larger homogeneity reached by separating RES4 by HPSEC elution and allowed to recognize that nominally larger, lipid-rich size fractions had significantly lower constant diffusivity, likely due to the formation of stable supramolecular associations, as promoted by hydrophobic interactions among alkyl chains. Instead, the diffusivity gradually increased in smaller-sized fractions, which contained a larger amount of aromatic and hydroxyaliphatic peaks and are not able to promote the formation of large suprastructures as much as alkyl-rich compounds (Nebbioso et al. 2014b).

Humin (HU) from a volcanic soil was characterized by Humeomics before and after a HF treatment, which was used to destroy the clay mineral components (Nebbioso et al. 2015). The obtained fractions were studied by GC-MS, thermochemolysis GC-MS and ^{13}C CPMAS NMR. Both weight and chromatographic yields were greater for the mineral-deprived HU than for bulk HU, hence, allowing better molecular identification in the Humeome of the former. Saturated and unsaturated alkanoic, ω-alkanedioic, hydroxyalkanoic acids, alkanols and hydrocarbons were identified in both HU samples. The large content of odd-numbered n-alkanoic acids found in ORG1 suggested an accumulation of free microbial metabolites, whereas plant-derived (even-numbered) alkanoic acids were predominantly recognized in both ORG2 and ORG3 fractions, implying an interaction of such compounds with the HU stronger than for microbial metabolites. Instead, unsaturated, n-alkanedioic and hydroxyalkanoic acids were noted only after hydrolysis of complex esters, i.e. in the ORG 3 fraction (Nebbioso et al. 2015). In line with previously reported findings for HA, the amount of aromatic compounds in HU residues progressively increased with Humeomics steps, while that of alkyl and hydroxy-alkyl molecules was reduced. The AQU2 extracts, in fact, were mainly composed of alkyl aromatic and carbohydrate-like compounds, while aromatic moieties were predominant in the RES fractions. Moreover, a thorough similarity in the identified compounds in both HA and HU isolated from the same soil further indicated that the differences among the traditional humic fractions (FA, HA and HU) resided more in the supramolecular arrangement of the contained molecules than in their molecular composition (Nebbioso et al. 2015).

Altogether, the above-described results confirmed that Humeomics enables to reach a deep and exhaustive molecular characterization of HS, thus allowing the isolation of molecules bound to the humic suprastructures by either weak dispersive forces or ester and ether linkages, without breaking carbon–carbon bonds. The progressive extraction of single humic molecules and their structural identification by cutting-edge physical–chemical analytical tools effectively advance the knowledge of the molecular composition of Humeome. In addition, Humeomics procedure may help to develop novel realistic computational models for Humeome conformation, in order to further relate its molecular structure to its biological activity towards both plants and microbes, thereby unravelling its environmental reactivity (Schaumann and Thiele-Bruhn 2011; Orsi 2014).

7.3 The Role of the Humeome in Soil Functions

The inherent molecular complexity and widespread diffusion of the soil Humeome determine the environmental importance of HS as keystone of SOM functions. In agricultural and forestry ecosystems, humic components play a crossroad influence in all the physical, chemical and biological processes associated with soil microbial–plant interactions and overall soil fertility.

7.3.1 Effect on Plant Growth and Soil Biological Fertility

The soil Humeome is a determinant factor for the maintenance of soil processes associated with plant growth and crop productivity (Olaetxea et al. 2018). In fact, besides their role in soil nutrient dynamics, humic molecules are involved in the biostimulation of plant development, acting as bioeffectors in a large range of biochemical activities, physiological processes and plant–microorganisms interactions (Canellas and Olivares 2017; Cozzolino et al. 2013). Therefore, an increasing effort is devoted to understand the effects of HS on plant metabolism and physiology, in order to support innovative technologies enabling the sustainable management of agro-ecosystems (Canellas et al. 2011, 2015).

The most important a-specific influence of the Humeome on plant growth is referred to the "*nutritional-environmental pathways*" (Olaetxea et al. 2018). This is related to the improvement in nutrient availability due to the effect of organic matter on the physical and chemical processes of soil rhizosphere (porosity, aggregation-texture, respiration, water availability and diffusivity, pool of bioavailable nutrients).

HS are recognized as important intermediates in the biogeochemical cycle of nitrogen, phosphorus and sulphur, as well as reservoir of these essential nutrients (Kelley and Stevenson 1996; Magid et al. 1996). A well-known effect is the increase in P availability induced by the adsorption of dissolved humic components on mineral surfaces of acidic soils and the concomitant displacement of P from iron and aluminium hydroxide. Furthermore, HS may interact with phosphate anions through the formation of metal bridges and enhance P solubilization and uptake (Urrutia et al. 2014).

HS form stable and soluble complexes with micronutrients, such as Fe, Cu, Mn and Zn, thus increasing the potential bioavailability of these elements in alkaline and calcareous soils, where these metal ions are otherwise insoluble and largely unavailable for plants (Chen 1996). In this respect, a recent work used two humic acids from a volcanic soil (HAs) and leonardite (HAl) to synthesize insoluble complexes with iron and simulated the effect of plant exudates on iron solubilization in the rhizosphere (Nuzzo et al. 2018). This was achieved by displacing iron from the iron–humate complexes with a solution of either a microbial siderophore (deferoxamine mesylate DFOM) or a mixture of organic acids such as citric, oxalic, tartaric and ketoglutaric acids. The variation in iron extractability between the two iron–humates was attributed to both the humic molecular composition and the steric hindrance of the extractants. Iron was more easily released from complexes formed with the more flexible conformational structure of HAs rich in aliphatic carbon than the rigid conformation of iron complexes with HAl dominated by aromatic carbon. This work showed that iron–humate complexes can be used to enhance iron nutrition of plants, whose exudates rich in organic acids, together with microbial siderophores, can displace iron from complexes and enhance its solubility in the rhizosphere.

Concomitantly, the same process may solubilize humic matter from metal–humates, thereby enhancing the concentration of bioactive humic molecules in the soil solution. In fact, humic matter from different sources may improve plant nutrition by activating the biochemical functions involved in nutrient root uptake, transport and metabolic cycles in plants (Piccolo et al. 1992; Vaccaro et al. 2009; Canellas et al. 2011). These HS-mediated effects were observed at both transcriptional and post-transcriptional levels (nutrient root transporters, proton pumps and key enzymatic activities) (Trevisan et al. 2010; Canellas et al. 2015). Therefore, the metal complexation by humic molecules can be complementary to the biochemical pathways in exerting the HS effect on plant growth (Olaexeta et al. 2018). In fact, humic extracts from either leonardite or composts were found to influence the expression of genes encoding the Fe(III) chelate reductase (CsFRO1) and the Fe(II) root transporter (CsIRT1) (Aguirre et al. 2009), and thus up-regulate the molecular mechanisms involved in Fe root uptake (Canellas et al. 2015).

The soil Humeome is involved in the stimulation of N assimilation by plant tissues and in the subsequent modulation of amino acid metabolism at both molecular and physiological levels (Vaccaro et al. 2015). Furthermore, a soil humic acid and its size fractions were found to induce a partial relief from response to phosphate (Pi) starvation in the cell culture of tobacco plants, thus increasing total cell phosphate amount, ATP and glucose-6-phosphate levels, as well as the activity of acid phosphatases (Zancani et al. 2009). Soil addition with a humic acid from vermicompost modified the expression of the high-affinity P root transporter gene in tomato plants (*LePT2*) (Jindo et al. 2016). In fact, in respect of control supplied with a low inorganic Pi concentration, plants added with HA showed a more efficient assimilation of available P sources and an increase in global phosphate uptake. These results indicated that the soil Humeome is strategically involved in plant adaptation to P availability through the activation of the metabolic pathway associated with the utilization of available P at low concentrations. Moreover, these findings confirm the complementarity of indirect and direct effects of HS on plant development and the relationships between biological processes and molecular properties of the soil Humeome (Nardi et al. 2007; Cozzolino et al. 2016a, b).

A direct beneficial action of HS on plant growth is related to their interaction with cell membranes of root surfaces (Nardi et al. 2017; Olaetxea et al. 2018). In fact, HS affect both the whole root growth, expressed by dry matter production, and the emergence and elongation of lateral roots and root hairs (Trevisan et al. 2010; Canellas et al. 2012; Canellas and Olivares 2017). It is believed that the biostimulation properties of HS are related to the increase in both the expression and activity of the plasma membrane (PM) H$^+$-ATPase enzyme, with a consequent hormone-like induction similar to that of auxinic phytoregulators (Dobbss et al. 2010; Canellas et al. 2011; Scaglia et al. 2016; Monda et al. 2018). Recent studies have further evaluated the mechanisms involved in the HS-dependent root development (Ramos et al. 2015). Plant treatment with a humic acid from vermicompost induced the emergence of lateral root in rice seedlings and specific H^+ and Ca^{2+} fluxes in the root elongation zone as activated by the plasma membrane H^+-ATPase and the Ca^{2+}-dependent protein kinase (CDPK). The authors pointed out that the HA act as molecular elicitors of H^+ and Ca^{2+} fluxes, which seem to be the upstream of a complex CDPK cell-signalling cascade (Ramos et al. 2015). Currently, the HS enhancement of PM-H^+ ATPase functions observed in many crops is associated with the stimulation of integrated hormone crosstalk systems, which include, besides the auxin synthesis, the nitrous oxide (NO) and abscisic acid (ABA) metabolic pathways (Mora et al. 2010; Olaetxea et al. 2018). HS may also regulate the amount of reactive oxygen species (ROS) at the cellular level, as well as the expression of superoxide dismutase (SOD) genes in the cytosol (cCu/Zn-SOD1 and cCu/Zn-SOD2). This regulation of ROS (H_2O_2 and O_2^-) control by humic molecules occurs in the root regions involved in cell elongation and differentiation, thereby indicating a possible synergic effect of ROS in the HS-mediated root development (García et al. 2016).

Although the main HS action as bioeffectors is mediated by the interaction with root systems, the biostimulant activities of Humeome may affect the physiological processes linked with the development of shoot and leaves tissues in various crop species (Zancani et al. 2011; Canellas and Olivares 2014; Vaccaro et al. 2015). The application of soil humic acid and derived size fractions in nutrient solutions of maize seedlings showed a positive effect on plant metabolism, with a clear influence on the enzymatic steps of both glycolytic and Krebs biochemical cycles in plant leaves (Nardi et al. 2007). In distinct bioassay tests on maize, the use of different Humeome components isolated from various green composts revealed a steady positive correlation with the chlorophyll content of maize leaves combined with a significant increase in fresh and dry weight of the whole plant biomass (Monda et al. 2017; 2018). The increase in nitrate uptake observed in cucumber plantlets treated with a solution of humic acids was found to be coupled with an active translocation of cytokinins from root to shoot tissues and with a subsequent large mobilization of K, Ca and Mg in leaves (Mora et al. 2010; Rose et al. 2014). The HS-induced growth of shoots in cucumber seedlings was related to an increase in root hydraulic conductivity and K fluxes, determined by the ABA-dependent pathways that regulate plasma membrane aquaporin activity in roots (Olaetxea et al. 2015).

Most research works on HS bioactive properties are conducted in confined and simplified environments under controlled conditions and preliminary selected variables, thereby allowing the identification of specific biochemical and metabolic processes. On the other hand, the HS effects in real agroecosystems are probably the result of many interconnected biological and chemical processes and dependent on the simultaneous interaction of the dynamic supramolecular structure of the Humeome with soil microbiota and plant species (Nardi et al. 2017; Piccolo et al. 2018a). Part of the beneficial effects of soil humus on plants stems from the interaction between microbes and organic matter. Microbes behave as mediators of carbon and nutrient turnover of soil humus, while the Humeome affects the survival and efficiency of microorganisms (Insam 1996; Lovely et al. 1996). In fact, the organic humified components influence both composition and activity of microbial communities of agroecosystems (Puglisi et al. 2013; Cozzolino et al. 2016a, b) with specific alternative effects, favouring either the selective preservation of beneficial microorganisms (Martinez-Balmori et al. 2013; Canellas and Olivares 2017) or an effective suppressive capacity of pathogens for different crops (Pane et al. 2011, 2013).

Therefore, additional complementary approaches on the study of bioactive properties of humic substances should be focused on the relationships between Humeome characteristics, and biological functions and cropping systems (Puglisi and Trevisan 2012; Ventorino et al. 2012; Scotti et al. 2016). In a bi-annual field experiment, the modification of SOM composition and biological activities attained by soil treatments with green compost produced a combined biostimulation effect on tomato plants with a significant improvement in nutrient uptake, fruit yield and quality (Pane et al. 2015). Furthermore, the inoculation of humic extracts and selected beneficial microbial species may promote a recovery of biochemical efficiency in crops grown under biotic and abiotic stress (Aguiar et al. 2016; da Piedade Melo et al. 2017).

The combined application to soil of compost and plant growth-promoting bacteria was tested as valuable and sustainable alternative management to the conventional phosphate mineral fertilization methods. The effects of the inoculation with different microbial isolates on the growth and metabolic processes of maize plants were evaluated in soil treated either with mineral fertilizers (triple superphosphate and rock phosphate) or with different humified manure-based compost as P-fertilizers (Li et al. 2017; Vinci et al. 2018a, b). Plants inoculated with organic materials and microorganisms showed larger P and N contents and a more differentiated metabolome as compared to the treatments with inorganic fertilizers. The metabolomic analyses revealed a significant increase in glucose, fructose, alanine and GABA metabolites in maize leaves, thus suggesting an improved photosynthetic activity due to enhanced P and N uptake. The inoculation with humified organic P-fertilizers sustained plant growth and the activity of microbial species and appeared as an efficient alternative to mineral fertilizers to enhance nutrients uptake and ensured plant growth under deficient soil P availability (Li et al. 2017; Vinci et al. 2018a, b).

Although the specific mechanisms underlying the action of humic substances as bioeffectors have not yet been completely elucidated, it is ascertained that the molecular characteristics and the structural properties of the Humeome have an influence on the plant activity and soil biological functionalities (Cozzolino et al. 2016a; Piccolo et al. 2018a). The supramolecular conformation based on the dynamic interaction of hydrophobic and hydrophilic domains was found to determine the bioactive properties of humic molecules in either exogenous organic inputs or SOM pools (Piccolo et al. 1992, 2018a; Piccolo 2002; Canellas et al. 2012). The hydrophobic composition of humic superstructures protects from microbial degradation the bioactive low molecular weight molecules of SOM, such as lignin fragments, phenol derivatives, carbohydrates, peptides and aromatic acids (Piccolo 2012; Muscolo et al. 2013; Piccolo et al. 2018a). In fact, the large concentration of root exudates and microbial products in the rhizosphere favours the physical–chemical interactions with the metastable supramolecular structures (Canellas et al. 2008; Nardi et al. 2017). The flexible Humeome conformation may hence undergo a favourable thermodynamic rearrangement with a consequent release of the retained molecules which may exert their bioactive properties on root membranes and microbial cells (Piccolo 2002, 2012, 2016). The hydrophobic/hydrophilic ratio and the structural modifications induced by the physical–chemical interaction in the rhizosphere have been identified as the key structural processes of the bioactive properties of the Humeome from soil, sediments and recycled biomasses (Puglisi et al. 2008; Dobbss et al. 2010; Aguiar et al. 2013; Martinez-Balmori et al. 2014; Piccolo et al. 2018a).

The effects of a humic acid and its size fractions on plants carbon deposition and the structure of microbial communities were analysed in the soil rhizosphere of maize plants with rhizo-box systems, which allowed the simultaneous but distinct characterization of both bulk rhizosphere and plant roots exudates (Puglisi et al. 2009). Following the interaction with root exudation, all humic treatments were able to modify the structure of the rhizospheric microbial communities. However, while the small-sized and more bioavailable humified fractions provided a prompt response to the microbial activity, both the bulk HA and the most hydrophobic fractions required additional plant rhizodeposition before their bio-transformation released additional metabolic carbon.

The positive effect on maize development of different water-soluble Humeomes isolated from green composts was attributed to their significant hydrophobicity that favoured adhesion to the root surfaces. Then, exudation of organic acids from plant roots fosters the subsequent modification of the humic conformation and the concomitant release of bioactive molecules (Monda et al. 2017, 2018). However, in the same experiment, the large content of soluble phenolic compounds in the Humeome from tomato- and artichoke/fennel-based composts revealed either a small stimulation or even toxic effects on roots growth. The biostimulation was recovered at larger Humeome supply rates, and this was explained with a conformational tightening due to the hydrophobic components, whose closer intermolecular aggregation prevented the release of soluble toxic components (Monda et al. 2018).

In order to provide evidence for the Humeome structure–activity relationship, the effect of root exudation on the bulk Humeome superstructures was simulated with different mild chromatographic and physical–chemical fractionation procedures (Nardi et al. 2007; Puglisi et al. 2009; Vaccaro et al. 2009; Canellas et al. 2011, 2012). For all tested humic materials, results revealed that the more flexible the bulk Humeomes, the larger was the release of low molecular weight components and the larger the stimulation of biochemical and metabolic processes. The bioactivity of Humeome thus appears to be closely related to its molecular composition and consequent stability of its conformational structure in solution, whose dynamic interaction with rhizosphere determines the potential release of bioactive molecules that directly or indirectly influences plant growth.

The understanding of the bioactive properties of humic substances may support the adoption of alternative technologies to develop effective sustainable SOM management of agroecosystems. The current increasing demand of renewable energy based on the exploitation of biomasses, such as non-food crops, agro-industrial by-products and organic wastes, leads to the increased availability of secondary products. Those left over from biorefinery processes include large amounts of lignocellulose residues which are currently burnt to produce thermal energy, with a net loss of photosynthate organic carbon that may be rather employed in the sustainability of agricultural production.

To this purpose, a specific alkaline oxidative extraction was applied to lignocellulosic biorefinery wastes and provided significant yields of water-soluble derivatives enriched in lignin components (Savy and Piccolo 2014). The detailed characterization of these organic extracts revealed that they may act as humic-like substances (HULIS) which retain the molecular features and the bioactive properties of the natural Humeome (Savy et al. 2016b, 2017). Therefore, bioenergy crops and biorefinery wastes should be regarded as valuable source of natural plant biostimulants, either by direct extraction procedures (Savy et al. 2016a, 2017) or after their use as humic precursors in composting processes (Spaccini et al. 2019).

The water-soluble HULIS derivatives isolated from either lignocellulose biorefinery residues or lignosulphonate as wastes of paper mill industries were evaluated for their molecular characteristics and bioactivity on maize seedlings (Savy et al. 2017, 2018). All these materials showed a positive influence on plant growth, with different responses depending on specific structural composition and concentration rates. As found for the soil Humeome, the biostimulation effects were attributed to the humic-like conformational behaviour that determined the release of bioactive lignin-derived phenolic compounds.

A direct extraction of HULIS bioactive components from lignin-rich biomasses was applied to four non-food bioenergy crops: cardoon (CAR), eucalyptus (EUC), and two black poplars (RIP) and (LIM). After a detailed characterization of the molecular components of extracts, the potential bioactivity towards seed germination and early growth of maize plantlets was evaluated (Savy et al. 2015). These lignin-derived HULIS stimulated the maize seedling development, though to a different extent depending on the hydrophobic/hydrophilic properties of humic-like molecules. The most significant bioactivity, found in CAR-derived extracts followed by RIP and LIM treatments, was related to a large content of aliphatic OH groups, CO carbons and smallest hydrophobicity, as assessed by ^{31}P-NMR and ^{13}C CPMAS NMR spectroscopies. The poorest biostimulation of EUC HULIS was attributed to its smallest content of polar OH groups and largest hydrophobicity. Both these features may be conducive of a EUC conformational structure tight enough to prevent its alteration by organic acids exuded from vegetal tissues. Conversely, the more labile conformational arrangements of the other more hydrophilic lignin extracts promoted their bioactivity by releasing biologically active molecules upon the action with exuded organic acids.

In a corresponding study, water-soluble HULIS lignins were isolated from the most representative non-food bioenergy crops such as giant reed (AD) and miscanthus (MG) (Savy et al. 2016a, b). The extracts were characterized for their conformational structure and molecular composition employing size-exclusion chromatography and NMR spectroscopy. Their effects at different concentration rates on germination of maize seeds and growth of maize plantlets were assessed concomitantly in growth-chamber experiments. HULIS showed both humic-like supramolecular structures and different conformational stability and molecular composition. Their largest bioactivity was revealed at 10 and 50 ppm of OC concentrations, with a significant increase in roots and coleoptiles length of maize seedlings, as well as of total shoot and root dry weights and root length of maize plantlets. Again, the differences in bioactivity of HULIS from AD and MG were attributed to their conformational stabilities and content of amphiphilic molecules, which may control both the adhesion to plant roots and the release of bioactive molecules upon interactions with plant-exuded organic acids.

7.3.2   Effects on Soil Stability

The physical properties of soil closely depend on the interaction with the soil Humeome. In fact, humus is involved in soil aggregation processes at different hierarchical levels, from the formation of organo-mineral complexes up to the stabilization of large size aggregates (Oades and Waters 1991). The humic molecules are the basic cementing agent of stable microaggregates which represent the resilient core of soil aggregate dynamics (Piccolo and Mbagwu 1999; Six et al. 2004). Humic substances promote an effective association of fine soil particles into larger aggregates, based on their hydrophobic components which exert a more permanent physical stabilization than the temporary and transient aggregation provided by fresh and biolabile plant inputs and microbial exudates (Piccolo and Mbagwu 1999; Spaccini and Piccolo 2012; Sarker et al. 2018; Piccolo et al. 2018a). The water-soluble components of the Humeome can perform a variable range of reversible chemical interactions with polar or charged mineral surface, like electrostatic interactions, ligand exchanges, cation bridges, hydrogen bonding, etc. (Nebbioso and Piccolo 2013; Paul 2016). The divalent Ca^{2+} and Mg^{2+} cations are believed to play a role for the initial mutual approaches of organic and mineral surfaces of clay–Humeome complexes in neutral and alkaline soils (Sutton and Sposito 2006), whereas the adsorption of HS occurs on the surface of iron and aluminium oxides in acidic soils (Oades 1988; Song et al. 2013).

The interaction of hydrophobic molecules with inorganic components enables a long-term stabilization to the soil structure (Piccolo and Mbagwu 1999). The lower affinity of dissolved hydrophobic compounds for the soil aqueous solution enhances the chemical potential towards the molecular adsorption on the available apolar and low-charged mineral surfaces (Schwarzenbach et al. 2003a; Feng et al. 2005; Nebbioso et al. 2015). This thermodynamic process is further increased by the small volume of the soil porous system, the effect of ionic strength of dissolved salts (e.g. "salting out" effect) and the presence of soluble organic components (Schwarzenbach et al. 2003a). The progressive accumulation of humic molecules constitutes an hydrophobic environment in soil that becomes tightly impermeable to the soil aqueous solution irrespective to pH values (Piccolo and Mbagwu 1999; Piccolo 2012; Masoom et al. 2016). This process leads to the progressive decrease in wettability of inorganic surfaces, thereby increasing soil aggregate stability. This organic coating may thus strengthen the association between mineral particles and contribute to the stability of microaggregates by favouring interparticle aggregation by hydrophobic interactions (Piccolo et al. 2018a). Conversely, the highly hydrophilic biomolecules (carbohydrates and peptides) are more likely to partition into the aqueous soil solution, where they are rapidly degraded by microorganisms, thereby providing at best only a temporary effect on soil structural stability (Piccolo and Mbagwu 1999; Six et al. 2004).

The characterization of organic compounds associated with stable microaggregates in different soil types and land uses revealed the steady selective accumulation of alkyl hydrophobic molecules (Baldock and Skjiemstad 2000; Piccolo et al. 2005a; Mao et al. 2007). The relation between soil aggregate stability and hydrophobic humic components was highlighted in heavily weathered African soils under original forest and agricultural management (Spaccini et al. 2002a, 2004). The aggregate size fractions derived from forest soils were characterized by a large structural stability and revealed an incorporation of highly hydrophobic humic materials (Piccolo et al. 2005a; Spaccini et al. 2006). The shift of land use with deforestation and cultivation produced a decrease in the hydrophobic portion of the soil Humeome with a concomitant progressive loss of large-sized water-stable aggregates and of overall physical stability.

7.3.3   Effect on Carbon Sequestration

The soil Humeome represents the most abundant SOC pool of forest and agricultural soils, acting as the bulk compartment for the incorporation, accumulation and stabilization of OM inputs (Piccolo 2012; Piccolo et al. 2018a). Despite the wide combination of ecological (soil type and texture, biological components, climate, vegetation) and anthropic (land use, management, cropping systems) that affects SOM dynamics, the soil Humeome from different ecosystems reveals comparable molecular characteristics and reactivity (Piccolo 1996; Drosos et al. 2017, 2018). This finding suggests the occurrence of general mechanisms which control the decomposition/accumulation reactions involved in the Humeome formation (Piccolo 2002; Piccolo et al. 2018a). The selective incorporation of less decomposable or persistent hydrophobic compounds, derived from the biotic and abiotic transformation of biomolecules released by dead cells, may be conceived as the driving stage underlying the humification process (Winkler et al. 2005; Lorenz et al. 2007; Nebbioso et al. 2015; Piccolo et al. 2018a). As noted above, the poorly soluble apolar substances undergo a favourable partition from the aqueous soil solution to the solid phase (Piccolo 2002; Schwarzenbach 2003a; Piccolo et al. 2018a). Organic molecules are hence adsorbed and retained on the surface of fine secondary minerals by non-covalent dispersive forces (Kleber et al. 2007; Paul 2016). However, although the adsorption is mainly based on weak reversible interactions, the hydrophobic effect strengthens the progressive accumulation of dissolved molecules onto the already existing adsorbed humic layers (Piccolo 2002; Piccolo et al. 2018a). The separation from the aqueous environment thus increases the thermodynamic barrier required for the biological oxidation and decomposition of organic components. The chemical affinity with dissolved polar organic molecules in soil solution drives their progressive incorporation into a multilayer superstructure composed by contiguous hydrophilic/hydrophobic domains in the close proximity of soil aggregates. Hence, the SOC accumulation and long-term persistence are accounted to the chemical protection that excludes water and thus preserves the organic layers by microbial activity (Piccolo 2016; Masoom et al. 2016; Piccolo et al. 2018a). The wet and drying cycles of soil aggregates further contribute to separate hydrophobic compounds from the soil solution and soil microbes, thereby increasing the mean residence time of Humeome components. This organic barrier may hence provide a biochemical hindrance to microbial decomposition and develop a dynamic mechanism based on the hydrophobic protection of biolabile organic compounds (Spaccini et al. 2002a). This process thus promotes the stable incorporation of potential bioavailable compounds, such as polysaccharides, peptides, P-containing compounds, deriving from either OM inputs, crop residues, roots exudation or microbial biomass (Piccolo 2012, 2016; Piccolo et al. 2018a).

The role of hydrophobic protection in the processes of SOC incorporation and stabilization has been tested in dedicated laboratory experiments. In a short-term incubation test, a ^{13}C-labelled 2-decanol was added to soil either alone or in combination with solutions of two different humic acids (HA) from lignite and compost. At defined sampling periods, the residual amount of labile alcohol was carefully measured in both bulk samples and soil size fractions by GC-IRMS (isotopic ratio mass spectrometry) analyses (Spaccini et al. 2002b). Results revealed a rapid loss of labelled carbon in control test with the biolabile 2-decanol alone, while a continuous preservation of ^{13}C–OC was found in soil samples added with exogenous HA. The larger the hydrophobic character of HA, the slower and smaller was the mineralization of the ^{13}C–OC. It was pointed out that while the unprotected bioavailable alcohol underwent a fast microbial decomposition, the partition in the hydrophobic domains of humic matter reduced the mineralization of the labelled compound. Moreover, at the last sampling time, the largest amount of residual labelled alcohol was found in soil that had received HA additions, thereby confirming the effective incorporation and preservation of labile component in the soil Humeome (Spaccini et al. 2002b).

These findings were further evaluated in one-year long experiment, where two soils with contrasting textural composition were treated with mature compost (Cmp), humic acids isolated from compost (HA-C) and lignite (HA-L), and a polysaccharide gum (AG). Compost and humic fractions were added to soils either before or after amendment with two different doses of polysaccharidic gum, in order to verify the effect of different humified materials on the mineralization of the biolabile gum (Piccolo et al. 2004). The variation in soil TOC content was monitored monthly for each soil treatment. Notwithstanding the comparable initial TOC content, an effective stabilization of labile OC was found, for both soil types treated with bulk compost and HA-C and HA-L mixed with polysaccharides, irrespective of the timing and doses of the AG additions. As compared to control samples that only received the labile AG alone, the mixing with humified materials revealed a progressive decrease in OC losses at increasing incubation time, with a final larger OC preservation in the finer textured silty loamy soil (Piccolo et al. 2004). These results confirm that humic materials may protect from biodegradation the biolabile molecules in soil by incorporation into the stable hydrophobic domains of the Humeome.

The findings of these laboratory experiments allowed to identify the processes involved in SOC accumulation. However, the relation between the molecular recalcitrance of the Humeome and SOC stabilization was also validated in following the OM transformation during composting and SOM dynamics in fields. In fact, the increase in biochemical stability associated with the humification stage of mature composts corresponded to the progressive preservation of hydrophobic molecules (Spaccini and Piccolo 2007, 2009; Martinez-Balmori et al. 2014). Moreover, the incorporation of apolar lipid and lignin plant derivatives was found to be associated with the increase in OC content in both tropical and temperate forest and agroecosystems, under long-term SOM management practices (Spaccini et al. 2000, 2009, 2013; Zhou et al. 2010; Song et al. 2013; Scotti et al. 2016).

Similar indications were found in long-term experiments conducted on weathered soils of Ethiopia and Nigeria, where SOC dynamics were evaluated following the vegetation shift from forest to cultivated soils. In these experiments, the soil TOC content, the NMR spectra of humic substances extracted from treated soils and the changes in ^{13}C–OC-labelled SOM due to replacement of C3 natural vegetation with C4 crops were followed (Piccolo et al. 2005a; Spaccini et al. 2006). In such highly weathered tropical conditions, despite the incorporation of recent C4 plant residues, 5 years of intensive cropping determined a decrease in SOC content in cultivated plots ranging from 46 to the 75% of the original forest soils. The molecular characterization of humic substances showed a clear relationship between the decrease in SOC content and the progressive lowering of hydrophobicity of the Humeome in cultivated plots. The authors pointed out that the reduced OM incorporation and increased mineralization of biolabile compounds were caused by the progressive loss of protective properties by hydrophobic compounds which were originally accumulated in soil under the pristine forests.

The SOC hydrophobic protection by the Humeome was verified in agricultural soils of temperate Mediterranean region for maize cropping systems (Spaccini and Piccolo 2012). In a 3-year experiment at three different field sites, the soil amendment with two annual rates of humified mature compost (2.7 and 5.4 OC ha^{-1}) was compared with conventional SOM management represented by minimum tillage and green manuring in soils undergone common ploughing and mineral fertilization. After 4 years, the amount of TOC in both bulk soils and water-stable aggregates of ploughed horizons (0–30 cm) did not reveal any significant difference between control and soils under green manuring. The soil treatment with minimum tillage showed annual and site-specific response with short-term and even negative effect on OC accumulation. Conversely, the soil amendment with hydrophobic mature composts produced a significant stabilization of SOM, revealing a persistent incorporation of SOC in both bulk samples and soil aggregates. Depending on the soil bulk density of each experimental site, field plots added with compost maintained a significant amount of added OC with compost, ranging from the 52 to 63% and from the 50 to 80%, for the low and high doses of amended compost, respectively.

These figures corresponded to a stable OC incorporation of 2.1–4.1 Mg ha^{-1} y^{-1} besides the OC added with compost. Moreover, ^{13}C CPMAS NMR spectral data on humus extracted from the treated soils, combined with multivariate analyses, revealed that compost-amended soils progressively incorporated alkyl and aromatic hydrophobic components into the soil Humeome during the experimental time (Spaccini and Piccolo 2012). These findings further highlight that the humification in soil proceeds by an increasing accumulation of less decomposable hydrophobic material that enables the progressive entrapment and protection of the biolabile compounds exuded by root or released by degrading crop residues.

This concept of humification has been recently adopted to develop an innovative model depiction of SOC dynamics, based on the effect of molecular recalcitrance of soil Humeome components on both SOC incorporation and physical aggregation mechanisms (Mazzoleni et al. 2012). The conceptual framework of the soil organic matter dynamics (SOMDY) model combines the molecular characterization of SOM and of exogenous OM inputs, as determined by ^{13}C CPMAS NMR, with their association with soil mineral particles. The vector descriptors of OC dynamics accurately anticipate both the decomposition/stabilization behaviours of a wide range of OM inputs and their effect on the soil aggregation processes (Incerti et al. 2017; Sarker et al. 2018). The close relationship between measured and simulated data obtained by the model confirmed that the molecular composition of the soil Humeome plays an important role in SOC stabilization and aggregation processes, being more predictive of SOM dynamics than the classical but less specific elemental features (e.g. C/N, lignin/N ratios) used in the current modelling approaches.

7.3.4 Effect on Inorganic and Organic Pollutants in Soils

The Humeome is an important environmental "buffer" that controls the fate of pollutants in soils, sediments and aqueous systems (Kozak 1996; Sannino and Piccolo 2013). The interactions promoted by humic fractions strongly determine the mobility and the bioavailability of xenobiotic compounds (Van Lipczynska-Kochany 2018). In fact, the molecular features and flexible conformational properties of humic superstructures are at the basis of a number of physical–chemical reactions with inorganic and organic chemicals.

The polar functional groups of humic molecules can complex the soluble and exchangeable forms of heavy metals, decrease their adsorption on mineral surface and counteract their negative effect on soil biological properties (Piccolo 1989; Zheng et al. 2018). Moreover, the formation of metal–humic complexes modifies the temporary bioavailability of less soluble heavy metals, thus preventing their transformation into insoluble species and improving the potential removal of these metal species (Halim et al. 2003).

The Humeome may also affect the adsorption, retention and biodegradation processes of organic pollutants. The same weak forces that stabilize the humic supramolecular conformations facilitate the incorporation of exogenous xenobiotic materials. The various organic pollutants may bind to humic molecules through several mechanisms, including hydrogen bonding, van der Waals forces, charge transfer and hydrophobic $\pi-\pi$ bonds (Schwarzenbach et al. 2003b). The chemical affinity with humic molecules and the poorly soluble organic pollutants fosters their selective incorporation and subsequent retention in the hydrophobic domains of the Humeome superstructures.

These thermodynamic processes driving the humic–pollutants interactions have been clearly highlighted by specific studies with NMR spectroscopy (Šmejkalová and Piccolo 2008a, b; Šmejkalová et al. 2009). NMR measurements by spin–lattice (T1) and spin–spin (T2) relaxation times and diffusion-ordered spectroscopy (DOSY) were applied to investigate the association of 2,4-dichlorophenol (DCP) with fulvic acid and two humic acids of different conformational structure. Results indicated that the larger the hydrophobic character of dissolved humic substances, the greater was the non-covalent binding of the pollutant. In fact, DCP became increasingly associated in host–guest complexes with the stable hydrophobic domains of dissolved humic matter. Moreover, the chemical shift variations for the three DCP aromatic protons indicated $\pi-\pi$ interactions, rather than H-bonding, as the main driving force for non-covalent association between DCP and dissolved humic substances (Šmejkalová and Piccolo 2008a, b).

By similar NMR approaches, the effect of different chemical structures of organic pollutants on the interaction with HS was followed. Non-substituted (phenol P) and halogenated phenols (2,4-dichlorophenol (DCP); 2,4,6-trichlorophenol (TCP); and 2,4,6-trifluorophenol (TFP)) were solubilized in the presence of different concentration of dissolved humic acid (Smejkalova et al. 2009). The progressive decrease in molecular mobility of dissolved pollutants at increasing HA concentration indicated the formation of stable non-covalent interactions. Binding increased by lowering solution pH value, thus revealing a dependence on the protonated forms of phenols in solution. However, it was found that the hydrophobicity conferred to phenols by chlorine atoms on aromatic rings was a determinant variable for the repartition within the HA hydrophobic domains. Calculated values of Gibbs free energy of transfer and related binding constants (K_a) revealed that the more hydrophobic and less soluble the chlorinated phenols, the more extensive was the binding to humic fractions, even at low HA concentration.

Also, the hydrophilic fraction of the Humeome becomes responsible for the associations with water-soluble organic pollutants. The association of glyphosate (N-phosphonomethylglycine) herbicide (GLY) with soluble fulvic acids (FA) and humic acids (HA) at different pH values was studied by ^{1}H- and ^{31}P-NMR spectroscopy (Mazzei and Piccolo 2012). Both the progressive broadening of GLY signals and the modification or NMR relaxation times with ever larger humic concentration indicated the occurrence of weak non-covalent interactions between GLY and humic molecules. Spectral data obtained at different pH values (5.2 and 7) showed a stronger binding with hydrophilic FA than with more non-polar HA and more extensively at pH value of 5.2 than at pH value of 7, thereby suggesting formation of hydrogen bonds between the carboxyl and phosphonate groups of GLY and the protonated oxygen functions in humic matter. The careful measurement of association constants (Ka) and Gibbs free energies of transfer obtained by diffusion-ordered NMR spectroscopy confirmed the thermodynamically favoured non-covalent association of GLY with the functional groups of FAs at pH 5.2 (Mazzei and Piccolo 2012).

The range of physical–chemical interactions between the Humeome and various pollutants suggests that humic substances may play an innovative eco-friendly role in developing a sustainable management of polluted soils (Sannino and Piccolo 2013). This may go as far as counteracting the effect of toxic soil pollutants on plant growth. A recent work combined the application of humic acid with arbuscular mycorrhizal fungi (AMF) in a Hg-polluted sandy soil cultivated with lettuce in order to verify the effects on the Hg tolerance of lettuce plants (Cozzolino et al. 2016b). It was found that plant biomass was significantly increased by the combined effect of AMF and humic acid treatments. Addition of humic matter to soil effectively complexed Hg and thus favoured the AMF capacity to improve plant nutrition, by enhancing the pigment content in plant leaves and inhibiting both Hg uptake and Hg translocation from roots to shoots.

Water-soluble humic substances may act as cheap and eco-friendly alternative to natural surfactants in the removal of inorganic and organic xenobiotics from soil through the washing technology. In fact, the use of synthetic surfactants in cleaning polluted sites by soil washing should be discouraged due to their toxicity to the soil biological quality, whose reduction will prevent any further natural attenuation of pollutants. Conversely, dissolved humic substances are natural surfactants without bearing any toxicity to soil microbial biomass. Moreover, the heterogeneous molecular composition of the Humeome enables a simultaneous removal of different organic and inorganic pollutants, thereby simplifying the soil washing steps and concomitantly enriching the soil with carbon useful to microbial metabolic processes.

Humic substances were successfully applied to remove polychlorinated biphenyls from model soils (Fava and Piccolo 2002), to reduce sorption of organic pollutants on spiked soils, thereby enabling desorption-remediation processes of polyaromatic hydrocarbons (PAHs) (Conte et al. 2001) and heavy metals (Halim et al. 2003). A water-soluble HA from leonardite was used for the soil washing of two highly polluted soils from the Italian polluted site of the ACNA industry, and its removal efficiency was compared to that of two classical synthetic surfactants, sodium dodecylsulphate (SDS) and Triton X-100 (TX100) (Conte et al. 2005). The soil washing with the HA solution showed even a slightly greater removal of a wide range of organic pollutants (PAHs, monoaromatic halogenated and nitrogenated compounds, thiophenes, sulphones, biphenyls) than by conventional surfactants, thereby proving the large surfactant activity of humic matter. An advantage over the synthetic chemicals is that humic materials are natural organic compounds, which do not exert toxicity on the microbial biomass and may instead stimulate the biological properties of polluted soils and promote a subsequent biological attenuation and a restoration of the biological quality of cleaned soils.

7.4 Case Studies

7.4.1 Direct Application of Humeomics on Soils

The application of the Humeomics fractionation on humic matter has allowed significant advancements in understanding HS molecular composition. However, an ideal SOM fractionation method should also be applicable directly on soils. Hence, Drosos et al. (2017) recently subjected an agricultural sandy loam to a complete Humeomic fractionation protocol. The residual OM still present in soil after the sequential fractionation was removed by an alkaline extraction providing a final fraction referred to as RESOM. Concomitantly, SOM was extracted by the traditional method employing a solution of NaOH and $Na_4P_2O_7$ and referred to as eSOM. The humic molecules released in each fraction were characterized by NMR, GC-MS and LC ESI-Orbitrap-MS.

The unbound fraction (ORG1) showed a larger amount of long alkyl chains than the other fractions, mainly unsaturated, mono- or poly-hydroxylated compounds. The ORG2 fraction contained fatty acids and sugars, and a minority of phenolic acids, imines, esters, dicarboxylic acids and hydroxyacids. Conversely, most molecules in the AQU2 fraction were ascribed to O-alkyl-containing compounds, with a small amount of alkyl and aromatic moieties. Other AQU fractions characterized by Orbitrap-ESI-MS showed the presence of amides, benzoic acids, nitrogen and oxygen heterocyclic compounds, ethers and a minority of phenolic esters, phenolic acids, hydroxyl acids, sterols, esters and alcohols. Furthermore, evidence of the occurrence of adducts between water-soluble humic components and Fe ions was also observed. The complexes were taken into account by further numerical elaboration of the empirical formulae obtained by MS and found that one to three Fe ions were covalently bound to humic molecules (Drosos et al. 2017). The molecular composition of the different separates by Humeomics was found similar to that reported for humic extracts, except for a greater abundance of cyclic acids (Nebbioso and Piccolo 2012).

The molecular composition of the RESOM fraction was compared to that of the eSOM extract. The smaller content of O-alkyl carbons in NMR solid-state spectra of RESOM than in those of eSOM was taken as an evidence of the progressive removal of hydroxyl-containing compounds (e.g. alcohols, saccharides) during Humeomics fractionation. Conversely, the intense alkyl peaks still detected in NMR spectra of RESOM indicated a persistent content of alkyl compounds tightly bound to soil minerals. This corroborates the fundamental role of hydrophobic components in stabilizing the conformation of humic suprastructures and in favouring the accumulation of organic carbon in soil (Piccolo 2001, 2012; Nebbioso et al. 2014a). In fact, the recalcitrance of hydrophobic compounds to the extraction even by Humeomics indicates an ultimate chemical inaccessibility that is likely due to their intimate interaction with soil mineral components. These findings show that RESOM stabilization should be attributed to the adsorption of its hydrophobic components on the alumina-silicate

Fig. 7.3 Organic carbon (OC) yields of Humeomic fractions (ORG1-3, AQU2-4 and RESOM) versus eSOM, including the OC of the unextractable material in both cases (RES) and of the material lost after extraction (LOSS)

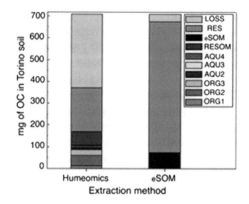

surfaces and to the formation of complexes between oxygen-containing hydrophilic groups and mineral iron.

Drosos et al. (2017) have also reported the mass yield (mg) of organic C (OC) obtained by either the traditional alkaline extraction (eSOM) or the Humeomics fractionation (Fig. 7.3). The traditional alkaline extraction was found significantly less efficient than the Humeomics in isolating humic compounds from soil. In fact, the authors pointed out a solubilization of 66% of the soil Humeome by Humeomics that was instead unextractable and, thus, unidentifiable by the traditional alkaline method (Drosos et al. 2017). Even though the C lost during eSOM extraction was smaller than for Humeomics, the unextractable OM in the case of eSOM was significantly greater (Fig. 7.3). The authors ascribed such a different solubilization of humic matter to the larger capacity of Humeomics to access humic molecules, which were progressively unmasked by first removing unbound organosoluble compounds and then the weakly and strongly bound organosoluble (ORG2, ORG3) and hydrosoluble (AQU2, AQU3) molecules after hydrolyses of esters of increasing stability, and, finally, the water-soluble AQU4 fraction released after breaking ether bonds.

This work highlighted the potential of Humeomics in unravelling the molecular nature of an amount of soil humic molecules far larger than that usually reached by traditional alkaline extraction of SOM and showed for the first time the molecular role of Fe in promoting the stabilization of OM in soil (Drosos et al. 2017). More-over, the reported high abundance of amines, amides and heterocyclic N-compounds, especially in AQU fractions, may contribute to shed light in the role of organic N-containing molecules in the reactivity of Humeome in soils and towards biota. Finally, results of Drosos et al. (2017) confirmed that the high-resolution Orbitrap-MS was more helpful in characterizing humic molecules than other analytical tools, such as low-resolution GC-MS or the poorly sensitive NMR spectroscopy (Saiz-Jimenez 1995; Šmejkalová et al. 2008a, b).

Humeomics was also exploited to assess and compare the molecular composition of the Humeome of two different grassland soils (Soil A and B) (Drosos et al. 2018b). All Humeomic organic and aqueous fractions, as well as RESOM and eSOM, were studied by HPSEC hyphenated to ESI-Orbitrap-MS. The spectrograms provided 175 empirical formulae for Soil A and 139 for Soil B, which were discussed in relation to their molecular size, unsaturation degree, oxygenation and the presence of N in the formula. Compounds containing only one N atom and showing no or low unsaturation were likely of microbial origin, while highly oxygenated molecules containing or not containing N atoms were mainly ascribed to saccharide moieties. Due to a predominant anoxicity (depletion of dissolved oxygen), Soil B contained a larger amount of nitrogenated compounds in the Humeome and a lower C/N ratio than Soil A. Conversely, the more oxic Soil A contained more highly oxygenated and unsaturated molecules than Soil B (Drosos et al. 2018b).

The observed empirical formulae were matched with their molecular structures, making use of ChemSpider and PubChem databases. Hence, the observed molecules were distributed into 16 specific chemical groups, whose distribution in the two soils was visually shown by Van Krevelen plots. By this approach, the organosoluble fractions in both soils were found to be dominated by aliphatic amides and saccharide ethers, while the hydrosoluble fractions comprised mainly aromatic amides, heterocyclic nitrogen compounds and saccharide ethers. Moreover, 66 molecules had been found common in both soils, with a main abundance of saccharide ethers, which were additionally bound to aromatic compounds in the water-soluble fractions. The study not only corroborated the potential of Humeomics in the clarification of the structure of N-containing compounds, but also highlighted the role of Humeomics in clarifying the mechanisms of long-term persistence of humic molecules in soil (Drosos et al. 2018b).

A similar Humeomic approach was adopted by Drosos and Piccolo (2018) to study the molecular features of the Humeome of a sandy loam agricultural soil that was cultivated with maize. Soil samples were collected after one and three years of maize cultivation under continuous conventional tillage and subjected to both the alkaline extraction and the Humeomic fractionation. As elsewhere, the extracts were characterized by GC-MS, HPSEC-ESI-Orbitrap-MS and elemental analysis. Twenty-six different classes of compounds were characterized by Humeomics, while only 11 classes were found in eSOM. In particular, 53 and 31 molecules were identified in eSOM of the first and third tillage year, whereas Humeomics allowed the structural assignment of 353 different molecules in the first year and 366 different molecules in the third year. While SOC extracted in eSOM was about 14.0% of total SOC in both years, the full Humeomic procedure allowed to extract 50.8% of total SOC after the first year (i.e. 3.65 times more C than for eSOM) (Drosos and Piccolo 2018). Furthermore, the Humeomic fractionation revealed both a progressive loss of hydrophobic protection in soil and a concomitant increase in SOM decomposition, due to a smaller ratio of organosoluble to hydrosoluble OC (from 0.77 to 0.59), passing from the first to the third year.

It was also found that several heterocyclic N-containing molecules constituted the main compound class in both eSOM and RESOM for both tillage years. Such molecules could have possibly derived from the methoxylation of tryptophan bound to iron hydroxide in soil (Nuzzo et al. 2013). Interestingly, most amides were found in both the first and third year Humeomes, even though their amount was reduced, thereby suggesting that such molecules have been lost with tillage. Conversely, the amines showed a greater resistance to tillage-induced transformation, since about the 99% of such compounds have been identified in the soil Humeome after both cultivation years (Drosos and Piccolo 2018). The overall recalcitrance of N-containing structures to soil tillage has been explained by the authors with the formation of covalent bonds between N and Fe in iron hydroxides. The Fe-bound molecules found in eSOM extracts were only slightly decreased from first to third year of tillage, whereas total Humeomics not only enabled the isolation of a larger number of these structures, but also revealed that their amount increased with tillage (Drosos and Piccolo 2018). Moreover, the quantitative increase from first to third year samples was 6% for heterocyclic N-containing molecules, 21% for amine and even 71% for phenolic ester. These values were in line with those reported for model systems based on Fe complexes with amines and heterocyclic N-containing compounds (Orlowska et al. 2017).

The Humeomic fractionation of the maize-cultivated soil also enabled to find an explanation for the results by Grignani et al. (2012), who reported that the biomass yields for maize grown on the soil fractionated by Humeomics by Drosos et al. (2018a, b) were reduced in the third year of tillage, more than in the first year. In fact, while the fertility loss could not be appreciated by changes in eSOM, Humeomics allowed to possibly relating this loss to a decrease in the overall aliphatic moieties of SOM (Drosos and Piccolo 2018). A decrease of hydrophobic carbon in soil is known to weaken the stability of soil aggregates, with a consequent decrease in the general soil quality (Piccolo and Mbagwu 1999; Piccolo et al. 2018a).

The advantage of the direct Humeomics application on soil is multiple. First of all, the fractionation of the soil Humeome is in itself a confirmation of the supramolecular nature of soil organic molecules, since they are progressively released from domains in which polar compounds (AQU fractions) are surrounded by or contiguous to hydrophobic structures (ORG fractions). These mixed hydrophilic/hydrophobic humic domains are not only stabilized by hydrogen, $\pi-\pi$ and metal-bridged bonds, but also by covalent bonds with either Fe or Al–Si components of oxides, hydroxides and clay minerals, thereby forming stable organo-mineral associations (Drosos et al. 2017). Moreover, Humeomics allows the characterization of more hydrophobic components of the Humeome, due to the lipid-rich ORG fractions, which are not solubilized in eSOM. The fact that most of ORG molecules are common among the ORG fractions becomes a further evidence of the supramolecular arrangement of the soil Humeome (Piccolo 2001; Piccolo et al. 2018a).

These considerations suggest that Humeomics, due to its capacity to show the changes in SOM composition with time or different experimental conditions, represents the most advanced tool for the ultimate assessment not only of the molecular composition of the soil Humeome, but also of its dynamics. For example, the molecular information provided by Humeomics may be used to introduce an index related to C stabilization in soil. In fact, since the lipidic compounds are thought to protect hydrophilic components from mineralization and microbial degradation (Piccolo et al. 2004; Spaccini et al. 2002b), the larger the organosoluble/hydrosoluble carbon ratio of the soil Humeome, the greater is the chemical protection of organic matter in a soil. Therefore, in the case of the agricultural soil studied by Drosos et al. (2017), a chemical protection ratio (CPR) has been developed (Eq. 7.1, Piccolo et al. 2018a).

$$\text{Extractable OM Chemical Protection Ratio (CPR)} : \frac{\sum ORG_i\, OC_{mg}}{RESOM\, OC_{mg} + \sum AQU_i\, OC_{mg}} \quad (7.1)$$

This CPR ratio based on the OC in ORG, AQU and RESOM fractions separated by Humeomics may be useful to evaluate the efficacy on SOM content of different soil management aimed to sequester C in soil. In fact, in the case of the work by Drosos and Piccolo (2018), an enhancement of the CPR value above 1.05 signified that the soil reached a more chemically protected status under a particular soil treatment. Conversely, a decrease in the CPR ratio would suggest a tendency of the soil to an advanced mineralization of SOC. It may be envisaged that similar alterations of the soil Humeome may be shown as a function of soil type, soil cropping and tillage systems. In fact, such management practices may affect the supramolecular structure of the soil Humeome and consequently alter its dynamics. Therefore, within the intensification of soil exploitation expected in the next decades with population growth, it becomes essential to rely on an advanced technique, such as Humeomics, that may efficiently provide insights into the relationship between the structure of the complex and heterogeneous soil Humeome and its functional reactivity in the environment (Piccolo et al. 2018a).

7.4.2 Carbon Sequestration by the in Situ Photocatalytic Polymerization of the Soil Humeome

The current view depicts the soil Humeome as supramolecular associations of relatively small heterogeneous molecules held together by weak dispersive forces. This novel paradigm of the chemical nature of soil humus allows conceiving innovative methodologies capable of promoting an effective SOC sequestration based on the inherent molecular characteristics of the soil Humeome. In this respect, an example of an innovative sustainable soil management towards an increased sequestration of SOC is represented by the implementation of a chemical technology aimed to bind together the different humic molecules by covalent bonds in larger oligomers and consequently increase the average molecular mass of the soil Humeome (Piccolo 2012; Piccolo et al. 2018a, b). This structural modification is hence expected to increase the internal energy of the Humeome and reduce or even inhibit the microbial mineralization of SOM.

The addition to solutions of humic molecules of an oxidative enzyme such as horseradish peroxidase, together with an oxidant such as hydrogen peroxide, induced a free radical-driven coupling reaction among humic aromatic molecules, with an effective oligomerization of the humic matter (Piccolo et al. 2000; Cozzolino and Piccolo 2002). A technological improvement was obtained by replacing the bio-labile enzyme with synthetic *biomimetic* catalysts such as metal porphyrins that mimic the activity of oxidative enzymes. To ensure water solubility and stable environmental reactivity, the porphyrin ring was chemically modified, by including electron-withdrawing dichloro-sulphonatophenyl groups. These water-soluble metal porphyrins were found to easily catalyse, under H_2O_2 oxidation, the covalent coupling of humic phenol components in greater molecular mass molecules (Piccolo et al. 2005b; Šmejkalová and Piccolo 2006; Šmejkalová et al. 2006).

Even more interesting was the finding that the formation of intermolecular C–C and C–O–C bonds among humic molecules catalysed by metal porphyrins could be equally achieved under photo-oxidation by simply exposing the reactive components to solar irradiation. The in situ catalysed photo-polymerization of SOM was shown in a laboratory experiment, whereby soils treated with iron porphyrin under solar light were found not only to significantly increase soil physical stability despite several cycles of wetting and drying but also to reduce microbial respiration and CO_2 emission to the atmosphere (Piccolo et al. 2011). Similar findings were shown by a mesocosm experiment with undisturbed soils treated with iron porphyrin and exposed to solar light (Gelsomino et al. 2010). These results indicated that the catalysed photo-polymerization of SOM enabled both an enhanced soil aggregation due to more extensive soil particles interactions and a reduced capacity of soil microorganisms not only to reach the larger humic molecules adsorbed on soil particles but also to overcome the increased internal energy of the newly formed covalent bonds.

The efficacy of the biomimetic catalysis on the photo-polymerization of humic molecules was further increased when a manganese porphyrin was immobilized through an imidazole spacer to the surface of common clay minerals such as kaolinite and montmorillonite (Nuzzo and Piccolo 2013; Nuzzo et al. 2013). The metal porphyrin immobilization on eco-compatible solid surfaces entailed a heterogeneous catalysis three times more efficient than the homogeneous catalysis by a water-soluble metal porphyrin. The enhanced efficiency of the heterogeneous catalysis was attributed to the larger persistence of the immobilized biomimetic catalyst on the soil surface, whereas the water-soluble metal porphyrins are easily eluviated down the soil profile, thus losing the capacity to benefit from the photo-oxidation that occurs at the soil surface (Nuzzo et al. 2016, 2017).

These positive findings were applied in real field conditions within an Italian national research project (MESCOSAGR) focused on the development of sustainable SOC sequestration methods in agricultural soils. The water-soluble iron porphyrin catalyst was amended to field soils under wheat cultivation at three sites along a north–south climate gradient in Italy (Spaccini and Piccolo 2012). After three years, the photo-polymerization of SOM promoted by the biomimetic catalyst increased the average TOC content in both bulk soils and their size separates from 2.24 to 3.90 t ha^{-1} y^{-1} in respect of control plots (Piccolo et al. 2018b). This exceptional extent of soil carbon sequestration was much larger than what is currently achieved by no-tillage practices which are reported to hardly exceed 0.5 tC ha^{-1} y^{-1} (Minasny et al. 2017). Since the experimental field plots were subjected to a conventional tillage system without any additional OM inputs, this work showed that the in situ catalysed photo-oxidative treatment effectively enhanced the chemical and biochemical stability of the soil Humeome, thus favouring a progressive SOC sequestration. Moreover, the objectively considerable amount of carbon sequestered yearly in this study by the applied photocatalytic technology was even achieved without either altering the soil biological quality or reducing the crop yields (Piccolo et al. 2018b). Fixation of SOC by amending soils with a photocatalyst, such as metal porphyrins, may be promising to establish an easy soil management practice that not only controls SOM dynamics and contributes to accumulate carbon stocks in agricultural soils, but also sustains the intensification of agriculture required to feed an increasing population.

References

Achard FK (1786). Chemische untersuchung des torfs. Crell's Chem Ann 2:391–403

Aguiar NO, Novotny EH, Oliveira AL, Rumjanek VM, Olivares FL, Canellas LP (2013) Prediction of humic acids bioactivity using spectroscopy and multivariate analysis. J Geochem Expl 129:95–102

Aguiar NO, Medici LO, Olivares FL, Dobbss LB, Torres-Netto A, Silva SF, Novotny EH, Canellas LP (2016) Metabolic profile and antioxidant responses during drought stress recovery in sugarcane treated with humic acids and endophytic diazotrophic bacteria. Ann Appl Biol 168(203–213):2016

Aguirre E, Leménager D, Bacaicoa E, Fuentes M, Baigorri R, Zamarreño AM, García-Mina JM (2009) The root application of a purified leonardite humic acid modifies the transcriptional regulation of the main physiological root responses to Fe deficiency in Fe-sufficient cucumber plants. Plant Physiol Biochem 47:215–223

Baalousha M, Motelica-Heino M, Galaup S, Le Coustumer P (2005) Supramolecular structure of humic acids by TEM with improved sample preparation and staining. Microsc Res Technol 66:299–306

Baalousha M, Motelica-Heino M, Le Coustumer P (2006) Conformation and size of humic substances: effects of major cation concentration and type, pH, salinity and residence time. Colloids Surf A 272:48–55

Baldock JA, Skjemstad JO (2000) Role of the soil matrix and minerals in protecting natural organic materials against biological attack. Org Geochem 31:697–710

Buurman P, van Lagen B, Piccolo A (2002) Increase in stability against thermal oxidation of soil humic substances as a result of self- association. Org Geochem 33:367–381

Canellas LP, Olivares FL (2014) Physiological responses to humic substances as plant growth promoter. Chem Biol Technol Agric 1:3

Canellas LP, Olivares FL (2017) Production of border cells and colonization of maize root tips by *Herbaspirillum seropedicae* are modulated by humic acid. Plant Soil 417:403–413

Canellas LP, Teixeira Junior LRL, Dobbs LB, Silva CA, Medici LO, Zandonadi DB, Façanha AR (2008) Humic acids cross-interactions with root and organic acids. Ann Appl Biol 153:157–166

Canellas LP, Dantas DJ, Aguiar NO, Peres LEP, Zsögön A, Olivares FL, Dobbs LB, Façanha AR, Nebbioso A, Piccolo A (2011) Probing the hormonal activity of fractionated molecular humic components in tomato auxin mutants. Ann Appl Biol 159:202–211

Canellas LP, Dobbs LB, Oliveira AL, Chagas JG, Aguiar NO, Rumjanek VM, Novotny EH, Olivares FL, Spaccini R, Piccolo A (2012) Chemical properties of humic matter as related to induction of plant lateral roots. Eur J Soil Sci 63:315–324

Canellas LP, Olivares FL, Aguiar N, Jones DL, Nebbioso A, Mazzei P, Piccolo A (2015) Humic and fulvic acids as biostimulants in horticulture. Sci Hortic 196:15–27

Chen Y (1996) Organic matter reactions involving micronutrients in soils and their effect on plants. In: Piccolo A (ed) Humic substances in terrestrial ecosystems. Elsevier, Amsterdam, pp 507–530

Chilom G, Bruns AS, Rice JA (2009) Aggregation of humic acids in solution: contribution of different fractions. Org Geochem 40:455–460

Conte P, Piccolo A (1999) Conformational arrangement of dissolved humic substances. Influence of solution composition on association of humic molecules. Environ Sci Technol 33:1682–1690

Conte P, Zena A, Pilidis G, Piccolo A (2001) Increased retention of Polyaromatic Hydrocarbons (PAH) in soils induced by soil treatment with humic substances. Environ Pollut 112:27–31

Conte P, Spaccini R, Piccolo A (2006) Advanced CPMAS-^{13}C NMR techniques for molecular characterization of size-separated fractions from a soil humic acid. Anal Bioanal Chem 386:382–390

Conte P, Spaccini R, Šmejkalová D, Nebbioso A, Piccolo A (2007) Spectroscopic and conformational properties of size-fractions separated from a lignite humic acid. Chemosphere 69:1032–1039

Cozzolino A, Piccolo A (2002) Polymerization of dissolved humic substances catalyzed by peroxidase. Effects of pH and humic composition. Org Geochem 33:281–294

Cozzolino A, Conte P, Piccolo A (2001) Conformational changes of humic substances induced by some hydroxy-, keto-, and sulfonic acids. Soil Biol Bochem 33:563–571

Cozzolino V, Di Meo V, Piccolo A (2013) Impact of arbuscular mycorrhizal fungi applications on maize production and soil phosphorus availability. J Geochem Explor 129:40–44

Cozzolino V, Di Meo V, Monda H, Spaccini R, Piccolo A (2016a) The molecular characteristics of compost affect plant growth, arbuscular mycorrhizal fungi, and soil microbial community composition. Biol Fertil Soils 52:15–29

Cozzolino V, De Martino A, Di Meo V, Salluzzo A, Piccolo A (2016b) Plant tolerance to mercury in a contaminated soil is enhanced by the combined effects of humic matter addition and inoculation with arbuscular mycorrhizal fungi. Environ Sci Pollut Res 23:11312–11322

da Piedade Melo A, Olivares FL, Médici LO, Dobbs LB, Torres-Netto A, Canellas LP (2017) Mixed rhizobia and *Herbaspirillum seropedicae* inoculations with humic acid-like substances improve water-stress recovery in common beans. Chem Biol Technol Agric 4:6. https://doi.org/10.1186/s40538-017-0090-z

Dobbss LB, Canellas PL, Lopes Olivares F, Aguiar ON, Peres LEP, Azevedo M, Spaccini R, Piccolo A, Façanha AR (2010) Bioactivity of chemically transformed humic matter from vermicompost on plant root growth. J Agric Food Chem 58:3681–3688

Drosos M, Piccolo A (2018) The molecular dynamics of soil humus as a function of tillage. Land Degr Develop 29:1792–1805. https://doi.org/10.1002/ldr.2989

Drosos M, Nebbioso A, Mazzei P, Vinci G, Spaccini R, Piccolo A (2017) A molecular zoom into soil Humeome by a direct sequential chemical fractionation of soil. Sci Total Environ 15:807–816

Drosos M, Nebbioso A, Piccolo A (2018a) Humeomics: a key to unravel the humusic pentagram. Appl Soil Ecol 123:513–516

Drosos M, Savy D, Spiteller M, Piccolo A (2018b) Structural characterization of carbon and nitrogen molecules in the Humeome of two different grassland soils. Chem Biol Technol Agric 5:14

Fava F, Piccolo A (2002) Effects of humic substances on the bioavailability and aerobic biodegradation of polychlorinated biphenyls in a model soil. Biotechnol. Bioeng 77(2):204–211

Feng XJ, Simpson AJ, Simpson MJ (2005) Chemical and mineralogical controls on humic acid sorption to clay mineral surfaces. Org Geochem 36:1553–1566

Fiorentino G, Spaccini R, Piccolo A (2006) Separation of molecular constituents from a humic acid by solid-phase extraction following a transesterification reaction. Talanta 68:1135–1142

García AC, Santos LA, Ambrósio de Souza LG, Tavares OCH, Zonta E, Gomes ETM, García-Mina JM, Berbara RLL (2016) Vermicompost humic acids modulate the accumulation and metabolism of ROS in rice plants. J Plant Physiol 192:56–63

Gelsomino A, Tortorella D, Cianci V, Petrovicová B, Sorgonà A, Piccolo A et al (2010) Effects of a biomimetic iron-porphyrin on soil respiration and maize root morphology as by a microcosm experiment. J Plant Nutr Soil Sci 173:399–406

Grignani C, Alluvione F, Bertora C, Zavattaro L, Fagano M, Fiorentino N, et al (2012) Field plots and crop yields under innovative methods of carbon sequestration in soil. In: Piccolo A (ed)

Carbon sequestration in agricultural soils: a multidisciplinary approach to innovative methods. Springer, Heidelberg, Berlin, pp 39–60

Halim M, Conte P, Piccolo A (2003) Potential availability of heavy metals to phytoextraction from contaminated soils induced by exogenous humic substances. Chemosphere 52:265–275

Hayes MHB (2009) Evolution of concepts of environmental natural nonliving organic matter. In: Senesi N, Xing B, Huang PM (eds) Biophysico-chemical processes involving natural nonliving organic matter in environmental systems. Wiley, New Jersey, pp 1–39

Hertkorn N, Benner R, Frommberger M, Schmitt-Kopplin P, Witt M, Kaiser K et al (2006) Characterization of a major refractory component of marine dissolved organic matter. Geochim Cosmochim Acta 70:2990–3010

Incerti G, Bonanomi G, Giannino F, Carteni F, Spaccini R, Mazzei P et al (2017) OMDY: a new model of organic matter decomposition based on biomolecular contentas assessed by 13C-CPMAS-NMR. Plant Soil 411:377–394

Insam H (1996) Microorganisms and humus in soils. In: Piccolo A (ed) Humic substances in terrestrial ecosystems. Elsevier, Amsterdam, The Netherlands, pp 361–406

Jindo K, Soares TS, Peres LEP, Azevedo IG, Aguiar NO, Mazzei P, Spaccini R, Piccolo A, Olivares FL, Canellas LP (2016) Phosphorus speciation and high-affinity transporters are influenced by humic substances. J Plant Nutr Soil Sci 179:206–214

Kelley KR, Stevenson FJ (1996) Organic forms of N in soil. In: Piccolo A (ed) Humic substances in terrestrial ecosystems. Elsevier, Amsterdam, The Netherlands, pp 407–427

Kleber M, Sollins P, Sutton R (2007) A conceptual model of organo-mineral interactions in soils: self-assembly of organic molecular fragments into zonal structures on mineral surfaces. Biogeochemistry 85:9–24

Köchy M, Hiederer R, Freibauer A (2015) Global distribution of soil organic carbon—Part 1: masses and frequency distributions of SOC stocks for the tropics, permafrost regions, wet-lands, and the world. Soil 1:351–365

Kozak J (1996) Soil organic matter as a factor influencing the fate of organic chemicals in the soil environment. In: Piccolo A (ed) Humic substances in terrestrial ecosystems. Elsevier, Amsterdam, The Netherlands, pp 652–664

Li M, Cozzolino V, Mazzei P, Drosos M, Monda H, Hu Z, Piccolo A (2017) Effects of microbial bioeffectors and P amendments on P forms in a maize cropped soil as evaluated by ^{31}P–NMR spectroscopy. Plant Soil 427:87–104. https://doi.org/10.1007/s11104-017-3405-8

Lipczynska-Kochany E (2018) Humic substances, their microbial interactions and effects on biological transformations of organic pollutants in water and soil: a review. Chemosphere 202:420–437

Lorenz K, Lal R, Preston CM, Nierop KGJ (2007) Strengthening the soil organic carbon pool by increasing contributions from recalcitrant aliphatic bio(macro)molecules. Geoderma 142:1–10

Lovley DR, Coates JD, Blunt-Harris EL, Phillips JP, Woodward JC (1996) Humic substances as electron acceptors for microbial respiration. Nature 282:445–448

Magid J, Tiessen H, Condron LM (1996) Dynamics of organic phosphorus in soils under natural and agricultural ecosystems. In: Piccolo A (ed) Humic substances in terrestrial ecosystems. Elsevier, Amsterdam, The Netherlands, pp 429–466

Maia CM, Piccolo A, Mangrich AS (2008) Molecular size distribution of compost-derived humates as a function of concentration and different counterions. Chemosphere 73:1162–1166

Mao J, Fang X, Schmidt-Rohr K, Carmo AM, Hundal LS, Thompson M-L (2007) Molecular-scale heterogeneity of humic acid in particle-size fractions of two Iowa soils. Geoderma 140:17–29

Martinez-Balmori D, Olivares FL, Spaccini R, Aguiar KP, Araújo MF, Aguiar NO, Guridi F, Canellas LP (2013) Molecular characteristics of vermicompost and their relationship to preservation of inoculated nitrogen-fixing bacteria. J Anal Appl Pyr 104:540–550

Martinez-Balmori D, Spaccini R, Aguiar NO, Novotny EH, Olivares FL, Canellas LP (2014) Molecular characteristics of humic acids isolated from vermicomposts and their relationship to bioactivity. J Agric Food Chem 62:11412–11419

Masoom H, Courtier-Murias D, Farooq H, Soong R, Kelleher BP, Zhang C et al (2016) Soil organic matter in its native state: unravelling the most complex biomaterial on earth. Environ Sci Technol 50:1670–1680

Mazzei P, Piccolo A (2012) Quantitative evaluation of noncovalent interactions between glyphosate and dissolved humic substances by NMR spectroscopy. Environ Sci Technol 46:5939–5946

Mazzoleni S, Bonanomi G, Giannino F, Incerti G, Piermatteo D, Spaccini R et al (2012) New modeling approach to describe and predict carbon sequestration dynamics in agricultural soils. In: Piccolo A (ed) Carbon sequestration in agricultural soils. Springer, Berlin, pp 291–307

Minasny B, Malone BP, McBratney AB, Angers DA, Arrouays D, Chambers A, Chaplot V, Chen Z-S, Cheng K, Das BS, Field DJ, Gimona A, Hedley CB, Hong SY, Mandal B, Marchant BP, Martin M, McConkey BG, Mulder VL, O'Rourke S, Richer-de-Forges AC, Odeh I, Padarian J, Paustian K, Pan G, Poggio L, Savin I, Stolbovoy V, Stockmann U, Sulaeman Y, Tsui C-C, Vågen T-G, van Wesemael B, Winowiecki L (2017) Soil carbon 4 per mille. Geoderma 292:59–86. https://doi.org/10.1016/j.geoderma.2017.01.002

Monda H, Cozzolino V, Vinci G, Spaccini R, Piccolo A (2017) Molecular characteristics of water-extractable organic matter from different composted biomasses and their effects on seed germination and early growth of maize. Sci Total Environ 590–591:40–49

Monda H, Cozzolino V, Vinci G, Drosos M, Savy D, Piccolo A (2018) Molecular composition of the Humeome extracted from different green composts and their biostimulation on early growth of maize. Plant Soil 429:407–424

Mora V, Bacaicoa E, Zamarreño AM, Aguirre E, Garnica M, Fuentes M, Garcia-Mina JM (2010) Action of humic acid on promotion of cucumber shoot growth involves nitrate-related changes associated with the root-to-shoot distribution of cytokinins, polyamines and mineral nutrients. J Plant Physiol 167:633–642

Muscolo A, Sidari M, Nardi R (2013) Humic substance: relationship between structure and activity. Deeper information suggests univocal findings. J Geochem Explor 129:57–63

Nardi S, Muscolo A, Vaccaro S, Baiano S, Spaccini R, Piccolo A (2007) Relationship between molecular characteristics of soil humic fractions and glycolytic pathway and krebs cycle in maize seedlings. Soil Biol Biochem 39:3138–3146

Nardi S, Ertani A, Francioso O (2017) Soil–root cross-talking: the role of humic substances. J Plant Nutr Soil Sci 180:5–13

Nebbioso A, Piccolo A (2009) Molecular rigidity and diffusivity of Al^{3+} and Ca^{2+} humates as revealed by NMR spectroscopy. Environ Sci Technol 43:2417–2424

Nebbioso A, Piccolo A (2011) Basis of a humeomics science: chemical fractionation and molecular characterization of humic biosuprastructures. Biomacromol 12:1187–1199

Nebbioso A, Piccolo A (2012) Advances in humeomics: enhanced structural identification of humic molecules after size fractionation of a soil humic acid. Anal Chim Acta 720:77–90

Nebbioso A, Piccolo A (2013) Molecular characterization of dissolved organic matter (DOM): a critical review. Anal Bioanal Chem 405:109–124

Nebbioso A, Piccolo A, Spiteller M (2010) Limitations of electrospray ionization in the analysis of a heterogeneous mixture of naturally occurring hydrophilic and hydrophobic compounds. Rapid Commun Mass Spectrom 24:3163–3170

Nebbioso A, Piccolo A, Lamshöft M, Spiteller M (2014a) Molecular characterization of an end-residue of humeomics applied to a soil humic acid. RSC Adv. 4:23658–23665

Nebbioso A, Mazzei P, Savy D (2014b) Reduced complexity of multidimensional and diffusion NMR spectra of soil humic fractions as simplified by humeomics. Chem Biol Technol Agric 1:24. https://doi.org/10.1186/540538-014-0024-4

Nebbioso A, Vinci G, Drosos M, Spaccini R, Piccolo A (2015) Unveiling the molecular composition of the unextractable soil organic fraction (humin) by humeomics. Biol Fertil Soils 51:443–451

Nuzzo A, Piccolo A (2013) Oxidative and photo-oxidative polymerization of humic suprastructures by heterogeneous biomimetic catalysis. Biomacromolecules 14:1645–1652

Nuzzo A, Sánchez A, Fontaine B, Piccolo A (2013) Conformational changes of dissolved humic and fulvic superstructures with progressive iron complexation. J Geochem Explor 129:1–5

Nuzzo A, Madonna E, Mazzei P, Spaccini R, Piccolo A (2016) *In situ* photo-polymerization of soil organic matter by heterogeneous nano-TiO_2 and biomimetic metal-porphyrin catalysts. Biol Fertil Soils 52:585–593

Nuzzo A, Spaccini R, Cozzolino V, Moschetti G, Piccolo A (2017) In situ polymerization of soil organic matter by oxidative biomimetic catalysis. Chem Biol Technol Agric 4:12

Nuzzo A, De Martino A, Di Meo V, Piccolo A (2018) Potential alteration of iron–humate complexes by plant root exudates and microbial siderophores. Chem Biol Technol Agric 5:19. https://doi.org/10.1186/s40538-018-0132-1

Oades JM (1988) The retention of organic matter in soils. Biogeochemistry 5:35–70

Oades JM, Waters AG (1991) Aggregate hierarchy in soils. Aust J Soil Res 29:825–828

Olaetxea M, Mora V, Bacaicoa E, Garnica M, Fuentes M, Casanova E, Zamarreño AM, Iriarte JC, Etayo D, Ederra I, Gonzalo R, Baigorri R, Gonzalo R, Garcia-Mina J-M (2015) Abscisic acid regulation of root hydraulic conductivity and aquaporin gene expression is crucial to the plant shoot growth enhancement caused by rhizosphere humic acids. Plant Physiol 169:2587–2596

Olaetxea M De, Hita D, Garcia A, Fuentes M, Baigorri R, Mora V, Garnica M, Urrutia O, Zamarreño AM, Berbara RL, Garcia-Mina JM (2018) Hypothetical framework integrating the main mechanisms involved in the promoting action of rhizospheric humic substances on plant root- and shoot-growth. Appl Soil Ecol 123:521–537

Orlowska E, Roller A, Pignitter M, Jirsa F, Krachler R, Kandioller W, Keppler BK (2017) Synthetic iron complexes as models for natural iron-humic compounds: synthesis, characterization and algal growth experiments. Sci Tot Environ 577:94–104

Orsi M (2014) Molecular dynamics simulation of humic substances. Chem Biol Technol Agric 1:10. https://doi.org/10.1186/S40538-014-0010-4

Pane C, Spaccini R, Piccolo A, Scala F, Bonanomi G (2011) Compost amendments enhance peat suppressiveness to *Pythium ultimum*, *Rhizoctonia solani* and *Sclerotinia minor*. Biol Control 56:115–124

Pane C, Piccolo A, Spaccini R, Celano G, Villecco D, Zaccardelli M (2013) Agricultural waste-based composts exhibiting suppressivity to diseases caused by the phytopathogenic soil-borne fungi Rhizoctonia solani and Sclerotinia minor. Appl Soil Ecol 65:43–51

Pane P, Celano G, Piccolo A, Villecco D, Spaccini R, Palese AM, Zaccardelli M (2015) Effects of on-farm composted tomato residues on soil biological activity and yields in a tomato cropping system. Chem Biol Technol Agric 2:4. https://doi.org/10.1186/S40538-014-0026-9

Paul EA (2016) The nature and dynamics of soil organic matter: Plant inputs, microbial transformations, and organic matter stabilization. Soil Biol Biochem 98:109–126

Peuravuori J (2005) NMR spectroscopy study of freshwater humic material in light of supramolecular assembly. Environ Sci Technol 39:5541–5549

Peuravuori J, Pihlaja K (2004) Preliminary study of lake dissolved organic matter in light of nanoscale supramolecular assembly. Environ Sci Technol 38:5958–5967

Peuravuori J, Bursakova P, Pihlaja K (2007) ESI-MS analyses of dissolved organic matter in light of supramolecular assembly. Anal Bioanal Chem 389:1559–1568

Piccolo A (1996) Humus and soil conservation. In: Piccolo A (ed) Humic substances in terrestrial ecosystems. Elsevier, Amsterdam, The Netherlands, pp 225–264

Piccolo A (1989) Reactivity of humic substances towards plant available heavy metals. Sci Total Environ 81/82:607–614

Piccolo A (2001) The supramolecular structure of humic substances. Soil Sci 166:810–832

Piccolo A (2002) The supramolecular structure of humic substances: a novel understanding of humus chemistry and implications in soil science. Adv Agr 75:57–134

Piccolo A (2012) The nature of soil organic matter and innovative soil managements to fight global changes and maintain agricultural productivity. In: Piccolo A (ed) Carbon sequestration in agricultural soils. Springer, Heidelberg, pp 1–19

Piccolo A (2016) *In memoriam* Prof. F. J. Stevenson and the question of humic substances in soil. Chem Biol Technol Agric 3:23. https://doi.org/10.1186//s40538-016-0076-2

Piccolo A, Mbagwu JSC (1999) Role of hydrophobic components of soil organic matter in soil aggregate stability. Soil Sci Soc Am J 63:1801–1810

Piccolo A, Spiteller M (2003) Electrospray ionization mass spectrometry of terrestrial humic substances and their size-fractions. Anal Bioanal Chem 377:1047–1059

Piccolo A, Nardi S, Concheri G (1992) Structural characteristics of humic substances as related to nitrate uptake and growth regulation in plant systems. Soil Biol Biochem 24:373–380

Piccolo A, Nardi S, Concheri G (1996) Macromolecular changes of soil humic substances induced by interactions with organic acids. Eur J Soil Sci 47:319–328

Piccolo A, Cozzolino A, Conte P, Spaccini R (2000) Polymerization of humic substances by an enzyme-catalyzed oxidative coupling. Naturwissenschaften 87:391–394

Piccolo A, Conte P, Cozzolino A (2001) Chromatographic and spectrophotometric properties of dissolved humic substances compared with macromolecular polymers. Soil Sci 166:174–185

Piccolo A, Conte P, Spaccini R, Chiarella M (2003) Effects of some dicarboxylic acids on the association of dissolved humic substances. Biol Fertil Soils 37:255–259

Piccolo A, Spaccini R, Nieder R, Richter J (2004) Sequestration of a biologically labile organic carbon in soils by humified organic matter. Clim Change 67:329–343

Piccolo A, Conte P, Spaccini R, Mbagwu JSC (2005a) Influence of land use on the humic substances of some tropical soils of Nigeria. Eur J Soil Sci 56:343–352

Piccolo A, Conte P, Tagliatesta P (2005b) Increased conformational rigidity of humic substances by oxidative biomimetic catalysis. Biomacromol 6:351–358

Piccolo A, Spiteller M, Nebbioso A (2010) Effects of sample properties and mass spectroscopic parameters on electrospray ionization mass spectra of size-fractions from a soil humic acid. Anal Bioanal Chem 397:3071–3078

Piccolo A, Spaccini R, Nebbioso A, Mazzei P (2011) Carbon sequestration in soil by in situ catalyzed photo-oxidative polymerization of soil organic matter. Environ Sci Technol 45:6697–6702

Piccolo A, Spaccini R, Drosos M, Cozzolino V (2018a) The molecular composition of humus carbon: recalcitrance and reactivity in soils. In: Garcia C, Nannipieri P, Hernandez T (eds) The future of soil carbon-its conservation and formation. Academic Press, San Diego, pp 87–124

Piccolo A, Spaccini R, Cozzolino V, Nuzzo A, Drosos M, Zavattaro L, Grignani C, Puglisi E, Trevisan M (2018b) Effective carbon sequestration in Italian agricultural soils by in situ polymerization of soil organic matter under biomimetic photocatalysis. Land Degrad Develop 29:485–494. https://doi.org/10.1002/ldr.2877

Puglisi E, Trevisan M (2012) Effects of methods of carbon sequestration in soil on biochemical indicators of soil quality. In: Piccolo A (ed) Carbon sequestration in agricultural soils. Springer, Heidelberg, pp 179–207

Puglisi E, Fragoulis G, Ricciuti P, Cappa F, Spaccini R, Piccolo A, Trevisan M, Crecchio C (2009) Effects of a humic acid and its size-fractions on the bacterial community of soil rhizosphere under maize (Zea mays L.). Chemosphere 77:829–837

Puglisi E, Pascazio S, Suciu N, Cattani I, Fait G, Spaccini R, Crecchio C, Piccolo A, Trevisan M (2013) Rhizosphere microbial diversity as influenced by humic substance amendments and chemical composition of rhizodeposits. J Geochem Explor 129:82–94

Ramos AC, Olivares FL, Silva LS, Aguiar NO, Canellas LP (2015) Humic matter elicits proton and calcium fluxes and signaling dependent on Ca^{2+}-dependent protein kinase (CDPK) at early stages of lateral plant root development. Chem Biol Technol Agric 2:4. https://doi.org/10.1186/s40538-014-0030-0

Rose MT, Patti AF, Little KR, Brown AL, Jackson WR, Cavagnaro TR (2014) A meta-analysis and review of plant-growth response to humic substances: practical implications for agriculture. Adv Agron 124:37–89

Saiz-Jimenez J (1995) Reactivity of the aliphatic humic moiety in analytical pyrolysis. Org Geochem 23:955–961

Sannino F, Piccolo A (2013) Effective remediation of contaminated soils by eco-compatible chemical, biological and biomimetic practices. In: Basile A, Piemonte V, de Falco M (eds) Sustainable development in chemical engineering: innovative technologies. Wiley, Chichester (UK), pp 267–296

Sarker TC, Incerti G, Spaccini R, Piccolo A, Mazzoleni S, Bonanomi G (2018) Linking organic matter chemistry with soil aggregate stability: Insight from ^{13}C NMR spectroscopy. Soil Biol Biochem 117:175–184

Savy D, Piccolo A (2014) Physical-chemical characteristics of lignins separated from biomasses for second-generation ethanol. Biomass Bioenergy 62:58–67

Savy D, Cozzolino V, Vinci G, Nebbioso A, Piccolo A (2015) Water-soluble lignins from different bioenergy crops stimulate the early development of maize (*Zea mays*, L.). Molecules 20:19958–19970

Savy D, Cozzolino V, Nebbioso A, Drosos M, Nuzzo A, Mazzei P, Piccolo A (2016a) Humic-like bioactivity on emergence and early growth of maize (*Zea mays* L.) of water-soluble lignins isolated from biomass for energy. Plant Soil 402:221–233

Savy D, Mazzei P, Nebbioso A, Drosos M, Nuzzo A, Cozzolino V, Spaccini R, Piccolo A (2016b) Molecular properties and functions of humic substances and humic-like substances (hulis) from biomass and their transformation products. In: Vaz S Jr (ed) Analytical techniques and methods for biomass. Springer, Chan, pp 85–114

Savy D, Mazzei P, Drosos M, Cozzolino V, Lama L, Piccolo A (2017) Molecular characterization of extracts from biorefinery wastes and evaluation of their plant biostimulation. ACS Sustain Chem Eng 5:9023–9031

Savy D, Cozzolino V, Drosos M, Mazzei P, Piccolo A (2018) Replacing calcium with ammonium counterion in lignosulfonates from paper mills affects their molecular properties and bioactivity. Sci Total Environ 645:411–418

Scaglia B, Nunes RR, Rezende MOO, Tambone F, Adani F (2016) Investigating organic molecules responsible of auxin-like activity of humic acid fraction extracted from vermicompost. Sci Total Environ 562:289–295

Schaumann GE, Thiele-Bruhn S (2011) Reprint of: molecular modeling of soil organic matter: squaring the circle? Geoderma 169:55–68

Schwarzenbach RP, Gschwend PM, Imboden DM (2003a) Sorption III: sorption processes involving inorganic surfaces. In: Environmental organic chemistry, 2nd edn. Wiley Interscience, pp 387–458

Schwarzenbach RP, Gschwend PM, Imboden DM (2003b) Sorption I: general introduction and sorption processes involving organic matter. In: Environmental organic chemistry, 2nd edn. Wiley Interscience, pp 275–330

Scotti R, Pane P, Spaccini R, Palese AM, Piccolo A, Celano G et al (2016) On-farm compost: a useful tool to improve soil quality under intensive farming systems. Appl Soil Ecol 107:13–23

Siéliéchi JM, Lartiges BS, Kayem GJ et al (2008) Changes in humic acid conformation during coagulation with ferric chloride: implications for drinking water treatment. Water Res 42:2111–2123

Simpson AJ (2002) Determining the molecular weight, aggregation, structures and interactions of natural organic matter using diffusion ordered spectroscopy. Magn Reson Chem 40:S72–S82

Six J, Bossuyt H, Degryze S, Denef K (2004) A hystory of research on the link between microaggregates, soil biota and soil organic matter dynamics. Soil Till Res 79:7–31

Šmejkalová D, Piccolo A (2006) Rates of oxidative coupling of humic phenolic monomers catalyzed by a biomimetic iron-porphyrin. Environ Sci Technol 40:1644–1649

Šmejkalová D, Piccolo A (2008a) Aggregation and disaggregation of humic supramolecular assemblies by NMR diffusion ordered spectroscopy (DOSY-NMR). Environ Sci Technol 42:699–706

Šmejkalová D, Piccolo A (2008b) Host-guest interactions between 2,4-dichlorophenol and humic substances as evaluated by ^{1}H NMR relaxation and diffusion ordered spectroscopy. Environ Sci Technol 42:8440–8445

Šmejkalová D, Piccolo A, Spiteller M (2006) Oligomerization of humic phenolic monomers by oxidative coupling under biomimetic catalysis. Environ Sci Technol 40:6955–6962

Šmejkalová D, Spaccini R, Fontaine B, Piccolo A (2009) Binding of phenol and differently halogenated phenols to dissolved humic matter as measured by NMR spectroscopy. Environ Sci Technol 43:5377–5382

Smith PM, Bustamante H, Ahammad H, Clark H, Dong EA, Elsiddig H, et al (2014) Agriculture, forestry and other land use (AFOLU). In: Edenhofer O, Pichs-Madruga R, Sokona Y, Farahani E, Kadner S, Seyboth K, Adler A, Baum I, Brunner S, Eickemeier P, Kriemann B, Savolainen J, Schlömer S, von Stechow C, Zwickel T, Minx JC (eds) Climate change 2014: mitigation of climate change. Contribution of working group III to the fifth assessment report of the intergovernmental panel on climate change. Cambridge University Press, Cam-bridge, United Kingdom and New York, NY, United States, pp. 816–887. Environ Pollut 241:265–271

Song XY, Spaccini R, Piccolo A, Pan GX (2013) Stabilization by hydrophobic protection as a molecular mechanism for organic carbon sequestration in maize amended rice paddy soils. Sci Total Environ 458:319–330

Spaccini R, Piccolo A (2007) Molecular characterization of compost at increasing stages of maturity. 2. thermochemolysis-GC–MS and ^{13}C-CPMAS-NMR spectroscopy. J Agric Food Chem 55:2303–2311

Spaccini R, Piccolo A (2009) Molecular characteristics of humic acids extracted from compost at increasing maturity stages. Soil Biol Biochem 41:1164–1172

Spaccini R, Piccolo A (2012) Carbon sequestration in soils by hydrophobic protection and in-situ catalyzed photo-polymerization of soil organic matter (SOM). Chemical and physical-chemical aspects of SOM in field plots. In: Piccolo A (ed) Carbon sequestration in agricultural soils. Springer, Heidelberg, pp 61–105

Spaccini R, Piccolo A, Haberhauer G, Gerzabek M (2000) Transformation of organic matter from maize residues into labile and humic fractions of three European soils as revealed by ^{13}C distribution and CPMAS-NMR spectra. Eur J Soil Sci 51:583–594

Spaccini R, Piccolo A, Mbagwu JSC, Igwe CA, Zena TA (2002a) Influence of the addition of organic residues on carbohydrate content and structural stability of some highland soils in Ethiopia. Soil Use Manag 18:404–411

Spaccini R, Piccolo A, Conte P, Haberhauer G, Gerzabek MH (2002b) Increased soil organic carbon sequestration through hydrophobic protection by humic substances. Soil Biol Biochem 34:1839–1851

Spaccini R, Mbagwu JSC, Igwe CA, Conte P, Piccolo A (2004) Carbohydrates and aggregation in lowland soils of Nigeria as influenced by organic inputs. Soil Tillage Res 75:161–172

Spaccini R, Mbagwu JSC, Conte P, Piccolo A (2006) Changes of humic substances characteristics from forested to cultivated soils in Ethiopia. Geoderma 132:9–19

Spaccini R, Sannino D, Piccolo A, Fagnano M (2009) Molecular changes in organic matter of a compost-amended soil. Eur J Soil Sci 60:287–296

Spaccini R, Song X-Y, Cozzolino V, Piccolo A (2013) Molecular evaluation of soil organic matter characteristics in three agricultural soils by improved off-line thermochemolysis: the effect of hydrofluoric acid demineralization treatment. Anal Chim Acta 802:46–55

Spaccini R, Cozzolino V, Di Meo V, Savy D, Drosos M, Piccolo A (2019) Bioactivity of humic substances and water extracts from compost made by ligno-cellulose wastes from biorefinery. Sci Total Environ 646:792–800

Stevenson FJ (1994) Humus chemistry: genesis, composition, reactions, 2nd ed, Wiley, New York

Sutton R, Sposito G (2006) Molecular simulation of humic substance–Ca-montmorillonite complexes. Geochim Cosmochim Acta 70:3566–3581

Trevisan S, Francioso O, Quaggiotti S, Nardi S (2010) Humic substances biological activity at the plant-soil interface. From environmental aspects to molecular factors. Plant Sign Behav 5:635–643

Urrutia O, Erro J, Guardado I, San Francisco S, Mandado M, Baigorri R, Yvin J-C, Garcia-Mina JM (2014) Physico-chemical characterization of humic-metal-phosphate complexes and their potential application to the manufacture of new types of phosphate-based fertilizers. J Plant Nutr Soil Sci 177:128–136

Vaccaro S, Muscolo A, Pizzeghello D, Spaccini R, Piccolo A, Nardi S (2009) Effect of a compost and its water-soluble fractions on key enzymes of nitrogen metabolism in maize seedlings. J Agric Food Chem 57:11267–11276

Vaccaro S, Ertani A, Nebbioso A, Muscolo A, Quaggiotti S, Piccolo A, Nardi S (2015) Humic substances stimulate maize nitrogen assimilation and amino acid metabolism at physiological and molecular level. Chem Biol Technol Agric 2:5

Varga B, Kiss G, Galambos I, Gelencser A, Hlavay J, Krivacsy Z (2000) Secondary structure of humic acids. Can micelle-like conformation be proved by aqueous size exclusion chromatography? Environ Sci Technol 34:3303–3306

Ventorino V, De Marco A, Pepe O, Virzo De Santo A, Moschetti G (2012) Impact of innovative agricultural practices of carbon sequestration on soil microbial community. In: Piccolo A (ed) Carbon sequestration in agricultural soils. Springer-Verlag, Heidelberg, pp 145–177

Vinci G, Cozzolino V, Mazzei P, Monda H, Spaccini R, Piccolo A (2018a) Effects of Bacillus amyloliquefaciens and organic and inorganic phosphate amendments on Maize plants as revealed by NMR and GC-MS based metabolomics. Plant Soil 429:1–14. https://doi.org/10.1007/s11104-018-3701-y

Vinci G, Cozzolino V, Mazzei P, Monda H, Spaccini R, Piccolo A (2018b) An alternative to mineral phosphorus fertilizers: The combined effects of *Trichoderma harzianum* and compost on *Zea mays*, as revealed by 1H NMR and GC-MS metabolomics. Plos One (in press). https://doi.org/10.1371/journal.pone.0209664

Winkler A, Haumaier L, Zech W (2005) Insoluble alkyl carbon components in soils derive mainly from cutin and suberin. Org Geochem 36:519–529

Woo DK, Quijano JC, Kumar P, Chaoka S, Bernacchi CJ (2014) Threshold dynamics in soil carbon storage for bioenergy crops. Environ Sci Technol 48:12090–12098

Wood S, Baudron F (2018) Soil organic matter underlies crop nutritional quality and productivity in smallholder agriculture. Agric Ecosyst Environ 266:100–108

Zancani M, Petrussa E, Krajňáková J, Casolo V, Spaccini R, Piccolo A, Macrì F, Vianello A (2009) Effect of humic acids on phosphate level and energetic metabolism of tobacco BY-2 suspension cell cultures. Env Exp Bot 65:287–295

Zancani M, Bertolini A, Petrussa E, Krajňáková J, Piccolo A, Spaccini R, Vianello A (2011) Fulvic acid affects proliferation and maturation phases in Abies cephalonica embryogenic cells. J Plant Physiol 168:1226–1233

Zheng Z, Zheng Y, Tian X, Yang Z, Jiang Y, Zhao F (2018) Interactions between iron mineral-humic complexes and hexavalent chromium and the corresponding bio-effects. Environ Pollut 241:265–271

Zhou P, Pan GX, Spaccini R, Piccolo A (2010) Molecular changes in particulate organic matter (POM) in a typical Chinese paddy soil under different long-term fertilizer treatments. Eur J Soil Sci 61:231–242

Chapter 8
Synthesis of New Agrochemicals

Paulo Marcos Donate and Daniel Frederico

Abstract The constantly growing world population calls for effective means of sustainable food production. This chapter will address some general aspects of the chemical action of products that protect crops against pest attack and the implications of these agrochemicals for the environment. The main types of active ingredients used in agrochemicals and their historical development will be concisely approached. The organic synthesis of original molecules and/or the structural modification of natural products that are already being used in agriculture will also be briefly discussed. This discussion will show some useful and versatile tools to obtain modern molecules with advantageous biological activities in agriculture. Given the need for further advances in crop protection, recently developed agrochemicals that can pave the way for future research, such as nanofertilizers, nanopesticides, and nanobiosensors, will be described. Finally, this chapter will discuss current challenges in the synthesis of agrochemicals, as well as cutting-edge developments in the design and synthesis of new agrochemicals, including the invention of more selective and environmentally friendly active ingredients, which will help to face the ongoing challenges of weed and pest resistance.

Keywords Agrochemical · Crop protection · Pesticide · Organic synthesis · Structural modification · Regulatory norms

P. M. Donate (✉)
Departamento de Química da Faculdade de Filosofia, Ciências e Letras, Universidade de São Paulo, Avenida Bandeirantes 3900, Ribeirão Preto, SP 14040-901, Brazil
e-mail: pmdonate@usp.br

D. Frederico
Dinagro Agropecuária Ltda., Via Anhanguera km 304, Ribeirão Preto, SP 14097-140, Brazil
e-mail: daniel.frederico@dinagro.com.br

© Springer Nature Switzerland AG 2019
S. Vaz Jr. (ed.), *Sustainable Agrochemistry*,
https://doi.org/10.1007/978-3-030-17891-8_8

8.1 Introduction

An agrochemical, a contraction of *agricultural chemical*, is a chemical product that protects crops against the attack and proliferation of pests such as insects, fungi, bacteria, viruses, mites, nematodes (parasites that attack plant roots), and weeds. Agrochemicals, or crop protection chemicals, include several groups of compounds, namely organochlorines, organophosphates, carbamates, pyrethroids, growth regulators, and neonicotinoids, which have been developed over the last decades. The sale of these products moves around US$50 billion per year (FAO Statistics 2017).

The large-scale use of agrochemicals is closely related to the evolution of agricultural production in numerous countries. For instance, the harsh winter period that helps to interrupt the pest cycle affecting agriculture in temperate zones is totally absent in tropical countries (Carvalho 2017). For this reason, over the last 100 years, pesticides and other agrochemicals have become an important component of agricultural systems worldwide and have contributed to raising crop yields and food production significantly (Alexandratos and Bruinsma 2012). Besides that, exponential human population growth has called for improved food production. The effects of climate change on agriculture have exacerbated the lack of food in many regions and have demanded further renewed efforts in food production (United Nations 2015).

Unfortunately, if on the one hand the use of agrochemicals has increased farming efficiency, on the other hand, it has raised great concern about environmental damage because agrochemicals pose higher risk of soil and water contamination. Over the last decades, agrochemical residues have been spread out in the environment, significantly polluting terrestrial ecosystems and poisoning human foods (Zhang et al. 2018). In addition, pesticide residues have repeatedly contaminated aquatic systems around the world, which has compromised aquatic food resources, fisheries, and aquaculture (European Environment Agency 2013; Khanna and Gupta 2018).

The adequate use of agrochemicals can effectively reduce the risks associated with them (World Health Organization 2012). Pesticides have been applied in agriculture in countless ways—the application modes vary from hand-operated spray conducted by workers on foot to truck and aerial spraying techniques (Figs. 8.1 and 8.2). However, handling agrochemicals without due care can also harm the health of workers and rural communities located near agricultural plantations. Various cases of intoxication among farmers, rural workers, and their families have occurred during pesticide applications and have been documented in reports on poisoning and on the effects of synthetic chemicals on the human health. Unintentional poisonings have killed an estimated 355,000 people worldwide each year, and such poisonings have been strongly related to excessive exposure and inappropriate use of toxic chemicals (Alavanja 2009; Alavanja and Bonner 2012). Moreover, dispersion of pesticide residues in the environment and mass extermination of non-human biota, like bees, birds, amphibians, fish, and small mammals, have been described (World Health Organization 2017).

Fig. 8.1 Rural producer applies agrochemicals in vegetable gardening with a manual backpack-type sprayer (from www.revistapesquisa.fapesp.br, No. 271, September 2018)

Fig. 8.2 Aerial application of pesticides in sugarcane plantation in the interior of São Paulo State, Brazil (from www.revistapesquisa.fapesp.br, No. 271, September 2018)

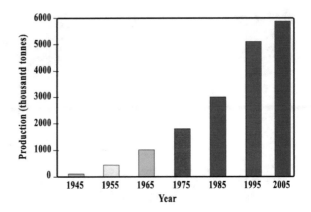

Fig. 8.3 World production of formulated agrochemicals based on FAO Statistics (from FAO Statistics 2017)

8.2   The Role that Agrochemicals Play in Agriculture

Agricultural production has increased markedly since the beginning of the twentieth century mainly to deal with demographic growth. The world population rose from 1.5 billion in 1900 to about 6.1 billion in 2000. At the current growth rates, the world population will be about 9.4–10 billion by 2050 (United Nations 2015). This means that food production will have to increase by at least twofold over the next 35 years to meet the food demand in 2050.

The rise in the world population along the twentieth century would not have been possible if there had not been a parallel growth in food production thanks to the application of chemical compounds. Organic fertilizers have been employed since the end of the nineteenth century, but mineral phosphate fertilizers started being introduced in the early years of the twentieth century, and their use has increased continuously (Gilland 2015). The application of phosphate compounds has led to an unprecedented improvement in agriculture productivity, the so-called Green Revolution, and cereal production has more than doubled per unit of surface area of cultivated land (Carvalho 2006). Human population growth and world production of phosphates for use as fertilizers have been significantly correlated over the last century (Roser and Ortiz-Ospina 2017, https://ourworldindata.org/world-population-growth/).

Since the 1940s, the introduction of synthetic crop protection chemicals has helped to increase food production further. The worldwide pesticide production has risen at a rate of about 11% per year, from 0.2 million tons in the 1950s to more than 5 million tons by 2000 (Fig. 8.3) (FAO Statistics 2017).

The use of agrochemicals has not been the same across the world: These chemicals are expensive (most of them are patented), agricultural labor costs are different, and specific pests exist in each climatic/geographic region. The Food and Agriculture Organization of the United Nations (FAO) has computed average application rates of agrochemicals per hectare of arable land, to find that Asia and some South American

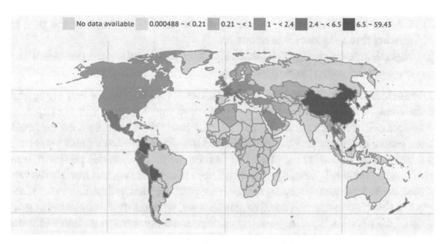

Fig. 8.4 Use of agrochemicals per hectare of arable land, kg ha^{-1}, in the years 2005–2009 (from FAO Statistical Yearbook 2013)

countries have the highest average values, 6.5–60 kg ha^{-1}. In North America and West Europe, herbicides have been intensively applied in agriculture and in urban areas over the last decades. In Asia, the use of herbicides has remained low and contrasts with the high use of insecticides (Fig. 8.4) (FAO Statistical Yearbook 2013).

Synthetic pesticides that had formerly been developed to control agriculture pests were also exhaustively used to control cattle ticks and human parasites in North America and Europe in the past. One example of such pesticides is dichlorodiphenyl-trichloroethane (DDT), whose insecticidal properties were discovered in 1939 by the Swiss entomologist Paul Müller. Although these former pesticides are now banned, they are still being used to preserve sundry fish in South Asia and are sometimes illegally applied to control malaria vectors and household pests in urban areas in the tropics (Taylor et al. 2003).

There are numerous criteria to classify pesticides including the type of pest they control, their action mode, their chemical structure, their toxic effect(s), and the hazards of using them (Hurst et al. 1994). The variety of plagues has determined the development of this wide diversity of pesticides. Most pesticides specifically act on a distinct type of pest. Concerning the type of pest they control, pesticides can be classified into insecticides (against insects), fungicides (against fungi), herbicides (against weeds), biocides (against a broad spectrum of organisms), and miscellaneous compounds (sterilants, sexual attractants, repellents, etc.).

Pesticides may destroy pests in different ways. The choice of pesticide will depend on their action mode. There are selective pesticides that control only one type of specific pest, while broader action pesticides control several plant diseases. Regarding the action mode, pesticides can kill pests in three main ways:

(1) Fumigants (which act as fumes or vapors at room temperature) eradicate the pest when they are breathed in as vapors.

(2) Contact pesticides (which attack by direct contact) exterminate the pest by passing through insect skin or cuticle.
(3) Ingestion pesticides (which attack after ingestion) destroy the pest when they are eaten by them (Hurst et al. 1994).

Pesticides may act as non-systemic pesticides at the place where they are spread, or they may act as systemic pesticides that move into the plant and consequently allow control of a larger area. When systemic pesticides are applied on the seeds, roots, stems, or leaves, they are absorbed and translocated to the various plant parts in amounts that are toxic to animals that feed on the plant. Systemic pesticides have advantages over non-systemic pesticides: They are selective against a particular type of pest, so they are less damaging to beneficial predatory and pollinating insects, and they are absorbed and translocated to plant tissues, which is useful to control pests in areas of the plant that ordinary contact pesticides cannot reach; e.g., bores of roots, stems, and fruits, and leaf miners (Hurst et al. 1994).

Pesticides destroy pests by interference at different levels. Some pesticides are involved in chitin metabolism; others block enzymes, respiration, and protein or hormone synthesis. They may also block nervous impulse transmission in pests. Compounds bearing different chemical groups have distinct toxic mechanisms and differ in the way they act on pest organisms. Organochlorine compounds (insecticides; e.g., aldrin, DDT, hexachlorocyclohexane, heptachlor, chlordane, and endosulfan) are generally very effective contact insecticides and are structurally related to steroid hormones that act on the respective hormone receptor (Tebourbi et al. 2011). Organophosphates (mostly insecticides; e.g., parathion, malathion, chlorpyrifos, diazinon, and dichlorvos) and carbamates (mostly herbicides and fungicides; e.g., aldicarb, carbofuran, ethienocarb, fenobucarb, and methomyl) act as acetylcholinesterase (AchE) inhibitors and disrupt nervous impulse transmission at the synaptic level. Pyrethroids (insecticides; e.g., cypermethrin, deltamethrin, esfenvalerate, and fenvalerate) act on the voltage-gated sodium channels in cell membranes, thereby impairing the Na^+ ion flux. Neonicotinoids (insecticides; e.g., acetamiprid, clothianidin, dinotefuran, and imidacloprid) act as agonists at the nicotinic acetylcholine receptors (nAChRs); they also act on the insect nervous system, which results in paralysis and death (Tomizawa and Casida 2005).

Unfortunately, the pesticide mechanism of toxic action is not restricted to target pests: Pesticides also exert toxicity on non-target, similar organisms, which harm biodiversity and the ecosystem. Organochlorine compounds heavily affect the top predators in terrestrial food chains, such as birds of prey. These compounds can accumulate in the adipose tissue of animals and humans and be transferred to newborns through milk fat, not to mention that they can act as endocrine disruptors (European Environment Agency 2013; Singh et al. 2016). Organophosphates are highly toxic to arthropods in general, including insects, shrimps, crabs, and other crustaceans. They are harmful to vertebrates, too. Pyrethroids also affect insects and vertebrates. Many other compounds, like herbicides, interfere in the mammalian central nervous system and excretory system, as well (Casida 2009).

Reports on environmental pollution and toxic effects on biota have prompted considerable efforts to design new chemicals and to improve pesticide formulations, application devices, and chemical delivery mechanisms (e.g., the use of degradable nanoparticles as a vehicle for pesticide delivery). These efforts are part of an attempt to reduce exposure to hazardous chemicals. Despite the countless issues regarding pesticide pollution, these compounds are still necessary. Thus, the search for new, more specific, and less toxic structures remains crucial.

8.3   The Use of Agrochemicals

The growing world population has demanded higher agricultural output (especially in terms of food production), but plant pathogens and pest problems have increased as a result (Peshin and Dhawan 2009). Crops are faced with 10,000 species of insects, 1500 plant diseases, 1800 kinds of weeds, and 1500 nematodes (microscopic soil worms), as well as various plant viruses (Stoner 2004; Govorushko 2018). To survive, plants need either natural or synthetic chemical defense. Most plants produce toxins, in smaller or larger amount, to protect themselves against pests (War et al. 2012). Curiously, an estimated 1500 mg of natural pesticides is consumed together with food each day. This quantity is 10,000 times larger than the quantity of synthetic pesticides that is eaten in the same period. Every plant species contains its own set of toxins. During a pest attack, plants can usually increase the amount of natural pesticides they produce to a dangerous level.

When it comes to natural pest control, the first option is to ensure that natural enemies of the pest exist in the growing environment, which can greatly reduce agricultural production losses. Weeds alone are estimated to promote losses of 34% of the world agricultural production (Oerke 2006). Another possibility for natural pest control involves the use of natural organic compounds directly extracted from microorganisms or vegetables and applied to the agricultural environment. Indeed, plant extracts have been employed as a strategy to control natural pests in agriculture since the eighteenth and nineteenth centuries (Ray 1991; Ujvary 2010). One example is the *Derris elliptica* (poison vine) extract containing the active ingredient rotenone. A further example is the *Chrysanthemum cinerariaefolim* (pyrethrum) flower extract, which displays insecticidal and repellent activities due to the presence of pyrethrins.

Another option to preserve plant health is to continue with the intensive use of agrochemicals, but this use should always be supported by research into more selective pesticides and improved application techniques. Some alternative nontoxic solutions have been proposed, as well. Despite scientific advances, clearly the use of pesticides in crop preservation cannot be avoided.

Pesticides have helped to maximize food and commodity production by helping to prevent losses during plant growth, harvest, transport, storage, and distribution. Nevertheless, according to statistics, 30–35% of the world production is lost each year in spite of modern agriculture techniques. Crop losses would double if pesticides were eliminated. In fact, pesticides have contributed to protecting crops and to improving production yields (Carvalho 2006).

The CAS Registry (www.cas.org) database provided by the American Chemical Society includes more than 130 million unique organic and inorganic chemical substances. More than 4000 new substances are added to this database every day. The number of chemicals has increased exponentially over the years; the average annual growth rate has been about 15% in the last decades (Binetti et al. 2008). A small fraction of these chemicals consist of pesticides. The US Pesticide Action Network database (PAN Pesticide Database 2016) lists 6400 pesticide active ingredients and their transformation products, as well as adjuvants and solvents used in pesticide products. The EU pesticide database contains more than 1300 compounds. Not all of these compounds have been approved for use, and about 700 registered chemicals have been employed as pesticides (Eurostat 2012).

The chemical structure of pesticides is very important: their action mode depends on it. Pesticides can prevent pests from spreading if they participate in metabolic processes that block a vital process in the pest organism, destroying it. Pesticides have complex chemical structures. Several types of skeletons and numerous functional groups have been encountered. Based on the chemical structure criterion, there are three classes of pesticides: inorganic, botanical, and synthetic organic compounds (Koehler and Belmont 1998). Figure 8.5 summarizes the history of the development of these agrochemicals.

Inorganic pesticides derive from naturally occurring minerals or chemical compounds. Inorganic insecticides were used mainly at the end of the nineteenth century. Many inorganic pesticides are relatively expensive and can control insects and other pests only moderately. Typical inorganic insecticides consist of heavy metal derivatives, mostly lead, mercury, and antimony. Fluoride, arsenate, sulfide and polysulfide, and borate have been used as anions. Most of these compounds are quite stable and usually accumulate in the environment. Some act as poisons (borates and boric acid) in the insect stomach. Others are considered contact poison (silica aerogel, diatomaceous earth, etc.) and are absorbed by the insect cuticle waxy layer (Koehler and Belmont 1998).

Plants produce botanical pesticides that are isolated from roots, stem, flowers, or seeds. Plant extracts display various properties including insecticide activity and toxicity to nematodes, mites, and other agricultural pests. They also possess antifungal, antiviral, and antibacterial actions (Singh and Saratchandra 2005; Elango et al. 2010; Khater 2012). On the basis of these observations, numerous natural compounds called botanicals have been used as insecticides. They offer the advantage of not being toxic to animals or humans. They do not accumulate along time; in fact, they are biodegradable. The main limitations to using them in large scale include the high cost of the raw materials, as well as the necessary extraction/purification and formulation steps and the difficulties in optimizing the quantity and quality of the active agents (Oliveira et al. 2018). Some examples of botanical pesticides are essential oils, carvacrol, citronellal, eugenol, geraniol, limonene, linalool, menthone, pyrethrins, rotenone, veratridine, ryanodine, and thymol, among others.

History of Agrochemicals

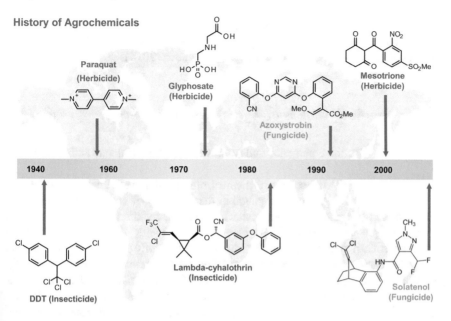

History of Agrochemicals

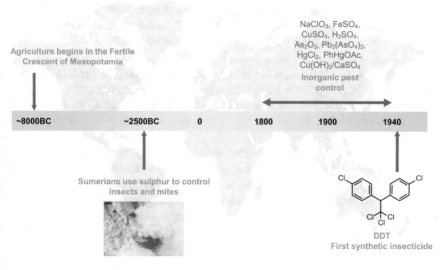

Fig. 8.5 Short history of the development of agrochemicals (from Syngenta 2014, available at http://www.oxfordsynthesiscdt.ox.ac.uk/resources/SBM-CDT-Agrochemistry.pdf)

Synthetic organic pesticides may have linear, carbocyclic, or heterocyclic structures. Chlorinated hydrocarbons, organophosphates, carbamates, and pyrethroids are the best-known synthetic organic pesticides (Stanescu 2014).

8.4 Organic Synthesis of Agrochemicals: A Brief Approach

Since 1950, with the emergence and improvement of nuclear magnetic resonance, mass spectrometry, and new chromatographic separation methods, organic synthesis has become strongly connected with medical, biological, and environmental sciences.

Whenever possible, chemists planning the synthesis of an agrochemical, drug, or any other molecule with biological activity should seek to use short and efficient routes, simple and accessible raw materials, asymmetric catalysts (to obtain an enantiomerically pure form), and environmentally friendly techniques. They should plan the reaction with a view to preventing the formation of isomers, and they should prioritize chemoselective and stereoselective processes. Reactions that promote simultaneous or sequential formation of chemical bonds in a single step are also relevant. In this context, one-pot, cascade, and multicomponent reactions are extremely important.

In a one-pot process, the raw material undergoes successive reactions in the same flask; the reagents are added in sequence. Product extraction and purification, which are often expensive and time-consuming, should only occur at the end of the synthesis. This saves time and solvent and reduces waste formation.

The fungicide fenpropimorph, which is a morpholine derivative compound used to control pests in cereal crops, has been prepared by the one-pot process shown in Fig. 8.6. Initially, compounds **1** and **2** are submitted to the Heck reaction conditions, to yield the aldehyde **3**. After total consumption of compound **1**, the morpholine derivative **4** is added to the same reaction flask, to produce fenpropimorph (Forsyth et al. 2006).

The (S)-(–)-enantiomer has greater fungicidal activity. However, in the industry, fenpropimorph is synthesized as a racemic mixture. About 40,000 tons of this fungicide are produced annually along with other similar molecules.

Cascade (or tandem or domino) reactions are processes during which several chemical bonds are sequentially produced without the need for isolating reaction intermediates, changing reaction conditions, or adding reagents. Cascade reactions have been used to synthesize the tetrahydroquinoline carbon skeleton of alkaloids iso-

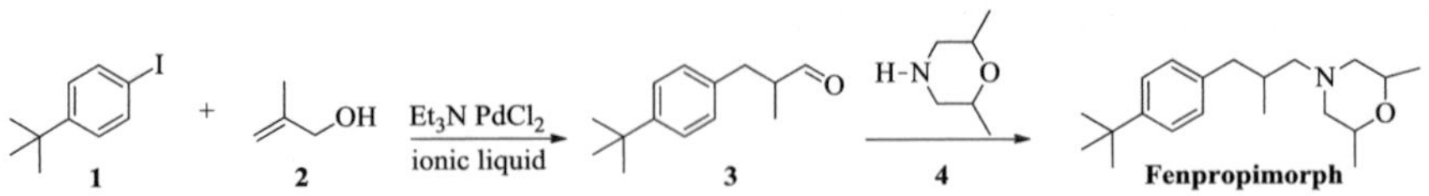

Fig. 8.6 Reaction scheme for one-pot production of the fungicide fenpropimorph

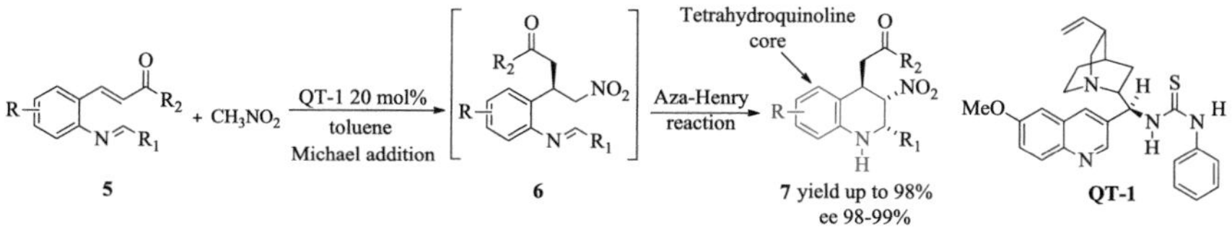

Fig. 8.7 Cascade reactions to prepare a tetrahydroquinoline nucleus

lated from the fungus *Penicillium simplicissimum* (Kusano et al. 2000), which have nematicidal activity in addition to other biological activities. The reaction shown in Fig. 8.7 is an organocatalyzed reaction that starts with Michael's addition of nitromethane to compound **5**, to give intermediate **6**. The latter intermediate is not isolated, but it reacts rapidly in a reaction of the aza-Henry type, to form the tetrahydroquinoline nucleus of compound **7** (Jia et al. 2011). In this case, the cascade reaction methodology is characterized by non-isolation of intermediate **6** and sequential formation of several C–C bonds. The bifunctional catalyst (Brφnsted acid/base) QT-1, whose structure is shown in Fig. 8.7, induces the final product stereochemistry.

Multicomponent reactions combine at least three reactants in the same pot to generate a product containing most (preferably all) atoms of the starting materials. Their atom economy, efficiency, mild conditions, high convergence, and concomitant step economy, in combination with their general compatibility with green solvents would justify a central place in the toolbox of sustainable synthetic methodologies (Cioc et al. 2014). Like the one-pot and cascade reactions, this process requires a smaller number of steps, thereby dismissing the need to purify and to isolate synthetic intermediates. This drastically reduces waste generation and allows for an ecologically and economically favorable process.

Compounds containing the bispyrazole-5-ol structure (Fig. 8.8) have been widely employed as fungicides (Singh and Singh 1991), insecticides, and pesticides (Londershausen 1996). The carbon skeleton of these compounds can be prepared by means of the multicomponent reaction protocol, by combining phenylhydrazine (**7**), ethyl acetoacetate (**8**), a substituted aromatic aldehyde (**9**), and cerium(IV) sulfate (as catalyst) in the absence of solvent (Hassankhani 2015). The phenyl hydrazone **10** and the arylidene pyrazolone **11** shown in Fig. 8.8 are not isolated because they are consumed in the subsequent steps of the process, thereby characterizing a multicomponent reaction. Fragments of all the starting materials used in the process have been observed in the structure of bispyrazole-5-ol.

The three models of organic synthesis presented above and their related benefits, such as reduced number of steps, lower waste generation, atom economy, solvent-free reactions or use of water as solvent, use of biocatalysts and/or non-metallic catalysts, and higher selectivity and efficiency, constitute the "ideal organic synthesis," or clean chemistry, and represent the current and future trends of this science that has been increasingly facing up to the serious environmental issues of today's world.

Nowadays, synthetic organic chemistry is used in practically all knowledge areas, including the production of polymers, dyes, food flavoring, detergents, and perfumes.

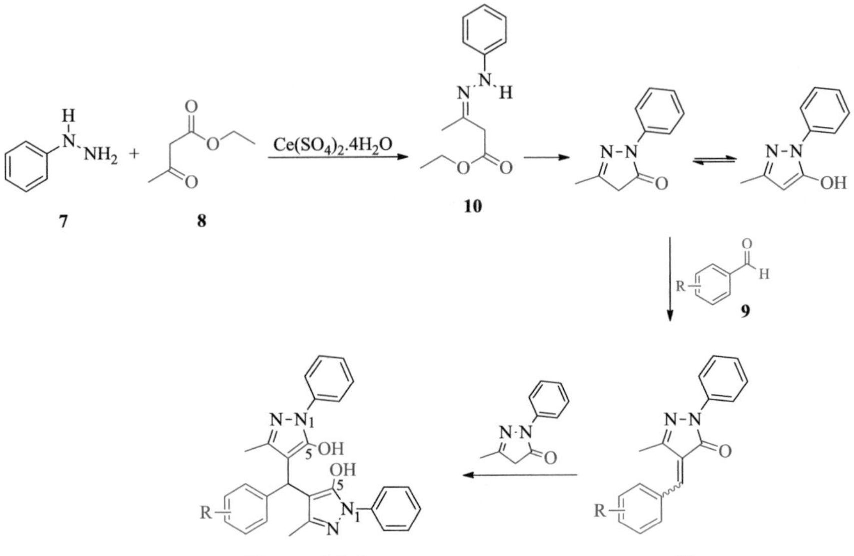

Fig. 8.8 Scheme used to produce bispyrazol-5-ol by multicomponent reaction

When it comes to producing pharmaceuticals and agrochemicals, synthetic organic chemistry is fundamental.

When a new molecule with potential agrochemical action is unveiled, no matter whether it was discovered by plant extraction, produced by microorganisms, or derived from some other natural product, a synthetic plan is usually devised to prepare this new active ingredient. Whenever possible, processes related to green chemistry should be adopted (Anastas and Warner 1998). To this end, the target molecule must be isolated and have its structure elucidated with respect to its spatial conformation and the absolute configuration of all its stereogenic centers, so that the synthesis process can accurately reproduce the structure of the biologically active compound.

The synthesis and use of new agrochemicals are always desirable because the constant application of the same molecules to control a certain pest leads the target organism to become resistant to the pesticide and may produce unwanted effects in non-target organisms (Oh et al. 2012). One way to overcome this problem is to modify the structure of the generally employed agrochemical in order to obtain a new molecule that is highly efficient toward the target species, has single action mode and low toxicity, and is easy to decompose.

The literature has described several syntheses of natural products that are used as agrochemicals. There are also descriptions of synthetic analogs that present similar (or even superior) biological activity to the activity of the natural product. Literature works have revealed the logic and creativity behind the synthetic route that is selected to prepare the carbon skeleton, to form heterocyclic systems, and to induce the asymmetry that will provide pure enantiomers.

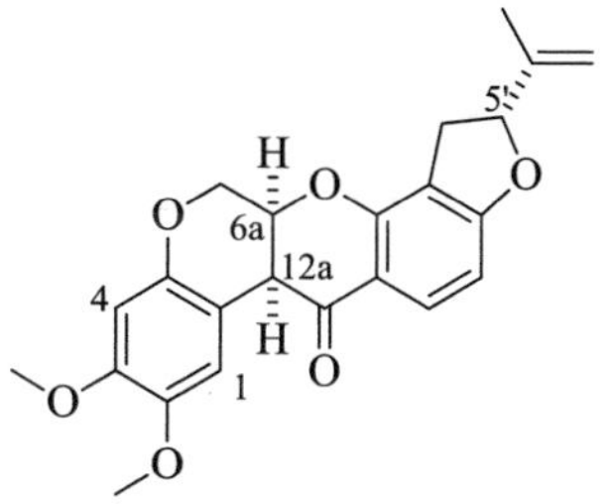

Fig. 8.9 Structure of 6aS,12aS,5'R-(−)-rotenone

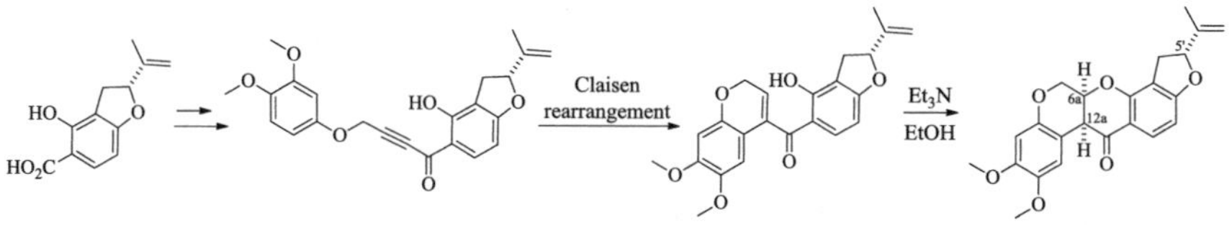

Fig. 8.10 Scheme used to synthesize 6aS,12aS,5'R-(−)-rotenone showing a stereospecific cyclization as the key step

The classic synthesis of (−)-rotenone is an example of stereochemical control during the preparation of a biologically active natural product (Sasaki and Yamashita 1979). (−)-Rotenone (Fig. 8.9) is a natural agrochemical extracted from the roots of *D. elliptica* (poison vine) or *Lonchocarpus utilis* (common lancepod), which grow in South America, Asia, and Australia. Rotenone has been used as an insecticide to fight against parasites in urban areas and in organic agriculture. Remarkably, rotenone can control undesirable fish in certain aquatic environments, so it is classified as piscicide (Solomon et al. 2010).

The (−)-rotenone structure has absolute configuration 6aS,12aS,5'R in the stereogenic centers, in a carbon skeleton with *cis* junction of the tetrahydrochromeno rings [3,4-*b*].

(−)-Rotenone has been synthesized from enantiomerically pure (R)-(−)-tubaic acid, to give the acetylenic intermediate **12** after four steps (Fig. 8.10). When submitted to Claisen rearrangement conditions, compound **12** gives compound **13**, which, after treatment with triethylamine in ethanol, undergoes a stereospecific cyclization controlled by the presence of the stereogenic center defined at C-5'.

This example of 6aS,12aS,5'R-(−)-rotenone synthesis is only one of the several synthesis (total or partial) procedures described for this molecule (Miyano et al. 1960; Miyano 1965; Fukui et al. 1967; Begley et al. 1989; Crombie et al. 1973). Each of the described methods aims to obtain the substance with the highest overall yield and the fewest synthetic steps possible. The need to reduce the number of steps of this synthesis and many other syntheses remains one of the main objectives of synthetic organic chemistry, which is still mainly concerned with the search for the best synthetic route to prepare a specific target molecule.

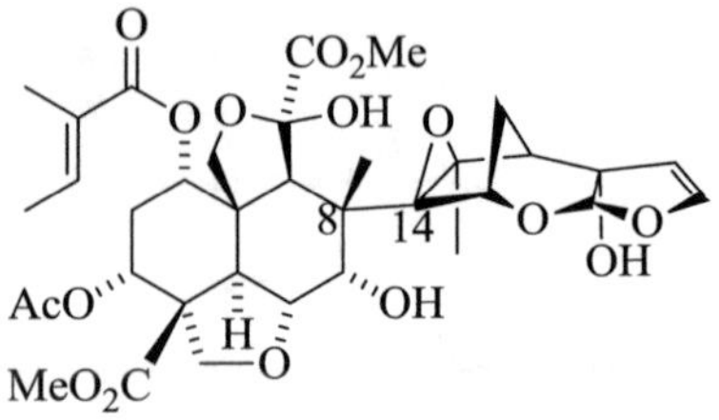

Fig. 8.11 Structure of azadirachtin

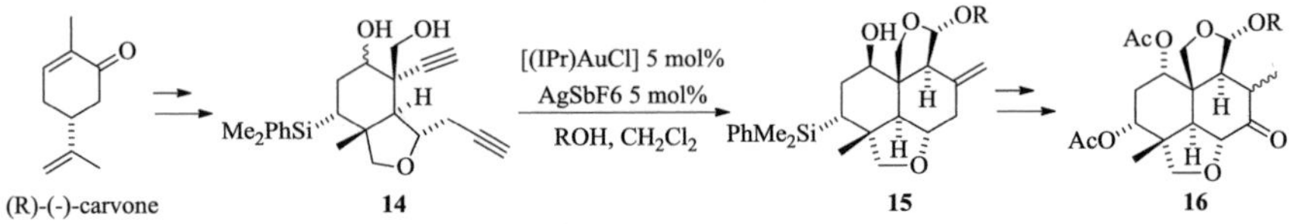

Fig. 8.12 Scheme used to obtain an azadirachtin fragment

Azadirachtin is another historically important and extremely structurally complex molecule (Fig. 8.11). It occurs in the extract of the neem tree *Azadiracta indica* (A. Juss). For centuries and especially in India, this tree has been used because of its medicinal properties. The extract of this plant displays acaricidal, fungicidal, repellent, nematicidal, and growth regulating actions. The azadirachtin molecule was isolated in 1968 (Butterworth and Morgan 1968) and had its structure elucidated in 1987 (Turner et al. 1987; Bilton et al. 1987; Kraus et al. 1987). The structure of azadirachtin has 16 stereogenic centers, seven of which are tetrasubstituted; there are also a number of oxygenated organic functions. The total synthesis of this molecule took 22 years and counted on the efforts of more than 40 chemical researchers who, in 2007, obtained azadirachtin after 64 synthetic steps (Veitch et al. 2007).

The azadirachtin molecule is sensitive to acid and alkaline conditions and is also photosensitive, which makes its structure prone to rearrangement reactions. These characteristics have probably frustrated several synthesis plans.

The structural complexity of azadirachtin has led researchers to propose numerous synthetic methodologies to construct its various fragments, which may be later used in its total synthesis. One of the synthetic proposals involves highly enantioselective and diastereoselective construction of the four-ring system through cascade reactions of diines catalyzed by palladium and gold (Shi et al. 2016). In this case, the monoterpene (*R*)-(–)-carvone is used as the starting material to prepare the diine **14**, which is achieved after 13 synthetic steps (Fig. 8.12). Treatment of intermediate **14** with gold salt gives the tetracyclic compound **15**, which is converted to the azadirachtin fragment **16** after further eight synthetic steps.

The literature contains further examples of schemes used to prepare azadirachtin fragments (Nishikimi et al. 1989; Henry and Fraser-Reid 1994; Ishihara et al. 1999a, b, 2003; Fukuzaki et al. 2002; Durand-Reville et al. 2002; Sakurai and Tanino 2015; Tan et al. 2015, and in the references cited in these articles).

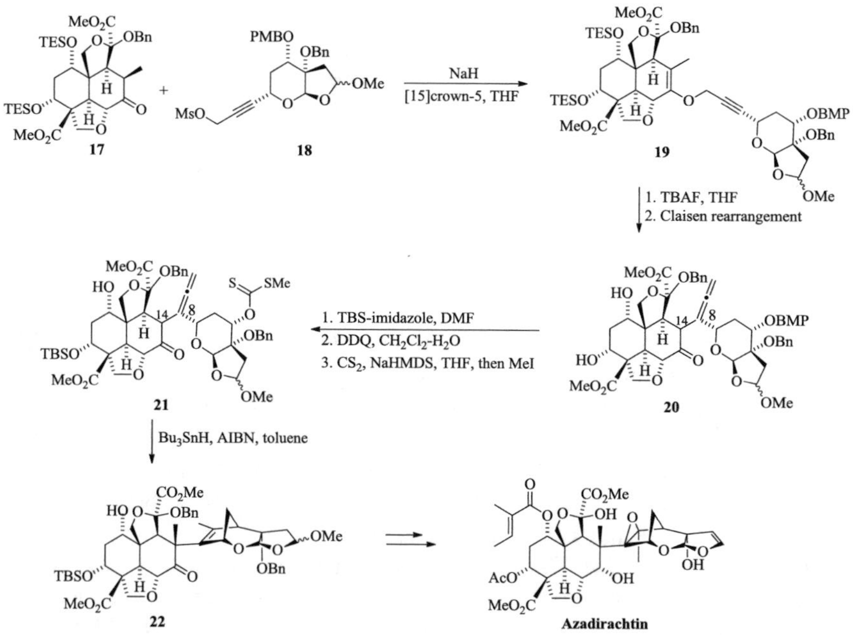

Fig. 8.13 Part of the strategy used to synthesize azadirachtin where C14–C8 bond formation occurs through a Claisen rearrangement

The total synthesis of azadirachtin performed by Ley et al. (Fig. 8.13) entails an O-alkylation reaction between compounds **17** and **18** (via keto-enolic equilibrium), to produce compound **19**. After deprotection of the two silicon ethers, a key step takes place: A Claisen rearrangement generates the sterically hindered bond between C14 and C8. Note that previous works had not reported good results (Anderson and Ley 1990; Ley 2005). The alternative synthetic route not only minimizes the steric hindrance, to afford compound **20**, but also gives the allene that is necessary for the subsequent radical cyclization reaction and the consequent production of intermediate **22** from **21** (Fig. 8.13). The natural product azadirachtin is obtained from compound **22** after five further synthetic steps (Veitch et al. 2007).

This brief description of the total synthesis of a greatly complex natural product shows that challenges encountered over the years have produced a wealth of information on the chemical and biological properties of azadirachtin. This total synthesis serves as a reference for researchers to develop new useful compounds for pest control.

The agricultural environment is actually a dynamic system: New insect species and new increasingly agrochemical-resistant types of weeds and plant pathogens are constantly arising. This situation requires constant changes in pest control methods, including the preparation and the use of new agrochemicals.

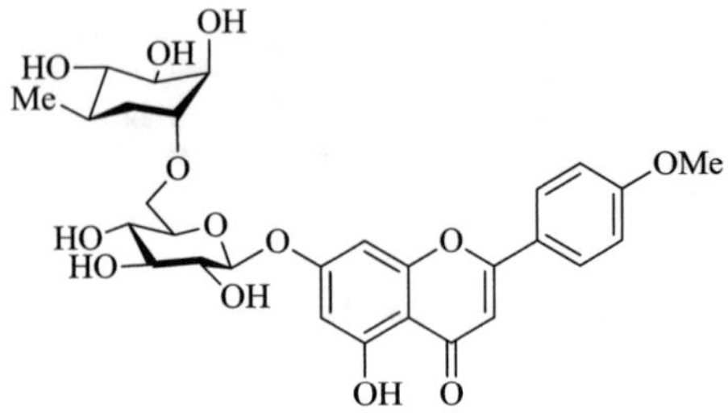

Fig. 8.14 Structure of the flavonoid linarin

To elaborate and to commercialize a new compound that can be potentially applied in the agricultural field, regulatory factors are extremely relevant because they establish restrictions for the registration of new molecules (FAO 2006; Sparks and Lorsbach 2017). These restrictions are related to environmental issues and demand that new molecules offer reduced risk of generating persistent residues. The way the new molecules affect non-target organisms and the toxicological profile of these new molecules has to be studied. This set of constraints makes the development of a new agrochemical that is more efficient, selective, and environmentally friendly a challenging mission. The time elapsed between the preparation of a new active compound and its actual use in agriculture is approximately 11 years (McDougall 2016).

Regulatory barriers do exist. Nonetheless, the development of new agrochemicals is increasingly urgent. When a new biologically active natural product is launched into the market, it is generally used as an isolated crude extract. If the product is economically viable, it can be synthesized on an industrial scale.

The flavonoid linarin (Fig. 8.14) is a natural product that can regulate plant growth. This compound is the main constituent of the organic extract of *Zanthoxylum affine*, a shrub that grows in tropical regions and semidesert zones (Rios et al. 2018).

Literature studies (Rios et al. 2018) have shown that linarin presents pre- and post-emergent actions on the dicotyledonous plant *Lactuca sativa* (lettuce). In the case of pre-emergent phytotoxic agents, the crude organic extracts can inhibit germination and residual growth (root and stem elongation). As for post-emergent agents, they can inhibit dry biomass. Linarin isolated from the crude extract acts as a pre-emergent herbicide that inhibits germination, seed respiration, residual bush growth, and, notably, root hair development. All these effects probably stem from changes in transcription factors and plant hormones. Furthermore, linarin inhibits ATP synthesis and the electron transport chain in isolated spinach chloroplasts; in other words, it behaves as a Hill reaction inhibitor. Therefore, linarin may be used as a lead for a potential green herbicide with different targets (Rios et al. 2018).

Like linarin, other natural products extracted from plants possess biological activities that make them useful in agriculture; e.g., essential oils extracted from aerial parts of plants have extensive use as antiseptic, antimicrobials, and insect repellents and can be applied in the medicinal and perfumery areas. Essential oils generally consist of a complex mixture of mono- and sesquiterpenes and of other oxygenated organic compounds produced as secondary plant metabolites. These oils are readily extracted, biodegradable, and non-persistent in soil and water; moreover, they exert little or no toxicity against vertebrates (Batish et al. 2008; Nerio et al. 2010). The barriers posed by regulatory factors make essential oils potential candidates for further studies designed to promote their wide use, mainly in weed control (Singh et al. 2003; Batish et al. 2004, 2007) and pest management (Isman 2000; Pawar and Thaker 2006; Abad et al. 2007).

Essential oils are an excellent example of natural products that can be directly used in the form of a mixture and act against various types of mites or function as insect repellent (Gillij et al. 2008). Although the biological activities of essential oils are attributed to some particular organic compounds, there is synergism among all the mixture constituents, which results in higher biological effect as compared to the action of an isolated component (Hummelbrunner and Isman 2001; Gillij et al. 2008; Gnankiné and Bassolé 2017). The biological activities of essential oils extracted directly from plants have proven much more effective as compared to the biological activities of synthetic combinations containing the main essential oil components only. These results have indicated that the presence of organic compounds at lower ratios in the mixture is important to confer bioactivity to the natural essential oils (Omolo et al. 2004). In addition, essential oil delivery nanosystems can be used against insects and as broad-spectrum bactericide/fungicide, which are very attractive alternatives to current strategies for crop protection (González et al. 2017; Oliveira et al. 2018).

A new agrochemical compound can also be prepared by modifying the structure of a natural active compound. This modification produces a semi-synthetic material. The synthesis of a natural product derivative generally involves the preparation of various compounds that are analogous to the starting material for further evaluation of the new biological activity. The results obtained with the analogous compounds allow the structure–activity relationship (SAR) to be assessed and a new series of agrochemicals to be designed.

The alkaloid sanguinarine (Fig. 8.15), which bears a benzo[c]phenanthridine structure, is an example of natural product that can undergo structural modification. This substance has many biological activities, including insecticide and acaricide actions. It is also an efficient termiticide against *Tetranychus urticae* (red and two-spotted spider mite) in apples (Lv et al. 2018). Structural modification of this molecule on the carbon adjacent to the quaternary nitrogen atom produces a neutral pseudo base (Dostál et al. 1996).

Alcohols and amines with different substituent groups, besides carbanions generated in beta-dicarbonyl compounds and malononitrile, are the nucleophilic species that have been used in these reactions. In this way, 32 sanguinarine analogs have been synthesized and had their biological activity evaluated against the fungi *Alternaria solani, Gibberella zeae, Rhizoctonia solani, Fusarium oxysporum,* and *Cercospora arachidicola* as compared to the fungicide tebuconazole, used as positive control. These sanguinarine synthetic analogs display moderate activities against all the fungi. Thirteen derivatives possess good effective concentration (EC_{50}), between 1.0 and 4.4 μg mL^{-1}, against *R. solani*, whereas sanguinarine has EC_{50} of 11.6 μg mL^{-1}. Structural modification of sanguinarine improves its potency by more than 11-fold

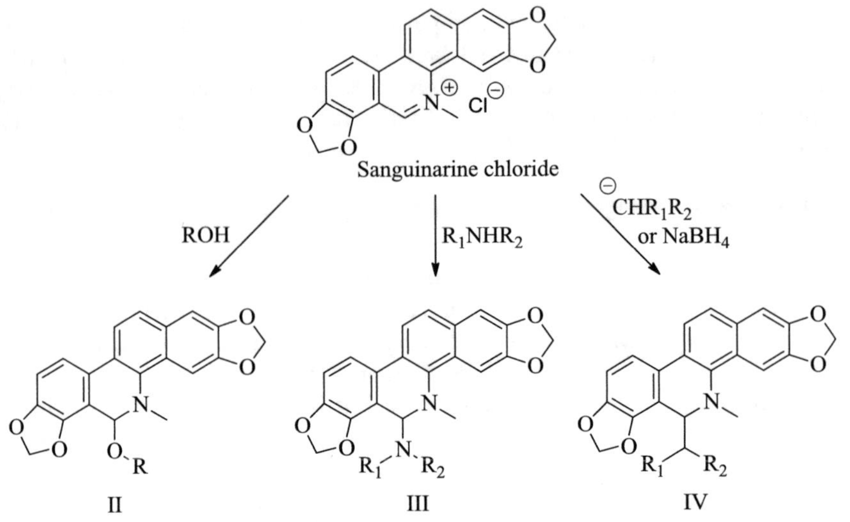

II-a - II-l

a R = CH_3
b R = CH_2CH_3
c R = n-Pr
d R = i-Pr
e R = n-Bu
f R = t-Bu
g R = i-Bu
h R = s-Bu
i R = n-Pent
j R = cyclopentyl
k R = n-hexyl
l R = cyclohexyl

III-a - III-p

a $R_1 = R_2 = CH_3$
b R_1 = H, $R_2 = CH_3$
c R_1 = H, R_2 = n-C_3H_7
d R_1 = H, R_2 = i-C_3H_7
e R_1 = H, R_2 = n-C_4H_9
f R_1 = H, R_2 = i-C_4H_9
g R_1+R_2 = -$(CH_2)_5$-
h R_1 = H, R_2 = cyclohexyl
i R_1 = H, R_2 = phenyl
j R_1 = H, R_2 = benzyl
k R_1 = H, R_2 = 4-Methylphenyl
l R_1 = H, R_2 = 3-Methylphenyl
m R_1 = H, R_2 = 2-Methylphenyl
n R_1 = H, R_2 = 4-Methoxybenzyl
o R_1 = H, R_2 = 2-Methoxybenzyl
p R_1 = H, R_2 = 2-Chlorobenzyl

IV-a - IV-c

a $R_1 = R_2$ = CN
b R_1 = H, $R_2 = COCH_3$
c R_1 = H, $R_2 = CO_2C_2H_5$
d dihydro sanguinarine

Fig. 8.15 Synthetic routes to obtain sanguinarine analogs (from Lv et al. 2018)

against *R. solani*. Two compounds obtained by reaction of sanguinarine with isopropanol and isopropylamine are highly toxic to *R. solani* and promote 100% inhibition when they are used at 50 μg mL^{-1}. These two compounds have approximately one-third of the potency of the commercial fungicide tebuconazole against *R. solani* in vitro. Therefore, they can be considered a lead structure for further design of agricultural fungicides (Lv et al. 2018).

The aforementioned example shows how organic synthesis is strongly associated with biology and environmental sciences. Examples of the structural modification

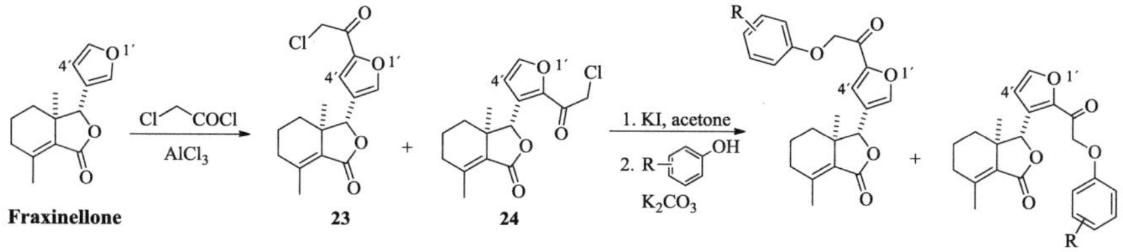

Fig. 8.16 Synthetic strategy to modify the fraxinellone structure

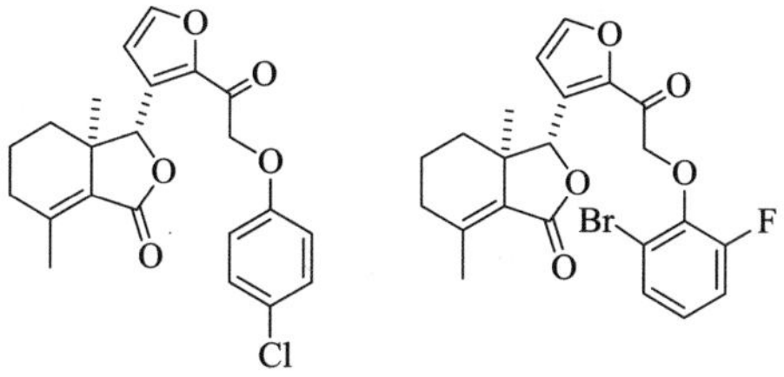

Fig. 8.17 Structure of fraxinellone phenolic derivatives with potential insecticidal activity

of bioactive molecules that will be presented further in this section reinforce this association and point out that organic synthesis is a strong, versatile, and essential tool to obtain new agrochemicals derived from natural products or not.

The natural product fraxinellone (Yang et al. 2018) has been structurally modified to combat the insect *Mythimna separata* (Walker) (oriental armyworm), a worldwide agricultural pest that can destroy crop fields. This insect has acquired resistance to the current insecticides because it has already been subjected to a large number of chemical compounds. Therefore, the development of new agrochemicals that can effectively combat this insect is an increasingly urgent and important action.

Fraxinellone (Fig. 8.16) is a degradation product of limonoids isolated from plants of the families *Meliaceae* and *Rutaceae*, such as *Dictamnus albus* (burning bush), *Dictamnus dasycarpus* (densefruit pittany), and *Melia azedarach* (chinaberry tree) (Zhang and Xu 2017). Fraxinellone has been structurally modified by electrophilic substitution in the furan ring, which produces two new chloroacetyl compounds **23** and **24**. After being submitted to the classic Finkelstein reaction (KI + acetone), both compounds **23** and **24** produce the corresponding phenolic ethers. In the latter reaction, the phenols are substituted with alkyl groups, halogen, and other oxygenated functions in the aromatic ring, to give an array of fraxinellone-derived compounds.

The insecticidal activity of the obtained compounds has been evaluated against the *M. separata* (Walker) pre-third-instar larvae in vivo as compared to the insecticide toosendanin, used as positive control. Compounds **25** and **26** (Fig. 8.17), which contain halogens in the aromatic ring, exhibit more potent insecticidal activity than toosendanin, and the final mortality rates are higher than 60% (Yang et al. 2018).

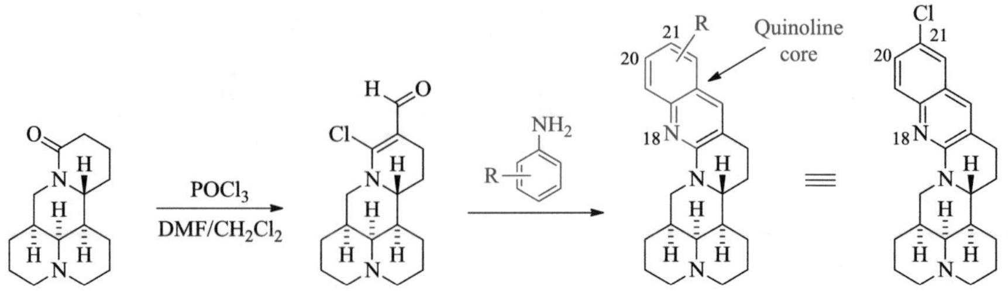

Fig. 8.18 Scheme used to prepare matrine derivatives by Vilsmeier–Haack reaction

A new agrochemical has been developed from the alkaloid matrine by using a procedure analogous to the one described above for fraxinellone (Huang et al. 2017, 2018). Matrine (Fig. 8.18) is a biorenewable quinolizidine alkaloid isolated from the roots of *Sophora flavescens* (Kushen) (bitter shen), which is widely distributed in Asia, Oceania, and the Pacific Islands. This natural product and its derivatives display a broad scope of biological properties such as anticancer, anti-inflammatory, and antiviral actions. Additionally, they are very powerful pesticides in the agricultural field.

As shown in Fig. 8.18, the key intermediate **27** (14-formyl-15-chloromatrine) is synthesized by the Vilsmeier-Haack reaction between matrine and DMF in the presence of POCl$_3$. Various derivatives like 14-formyl-15-aryloxymatrine have been easily prepared by reaction of intermediate **27** with the corresponding phenols in the presence of KOH (Huang et al. 2017). A series of quinolinomatrine derivatives have also been obtained by reacting compound **27** with different anilines (Huang et al. 2018).

The pesticide activity of the synthesized compounds has been assayed against three typically crop-threatening agricultural insect pests, *M. separata* (Walker) (oriental armyworm), *Plutella xylostella* (Linnaeus) (diamondback moth), and *Tetranychus cinnabarinus* (Boisduval) (spider mite). Infestations by these pests are very inconvenient and hard to control. Some interesting structure–activity relationship results have been obtained for the tested compounds. Compound **27** exhibits low insecticidal activity as compared to toosendanin, used as positive control. However, when the chlorine atom of compound **27** is replaced with aryloxy groups, the corresponding products have stronger insecticidal activity than matrine. Compared to matrine, all 14-formyl-15-aryloxymatrine derivatives inhibit the growth of early *M. separata* third-instar larvae more potently, and some compounds have more promising insecticidal activity than toosendanin (Huang et al. 2017).

Compared to spirodiclofen, as positive control, compound **28a** and other analogs have powerful acaricidal activity against *T. cinnabarinus* female adults. In addition, compared to toosendanin, as positive control, compound **28a** and its analogs exhibit potent insecticidal activity against *M. separata* pre-third-instar larvae, being compound **28a** (21-chloroquinolinomatrine) the most potent insecticidal and acaricidal

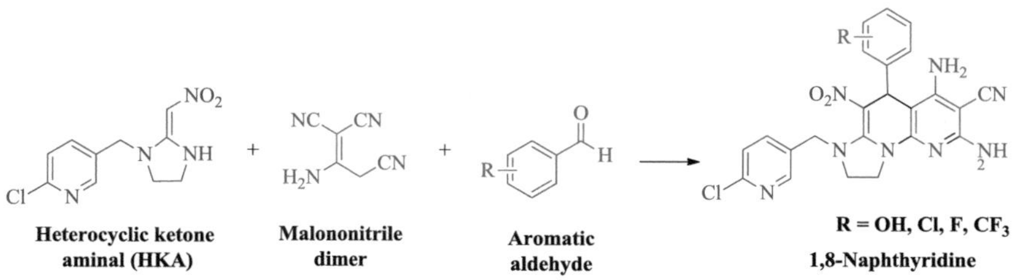

Fig. 8.19 Multicomponent reaction that forms the 1,8-naphthyridine skeleton

agent. Introduction of a chlorine atom at the C-21 position of quinolinomatrine is important for the insecticidal and acaricidal activities. These results provide the basis for further structural modifications in matrine to make it a more powerful acaricidal and insecticidal agent (Huang et al. 2018).

The preparation of neonicotinoids with a 1,8-naphthyridine nucleus is an excellent example of the application of multicomponent reactions to synthesize new compounds with insecticidal activity (Hou et al. 2017). The multicomponent reaction between a heterocyclic ketone aminal (HKA), the malononitrile dimer, and a suitably substituted aromatic aldehyde produces the 1,8-naphthyridine skeleton in a single step (Fig. 8.19).

Initially, the malononitrile dimer and the aromatic aldehyde undergo a Knovenagel reaction, to yield compound **29**, which in turn reacts with HKA in an aza-ene type of reaction, to afford compound **30** (Fig. 8.20). Compound **30** is an imine-enamine-like equilibrium that enables a complex double intramolecular cyclization sequence to occur until the 1,8-naphthyridine skeleton arises (Sun et al. 2015).

The assorted compounds obtained through this multicomponent reaction vary in terms of the substituent (R) of the aromatic aldehyde. The use of 4-fluor and 4-trifluoromethylbenzaldehyde produces 1,8-naphthyridine derivatives with excellent insecticidal activity against bean crop pests.

The multicomponent reaction process is extremely satisfactory during the search for a compound with biological activity against pests in agricultural crops: This process allows the environmental requirements of a potential large-scale production to be attained. In addition, it fulfills the strict regulatory norms that are necessary to register a new agrochemical.

Therefore, synthetic strategies that promote simultaneous formation of diversified carbon–carbon or carbon–heteroatom bonds during the synthesis of new agrochemical compounds are always beneficial. An example is the preparation of 1,3-oxazoline derivatives, which uses a cascade reaction process during which the heterocycle and the thioether function arise in only one synthetic step (Yu et al. 2015).

Etoxazole (Fig. 8.21) is the only commercially available acaricidal and insecticidal compound belonging to the chemical class of 2,4-diphenyl-1,3-oxazoline. Its biological activity inhibits the biosynthesis of chitin, an essential component of the insect cuticle and the fungal cell wall. Therefore, etoxazole presents specific action against target organisms and is practically harmless against non-target organisms,

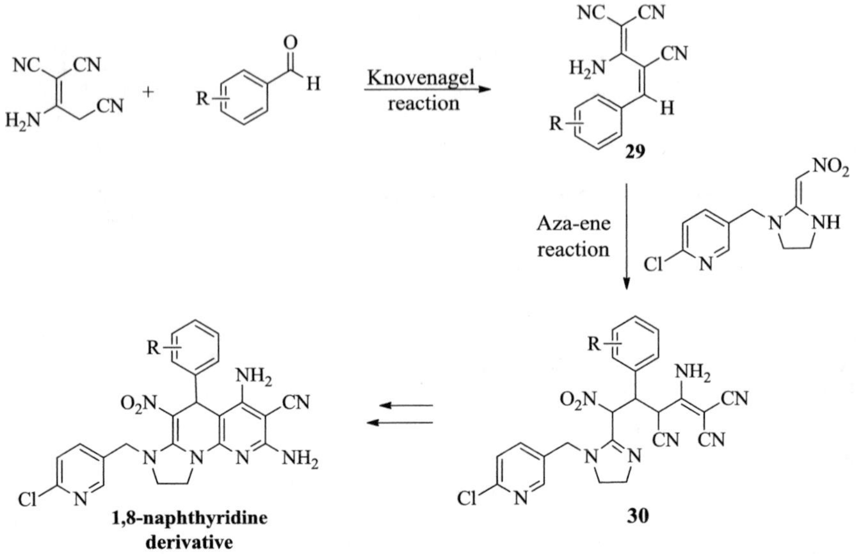

Fig. 8.20 Sequence of reactions that produce 1,8-naphthyridine derivatives (insecticidal activity was obtained when R = F and CF$_3$)

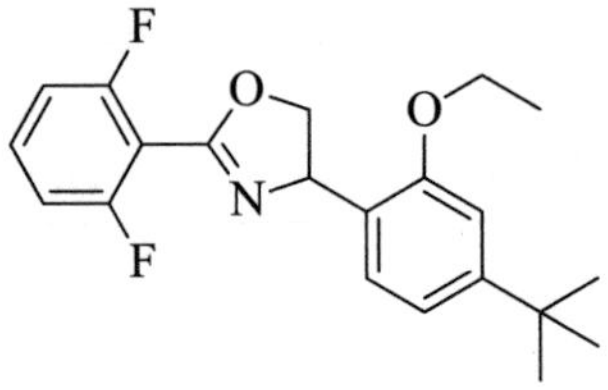

Fig. 8.21 Structure of etoxazole

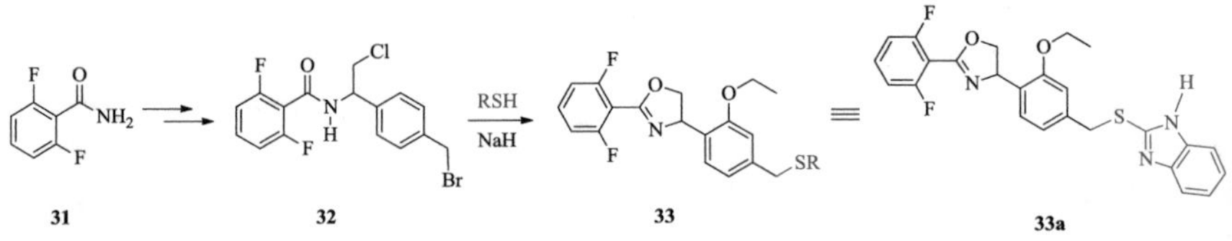

Fig. 8.22 Strategy for the synthesis of oxazole derivatives

such as mammals and higher plants (Oberlander and Silhacek 1998a, b; Merzendorfer 2013).

A large number of etoxazole derivatives have been prepared from 2,6-difluorobenzamide (**31**), which had been conveniently functionalized to give the key intermediate **32** (Fig. 8.22). The key step of this synthesis is to prepare the oxazoline ring and the thioether function in a single step by treatment of compound **32** with sodium hydride and a mercaptan derivative.

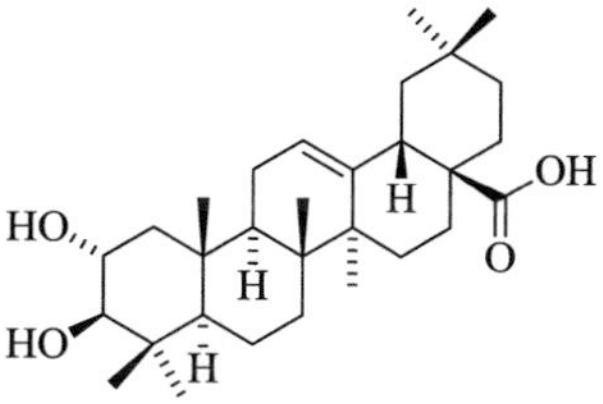

Fig. 8.23 Structure of maslinic acid

Compound **33a** exhibits excellent acaricidal activity against both *T. cinnabarinus* (carmine spider mite) eggs and larvae and against the mites *T. cinnabarinus* and *Polyphagotarsonemus latus* in the case of eggplant crops. The acaricidal activity of compound **33a** is superior to the activity of commercial etoxazole, used as positive control. No adverse effect on eggplant growth has been observed under any of the studied concentrations, indicating that this compound can be used as acaricidal (Yu et al. 2015).

In organic synthesis, cycloaddition reactions are the most often used to prepare carbonic and heterocyclic compounds (Carruthers 1990). These reactions generate structurally complex adducts, occur with total atom economy, and simultaneously produce two new sigma bonds (Brocksom et al. 2010). Such characteristics make cycloaddition reactions an excellent tool to synthesize new agrochemicals if we consider that they meet all the regulatory norms and the green chemistry requirements.

To obtain a new compound with potential agrochemical use, a research work has employed a [3+2] cycloaddition reaction (Huisgen 1961) as the key step to construct the 1,2,3-triazole heterocycle (Nejma et al. 2018) present in the structure of a compound derived from maslinic acid (Fig. 8.23).

Maslinic acid is a pentacyclic triterpene that is isolated mainly from the plant *Olea europaea* L. (olive). This acid has several activities related to the medicinal area (Montilla et al. 2003; Huang et al. 2011; Reyes-Zurita et al. 2011). To synthesize and to evaluate a potential herbicide derived from maslinic acid, the strategy shown in Fig. 8.24 has been used. In this reaction, the bond between maslinic acid and the phthalimide is established through a spacer of the 1,2,3-triazole type. To prepare the heterocycle, the [3+2] cycloaddition reaction catalyzed by a copper salt is performed between the terminal alkyne **34** and the azide **35**, to afford compound **36**.

The formation of 1,2,3-triazole by this procedure fits the "click chemistry," concept developed by Sharpless and collaborators to describe a thermodynamically favorable reaction that can connect two molecules in a simple and rapid way, with very high yields (Sharpless et al. 2001). In addition, "click chemistry" should be stereospecific and highly selective for a single product and should be carried out in the absence of solvent or in the presence of a green solvent (such as water) or an easily removable solvent. Finally, the product should not be difficult to isolate. By means of this methodology, the terminal alkyne **34** has been prepared directly from maslinic acid by reaction with propargylic bromide and sodium hydride, while the azide **35** has been synthesized from *N*-chloromethylphthalimide (Fig. 8.24) in three steps.

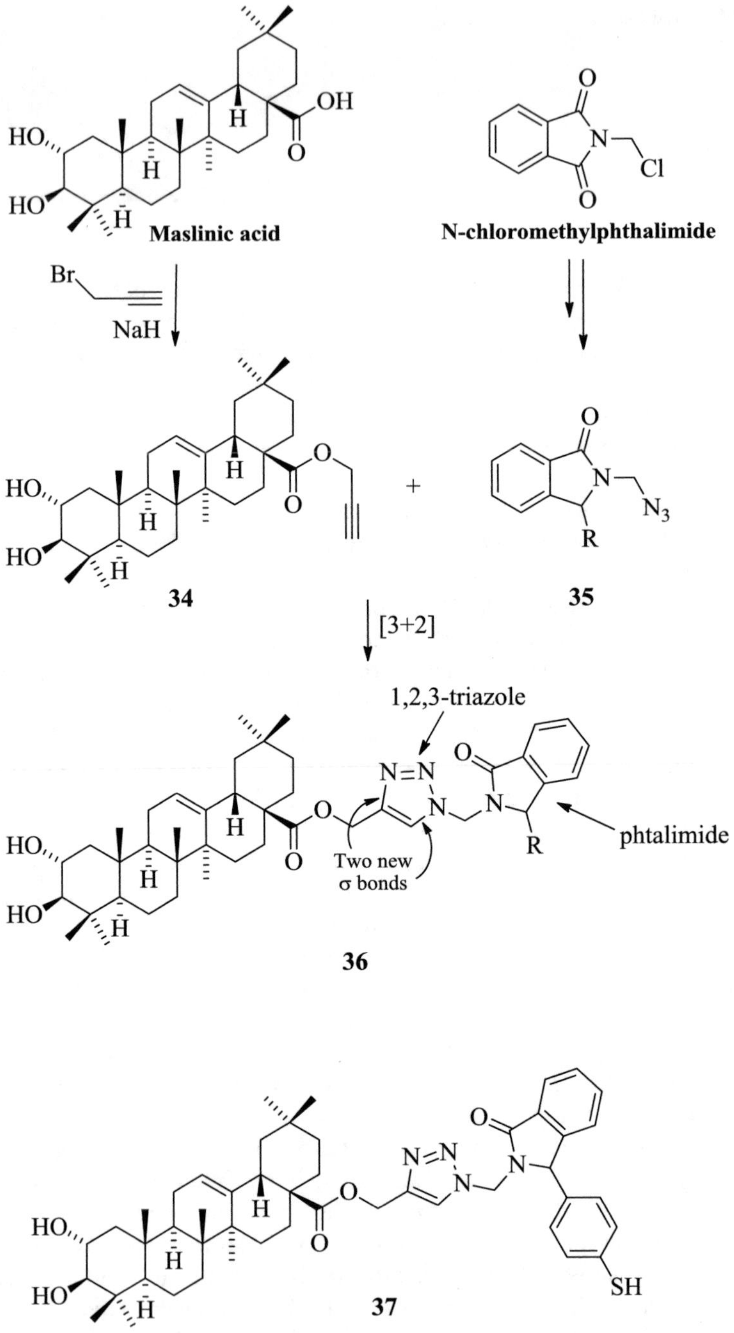

Fig. 8.24 Reaction scheme to obtain maslinic acid triazole derivatives

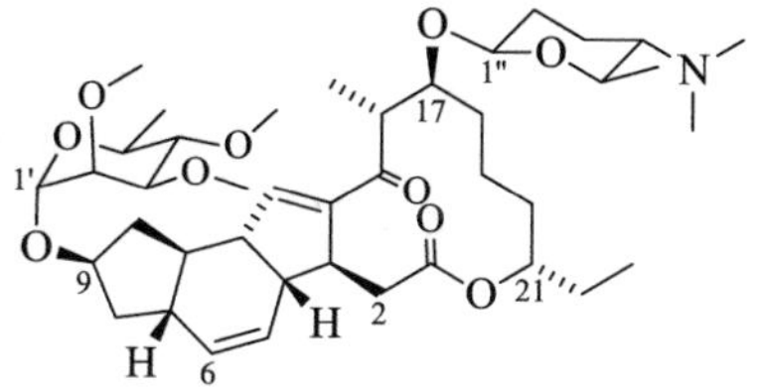

Fig. 8.25 Structure of (–)-spinosyn-A

The herbicidal activity of these compounds has been studied by observing seed germination and early growth stage inhibition in the plant *L. sativa* (lettuce). Free maslinic acid has very low inhibition percentage (5.5%). However, the '*R*' group present in compound **36** provides the new synthesized molecules with high herbicidal activity (from 91 to 100%). Compound **37**, which contains a thiol group at position 4 of the aromatic ring, affords the best result. Thus, structural modification involving 1,2,3-triazole as a spacer between maslinic acid and the phthalimide underlies the excellent herbicidal activity of the compounds. These structurally modified compounds could be used as a model to develop new and increasingly more active molecules.

Diels–Alder [4+2] is the best-known cycloaddition reaction used in organic synthesis. This reaction can be applied in the most varied situations, including enantioselective catalysis (Corey 2002), biosynthesis of natural products (Stocking and Williams 2003), and multicomponent reactions (Brocksom et al. 2007), apart from numerous other applications.

An interesting application of the Diels–Alder reaction is the synthesis of (–)-spinosyn-A (Fig. 8.25), a natural molecule that has two sugars attached to a tetracyclic nucleus containing a macrocyclic lactone. (–)-Spinosyn-A is produced by the bacterium *Saccharopolyspora spinos* and can effectively control insects, such as flies, caterpillars, and beetles that attack cotton crops, fruits, vegetables, lawns, and ornamental plants. This natural product acts fast and selectively against target pests without harming beneficial insects or predatory flies (Dagani 1999).

The Diels–Alder reaction is one of the steps during the biosynthesis of (–)-spinosyn-A and other natural products like lovastatin, macrophomate, solanapyrone-A, chaetoglobosin-A, and chlorothricin (Kelly 2008). Therefore, the synthetic strategy to prepare (–)-spinosyn-A is based on its biosynthesis (biomimetic synthesis). As shown in Fig. 8.26, a Horner–Wadsworth–Emmons reaction in compound **38** promotes macrocyclization and consequently forms intermediate **39**, which has not been isolated. Subsequently, a transannular Diels–Alder reaction occurs, which directly produces compound **40**. In this case, simultaneous formation of sigma bonds between C7–C11 and C4–C12 followed by two sequential reactions that form the macrocycle and the six-membered ring is an excellent example of cascade reactions. The subsequent synthetic step is a Morita–Baylis–Hillman reaction that provides compound **41**, which is conveniently

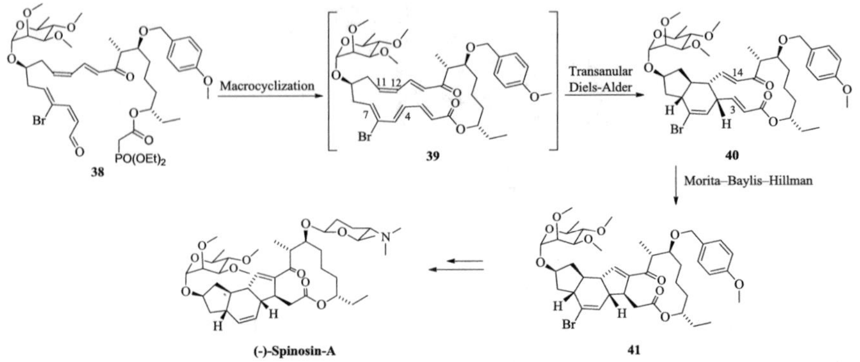

Fig. 8.26 Synthetic route to prepare (–)-spinosyn-A by biomimetic synthesis

converted into (–)-spinosyn-A. This synthesis of (–)-spinosyn-A has been performed in 31 steps with total yield of only 3% (Mergott et al. 2004).

As in the preceding syntheses of azadirachtin and (–)-rotenone, (–)-spinosyn-A preparation demonstrates the versatility of organic synthesis to construct natural products and derivatives for use as agrochemicals.

8.5   Recent Developments in Agrochemicals

A combination of different options may provide a balanced strategy to deal with the challenge of food production. For example, reduced food wastage and lower animal product consumption are desirable. Emerging innovative technologies like urban and/or vertical farmings may play an important role in this context: They may increase the amount of land available for plantation. Moreover, crop yields will have to enhance significantly in order to meet food requirements in 2050. This will demand further advances in crop protection, expanded use of genetically modified crops, and development of other modern technologies such as precision farming. Given all the scenarios for fostering improved agricultural harvest yields, crop protection products will remain essential.

Recently, a special volume of *Bioorganic & Medicinal Chemistry* has been entirely devoted to reviewing the recent developments in agrochemistry (Jeanmart 2016). The first article describes the synthesis of 30 agrochemicals that have received an international standardization organization (ISO) name over a period of five years (from January 2010 to December 2014). The purpose is to show the range and scope of chemical products that have been used to discover or to produce the latest active ingredients employed in the crop protection industry (Jeanmart et al. 2016).

Most agrochemicals listed in the Pesticide Manual have been discovered through screening programs based on trial-and-error testing. Current agrochemical discoveries have not taken much advantage of in silico original chemical compound iden-

herbicides   fungicide fluazinam

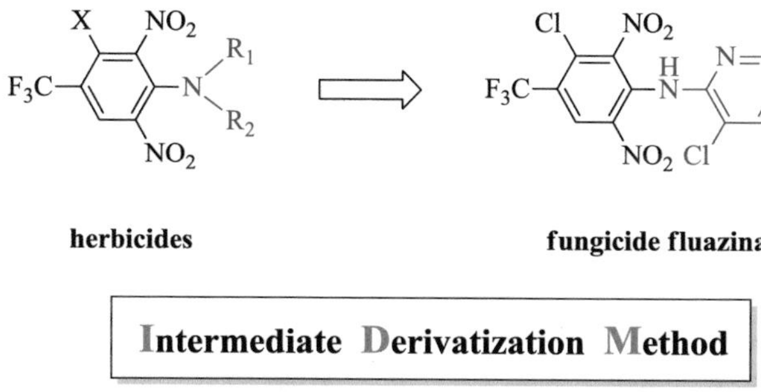

Intermediate Derivatization Method

Fig. 8.27 Obtaining agrochemicals containing a pyridine moiety by IDM (from Guan et al. 2016)

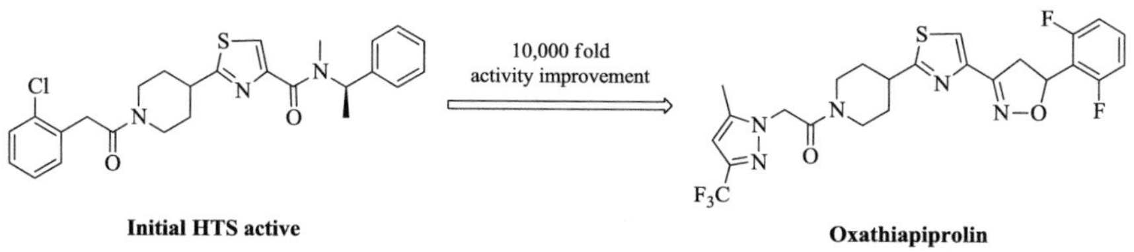

Initial HTS active   Oxathiapiprolin

Fig. 8.28 Synthesis of a new class of piperidinyl thiazole isoxazoline fungicides

tification/discovery techniques that are used in pharmaceutical research. Finding different methods that make the process of discovering novel lead compounds in the agrochemical field more efficient is important to shorten the time of research phases and to achieve the recent market requirements. Another review article has dealt with 18 representatives of known agrochemicals containing a pyridine moiety. The authors studied other derivatives that could have potential use as agrochemicals and extrapolated their findings from the perspective of *Intermediate Derivatization Methods* (IDM) in the hope that this approach will appeal to researchers engaged in the discovery of agrochemicals and/or pharmaceuticals (Fig. 8.27) (Guan et al. 2014, 2016).

The discovery, synthesis, optimization, and biological efficacy of oxathiapiprolin (Fig. 8.28), a new fungicide with a cutting-edge action mode, have also been presented (Pasteris et al. 2016). Oxathiapiprolin is the first member of a recent class of piperidinyl thiazole isoxazoline fungicides with exceptional activity against plant diseases caused by oomycete pathogens. Under field conditions, this molecule can control several oomycete species at rates as low as 12–30 g ha^{-1}. It acts by inhibiting a novel fungal target—an oxysterol-binding protein—thereby providing excellent protection, as well as curative and residual efficacy against key diseases of grapes, potatoes, and vegetables.

Broflanilides, which are novel *meta*-diamine insecticides (Fig. 8.29), act through a noncompetitive γ-aminobutyric acid (GABA) receptor antagonist. These compounds have high larvicidal activity against *Spodoptera litura* (tobacco cutworm or cotton leafworm) (Nakao and Banba 2016). The binding site of desmethyl-broflanilide is

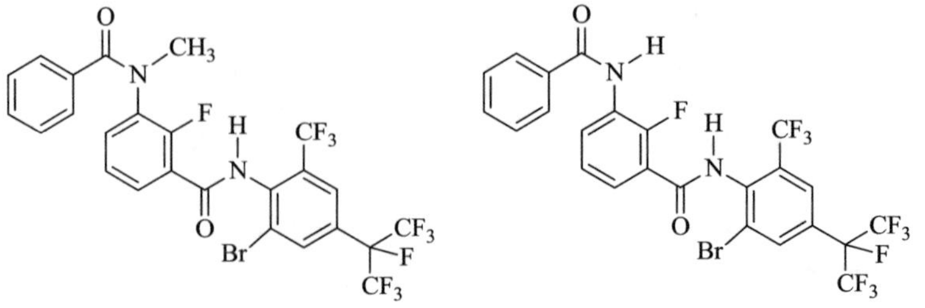

Broflanilide   **Desmethyl-broflanilide**

Fig. 8.29 Structure of broflanilides, new *meta*-diamine insecticides

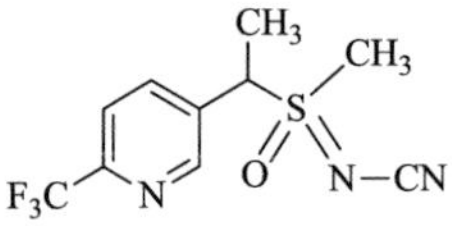

Sulfoxaflor
(Isoclast™ active)

Fig. 8.30 Structure of sulfoxaflor derivatives, a new sulfoximine class insecticide

distinct from the binding site of conventional noncompetitive antagonists such as fipronil.

Sap-feeding insect pests constitute a major insect pest complex that includes a range of aphids, whiteflies, planthoppers, and other insect species. Sulfoxaflor, a new sulfoximine class insecticide (Fig. 8.30), targets sap-feeding insect pests including pests that are resistant to numerous other classes of insecticides. A structure–activity relationship (SAR) investigation of sulfoximine insecticides has revealed that a 3-pyridyl ring and a methyl substituent on the methylene bridge linking the pyridine and the sulfoximine moiety are important to achieve strong activity against the green peach aphid (*Myzus persicae*). Model development has resulted in a highly predictive model for a set of 18 sulfoximines, including sulfoxaflor. The modeling is consistent with and helps to explain the highly optimized pyridine substitution pattern for sulfoxaflor (Loso et al. 2016).

There has been fundamental research toward the discovery of new leads that could help to discover original active ingredients for the crop protection market (Yang et al. 2016). A series of new pymetrozine analogs containing both methyl on the imine carbon and a phenoxy group in the pyridine ring have been designed and synthesized (Fig. 8.31). Their insecticidal activities against bean aphid (*Aphis craccivora*), mosquito larvae (*Culex pipiens pallens*), cotton bollworm (*Helicoverpa armigera*), corn borer (*Ostrinia nubilalis*), and oriental armyworm (*M. separata*) have been evaluated. The bioassay results indicate that most of the target compounds display good insecticidal activity against bean aphid. These compounds and known pymetrozine derivatives exhibit completely different structure–activity relationship. Introduction

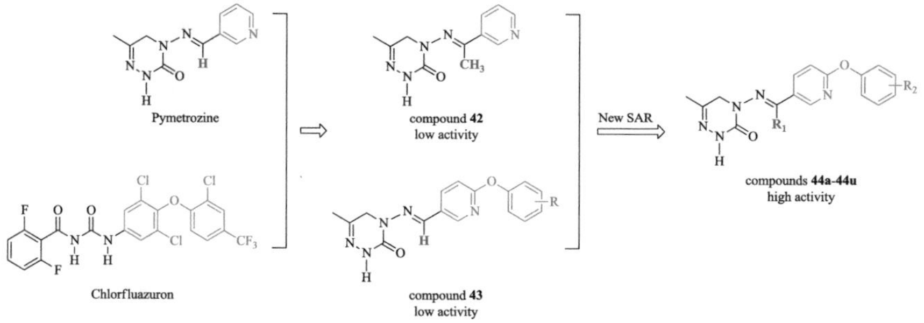

Fig. 8.31 Design of a series of new pymetrozine analogs (from Yang et al. 2016)

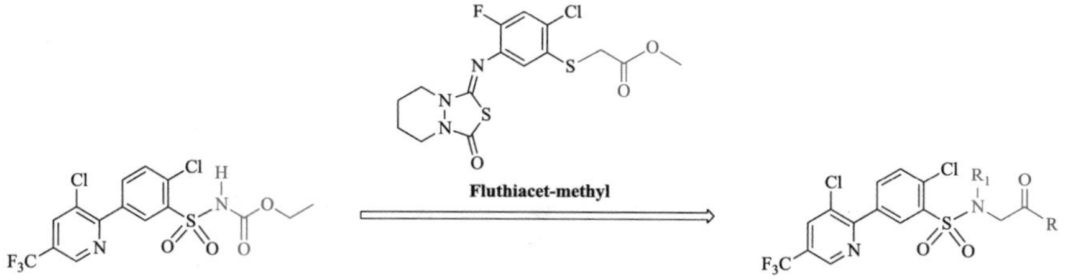

Fig. 8.32 Synthesis of new substituted 3-(pyridin-2-yl)benzenesulfonamide derivatives

of an alkyl group into the imine carbon could be detrimental to the activities. The results suggest that the methyl on the imine carbon and the phenoxy group at the pyridine ring might exert additive effects on aphicidal activity improvement. Moreover, some compounds containing an allyl group at the *para* position of the phenoxy group exhibit excellent insecticidal activity against mosquito larvae, lepidoptera pests cotton bollworm, corn borer, and oriental armyworm (Yang et al. 2016).

In an attempt to obtain novel candidate compounds for weed control, a series of substituted 3-(pyridin-2-yl)benzenesulfonamide derivatives have been designed and synthesized (Fig. 8.32), and their herbicidal activities have been evaluated in greenhouse tests (Xie et al. 2016). According to the weed controlling spectrum test, these compounds effectively control dayflower (*Commelina tuberosa*), bur beggarticks (*Bidens tripartita linn.*), youth-and-old age, cassia tora (*Cassiaobtusifolia* L.), velvet leaf, purslane (*Portulaca oleracea*), and false daisy (*Eclipta prostrata* L.). The best molecule is active at 37.5 g ha^{-1} against a wide range of weeds. In addition, a combination of some of these compounds and propanil could produce a synergistic effect and enhance the herbicidal activity. On the basis of the herbicidal activity assay in field test, these compounds could effectively control dayflower and nightshade (*Disambiguation*) with long-lasting persistence, indicating that they may be potent herbicidal agents (Xie et al. 2016).

Diphenylamine derivatives with good fungicidal, insecticidal, acaricidal, rodenticidal, and/or herbicidal activities have been reported. To find new lead compounds of this kind, a series of novel diphenylamine derivatives have been designed and syn-

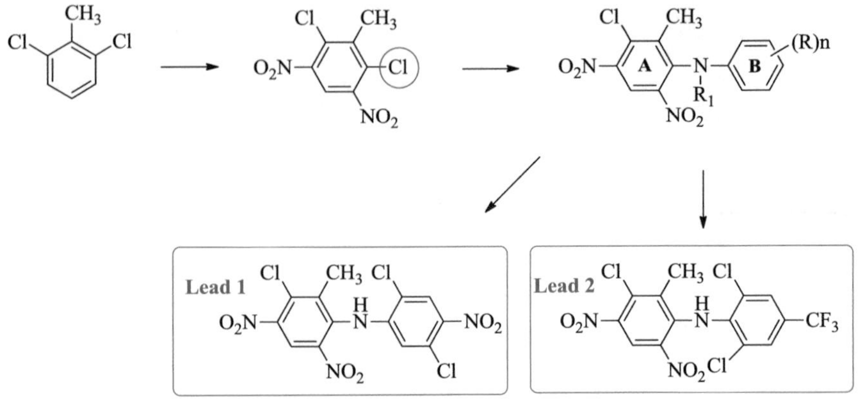

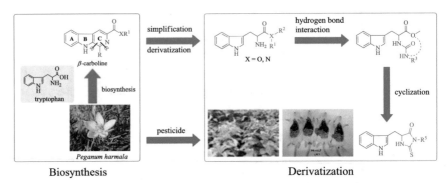

Fig. 8.33 Synthesis of a series of novel diphenylamine derivatives (from Li et al. 2016a)

Fig. 8.34 Synthesis of β-carboline alkaloid derivatives (from Huang et al. 2016)

thesized (Fig. 8.33). According to bioassays, some compounds substituted at 2,4,6-positions or 2,4,5-positions of phenyl ring B exhibit excellent fungicidal activities. The relationship between structure and fungicidal activities of these diphenylamine derivatives has been discussed, as well (Li et al. 2016a).

Tryptophan, the biosynthesis precursor of β-carboline alkaloids, and some derivatives have been synthesized (Fig. 8.34), and their biological activities and structure–activity relationships have been studied. The bioassay has shown that these compounds display good inhibitory activities against tobacco mosaic virus (TMV) both in vitro and in vivo, as well as antifungal and insecticidal activities. One of the synthesized compounds possesses anti-TMV activity in the field, as well as low acute oral toxicity (Huang et al. 2016).

Gossypol is part of the defense system of the cotton plant against pathogens and herbivorous insects. To discover gossypol analogs with broad spectrum and high activity, a series of gossypol alkylamine Schiff base, oxime, and hydrazone derivatives have been synthesized (Fig. 8.35) and bioassayed. On the basis of the

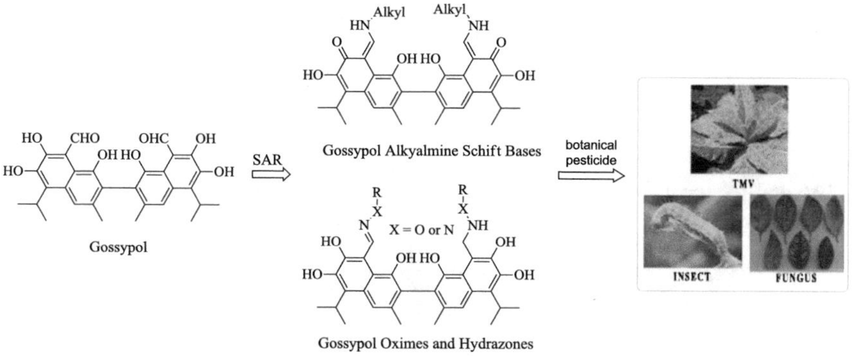

Fig. 8.35 Synthesis of a series of gossypol derivatives (from Li et al. 2016b)

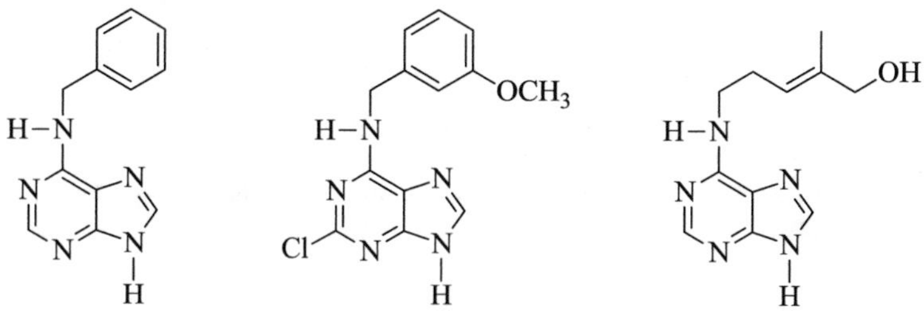

Fig. 8.36 Structures of some cytokinins used in the exogenous treatments in various field crops

biological results, most of these derivatives present higher anti-TMV activity than gossypol (Li et al. 2016b).

Cytokinins are plant hormones that regulate various aspects of plant growth and development, such as branching, senescence delay, nutrient remobilization, and flower and seed set control (Fig. 8.36). Therefore, cytokinins have become interesting substances in the search for potential agrochemicals. Since 1970, exogenous application of cytokinins has been tested in field conditions to improve the yields of important crops such as wheat, rice, maize, barley, and soybean worldwide. Despite extensive testing, so far cytokinins have not occupied a stable position among commercial plant growth regulators mainly because they have complex effects. A recent overview of the outcomes obtained during field experiments involving cytokinin exogenous treatments describes the application modes and points out the impacts on various field crops, vegetables, cotton, and fruit trees (Koprna et al. 2016).

Water availability dictates agricultural productivity, so drought is the major source of harvest losses worldwide. Abscisic acid is an apocarotenoid plant hormone that mediates responses to abiotic stress; this acid also modulates multiple growth and developmental processes (Dejonghe et al. 2018). Abscisic acid acts through a negative regulatory signaling module that is present in all land plant genomes sequenced to date. Abscisic acid levels rise in response to water deficit, and this acid modulates drought tolerance by reducing water consumption and inducing other drought-

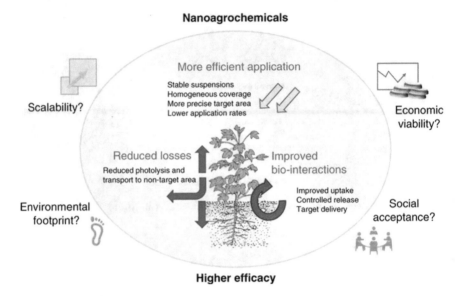

Fig. 8.37 Key drivers for nanotechnology application to improve the efficacy of agrochemicals (from Kah et al. 2018)

protective feedback. Recent identification of abscisic acid receptors, elucidation of their structures, and understanding of the core abscisic acid signaling network could serve as a blueprint for the development of future agrochemicals that can control plant water use and improve yield under water stress conditions (Helander et al. 2016). Chemical manipulation of plant water use has been reviewed, and it has been demonstrated that an existing agrochemical can be repurposed to control a certain trait. This is an advanced path for future research into crop protection.

Recently, research approaches have emerged to solve current challenges in agriculture. Researchers have attempted to design potential nanoscale pesticides, such as nanopesticides, which are a recent novel eco-friendly strategy to manage insect pests (Prasad et al. 2014; Nuruzzaman et al. 2016; Chhipa 2017; Bhan et al. 2018; Oliveira et al. 2018). Research into nanotechnology applications in agriculture, particularly in crop production, has received a lot of attention and generally aims to use resources more efficiently through the development of delivery nanosystems for agrochemicals (Rodrigues et al. 2017; Yin et al. 2018).

Most of the so-called nanopesticides and nanofertilizers proposed so far involve reformulation of registered active ingredients, to achieve improved performance as compared to the existing agents and to counteract the main drawbacks associated with current agrochemical products (Fig. 8.37) (Kah et al. 2018). Targeted delivery of the active ingredient to the pest and/or increased fertilizer efficacy is among the strategies that have been designed to aid maintenance or even to increase yields with significantly reduced application rates, while the negative impacts of agriculture on ecosystems and the human health are minimized.

Researchers have developed different types of nanopesticides like nanocapsulated formulations, nanoemulsion, nanogel, nanospheres, polymer matrix, yeast cells as matrix, and metal or metal oxide nanoparticles. Nanoencapsulation is a potent means to carry these nanopesticides to the target site. Aluminosilicate nanotube is an efficient nanomaterial. Insect hairs absorb aluminosilicate nanotubes spread on the surface of the plant, so insects consume the nanotubes filled with pesticides and die (Kumar et al. 2017).

Recent studies have shown that nanopesticides can reduce the toxic impact of chemical pesticides and can provide specific control of crop pest. Such studies are important for aiding the development of intelligent nanosystems that help to minimize the adverse problems of agriculture like environmental imbalance, food security issues, and food productivity concerns (Nuruzzaman et al. 2016; Chhipa 2017; Oliveira et al. 2018). These nanosystems can conduct controlled release of the active ingredient. They remain efficient for a long period and can help to solve the issues of eutrophication and residual pesticide accumulation. In addition, nanopesticides provide the active ingredient with improved solubility and stability for effective pest control (Chhipa 2017). The development of smart nanopesticides will offer countless solutions to the agrochemical industry, i.e., active ingredient solubility, stability, controlled release, and targeted delivery. However, lots of research is still required to understand the nanopesticide fate in the environment (Fraceto et al. 2016; Oliveira et al. 2018; Zhao et al. 2018).

There are some concerns about the altered profile of novel nanoagrochemicals, so these modern products must be critically evaluated against conventional analogs, and their associated benefits and risks must be assessed. Recently, several literature articles have critically determined the extent to which nanoagrochemicals differ from conventional products (Kah et al. 2018). These analyses have shown that nanoagrochemicals provide an average gain of about 20–30% in efficacy relative to conventional products. Nanoformulations may alter the environmental fate of agrochemicals, but changes may not necessarily translate into reduced environmental impact. Many studies lack nanospecific quality assurance and adequate controls. Currently, no comprehensive study has evaluated the efficacy and environmental impact of nanoagrochemicals under field conditions. This is a crucial knowledge gap, and it is essential that the benefits and recent risks that nanoagrochemicals pose are evaluated as compared to existing products.

8.6 Current Challenges in the Synthesis of Agrochemicals

Crop protection chemistry is a high-tech science that supports sustainable food production for a rapidly growing population. Cutting-edge developments in the design and synthesis of new agrochemicals have helped to create more selective and environmentally benign active ingredients to face the current challenges of weed and pest resistance, higher regulatory safety margins, and higher cost of goods.

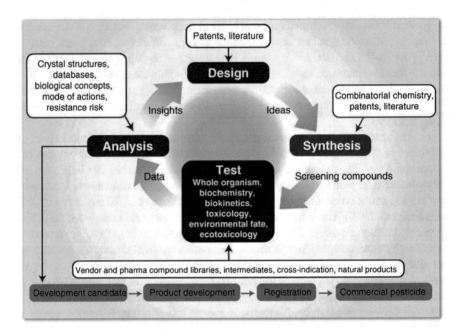

Fig. 8.38 Innovation and optimization process in crop protection research (from Lamberth et al. 2013)

Over the past 50 years, the amount of agrochemicals that is used to protect crops has decreased from 1 kg per ha to about 10 g per ha, that is, to only 1% of what used to be required. Therefore, there is a clear trend toward the use of molecules that combine low application rates with a more favorable toxicological profile. However, increased resistance of weeds and pests to some kinds of agrochemicals has stimulated the search for more selective, much safer, and cheaper chemicals (Baker and Umetsu 2001).

Research concerning the development of advanced agrochemicals has used a variety of recent chemical inputs taken from a series of areas in the vanguard of chemistry: natural products, combinatorial chemistry libraries, and compound collections from pharmaceutical and animal health companies, among others. The screening hits obtained from these sources are usually optimized through diverse rounds of a design–synthesis–test–analysis cycle, which is well represented in Fig. 8.38, obtained from Lamberth et al. (2013). This cycle closely resembles the design–synthesis–test–analysis cycle that is employed in the pharmaceutical industry. Agrochemical research typically verifies all compounds directly on the whole weed, fungus, or insect organism. However, molecular target-based approaches that use in vitro assays are becoming increasingly common.

Despite the various sources of new products for crop protection, the number of modern molecules introduced into the market has declined in recent years. This is because companies have decided that investment in the discovery, development, and launch of an original pesticide carries a high level of risk. Corporations have to translate the expectation or assumption of future needs into goals, and they must transform innovative discoveries into actionable projects. Much of that risk stems from two major factors: The time elapsed between goal development and eventual product launch is long, and markets around biological systems fluctuate unpredictably (Whitford et al. 2006). All of these challenges could be overcome through the application of advanced design strategies and state-of-the-art organic synthesis that can deliver solutions for the discovery of original agrochemicals.

According to Lamberth et al. (2013), modern agrochemicals and pharmaceuticals interact with their target receptors or enzymes via the same molecular recognition processes. In countless cases, a homologous enzyme/receptor is addressed, so it is not entirely surprising that one class of compounds can give rise to (different) active ingredients serving both industries. However, although bioavailability is vital to both pharmaceuticals and agrochemicals, the chemical environments that the active ingredients encounter on their way from the application site to the biochemical target are very different and generally require distinct physicochemical properties (Lindell et al. 2009; Schleifer 2007). For example, agrochemicals have a lower number of hydrogen bond donors (Tice 2001, 2002). In addition, for agrochemicals to meet the producer's need for the longest possible interval between spraying, they should have residual activity and an effect that lasts up to several weeks. The fact that most heterocycles in agrochemicals are heteroaromatics illustrates this point (Tice 2001, 2002). Likewise, the production cost of an agrochemical is generally much more constrained as compared to the cost of producing a pharmaceutical to match a medical need. Thus, selection of the most cost-effective molecule for a particular market, coupled with an inexpensive manufacturing route, plays an important role (Corsi and Lamberth 2015).

More than 70% of the active ingredients that have been introduced into the market within the past 30 years possess heterocyclic scaffolds (Lamberth and Dinges 2012), and a similar number of ingredients contain halogen substituents (Theodoridis 2006; Jeschke 2012). Around one-third of all agrochemicals are chiral compounds (Tambo and Bellus 1991; Kurihara and Miyamoto 1998; Jeschke 2018), and some of the most important and most abundantly manufactured agrochemicals have been introduced in an enantiomerically or diastereomerically enriched form. One example is the insecticide indoxacarb (**47**), whose abbreviated synthesis is shown in Fig. 8.39 (Lamberth et al. 2013).

The finding that only the (*S*)-enantiomer of the insecticide indoxacarb (**47**), which blocks the voltage-gated sodium channel of the target pests, possesses insecticidal activity has created the desire for an efficient enantioselective synthesis (Lamberth et al. 2013). Besides the complexity of the tricyclic core scaffold, tetrasubstitution at the chiral carbon center renders the synthesis especially challenging. This problem has been solved by asymmetric α-hydroxylation of the indanone ester **45** with *tert*-butyl hydroperoxide using cinchonine as a chiral base, which affords a 3:1 product ratio in favor of the biologically active (*S*)-enantiomer [50% enantiomeric excess (e.e.)]. Application of a chiral zirconium-based catalyst increases the enantioselec-

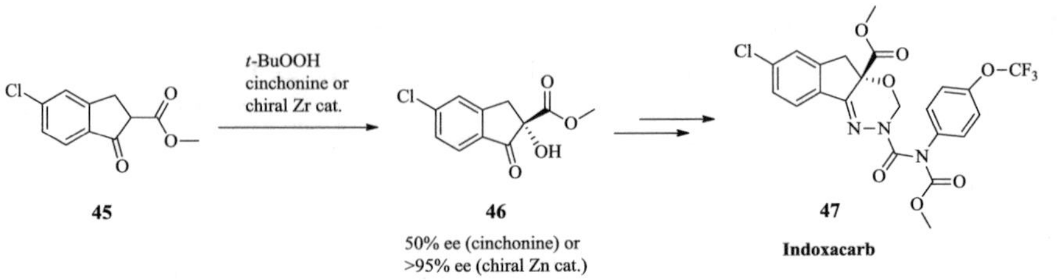

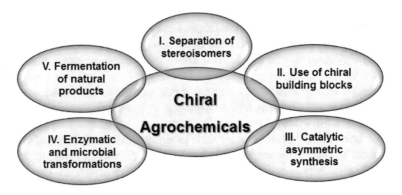

Fig. 8.39 Stereoselective synthesis of the insecticide indoxacarb

Fig. 8.40 Various methods to prepare chiral agrochemicals (from Jeschke 2018)

tivity to 95% e.e. In four further steps, the resulting 2-hydroxyindanone ester **46** is transformed into indoxacarb (**47**) (McCann et al. 2001, 2012).

In the past, selective preparation of enantiomers or their separation on an industrial scale was often difficult, inefficient, and expensive. However, in recent years, enormous progress has been made in asymmetric synthesis and catalysis, and stereoselective processes have become increasingly widespread in the preparation of modern agrochemicals. Today, various methods are available to prepare chiral agrochemicals (synthesized or resolved racemates in enantiopure compounds) and to analyze them with good resolution and sensitivity (Fig. 8.40) (Jeschke 2018).

Oxathiapiprolin, previously shown in Fig. 8.28 (Pasteris et al. 2016), is the first member of a novel class of piperidinyl thiazole isoxazoline fungicides that has been shown to have exceptionally high activity against diseases caused by *Oomycete* pathogens. Before oxathiapiprolin was discovered, a racemic mixture of (5*R*,*S*)-enantiomers could be obtained from assorted chiral key molecules that had been synthesized, and their fungicidal potency was evaluated. This can be exemplified by the preparation of the second lead structure (1*R*)-*N*-[(1*R*)-1-phenylethyl]-4-thiazole-carboxamide bearing a pyrazole acetamide fragment (the first lead bearing 2-chloro-phenyl acetamide), as outlined in Fig. 8.41 (Jeschke 2018). Whereas the (1*R*)-enantiomer at 10 mg L^{-1} provides excellent preventive and curative oomycete control, its corresponding (1*S*)-enantiomer is less active. Cyclizations near the chiral

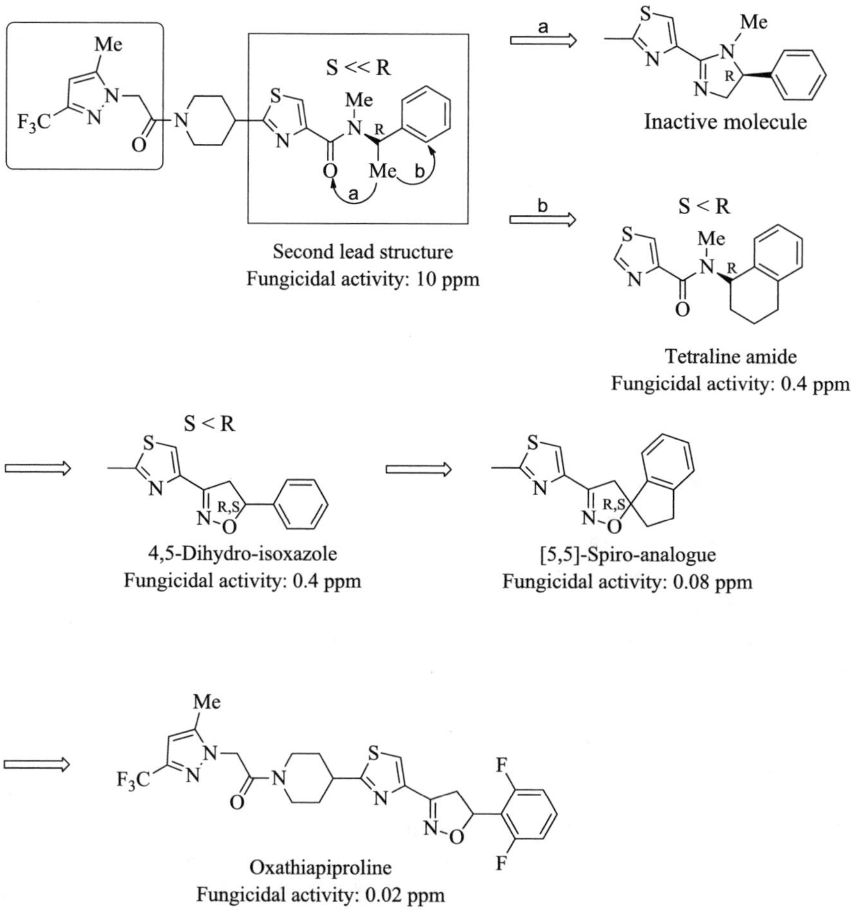

Fig. 8.41 Discovery of oxathiapiprolin based on the second lead structure, the tetraline amide, the 4,5-dihydro-isoxazole, and the 5,5-spiro analog (from Jeschke 2018)

center region have been explored in a program that investigates restricted conformations, e.g., cyclization of (±)-methyl (a) to the carbonyl oxygen position and (b) to the phenyl *ortho*-position.

Despite retaining the (1R)-configuration, the first modification leads to an inactive molecule. The second modification results in the 25-fold more active (1R)-[1,2,3,4-tetrahydro-1-naphthalenyl]-amide, but its (1S)-enantiomer is significantly less active. Surprisingly, at 0.4 ppm, the *racemic* (5R,S)-(4,5-dihydro-5-phenyl-3-isoxazolyl)-2-thiazolyl compound as benzyl amide bioisoster provides similar levels of both preventive and curative oomycete control, but the (5R)-enantiomer is more potent than the corresponding (5S)-enantiomer. The SAR depends on the 4,5-dihydro-isoxazoline moiety, so 5-substitution is essential for the high fungicidal activity and has been explored with a wide variety of substituents.

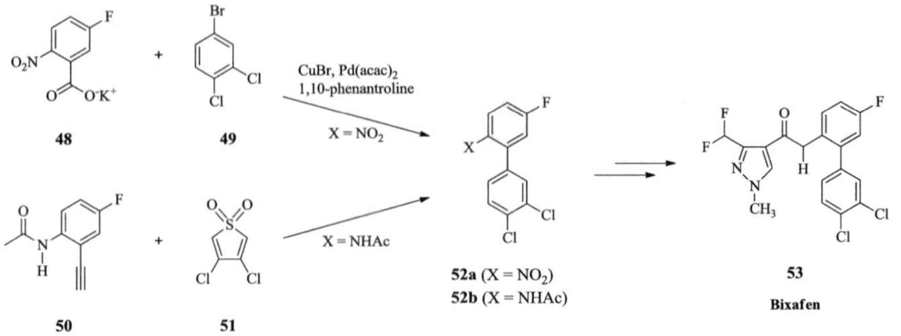

Fig. 8.42 Synthesis of the fungicide bixafen

Through admission of a second 5-substituent, the [5.5]-spiro analog contains an out-of-plane twisted phenyl ring that improves the activity five times. This twist can also be induced by 2,6-disubstituted phenyl rings on position 5 of (5R,S)-4,5-dihydro-isoxazoline, resulting in oxathiapiprolin, a second lead structure with 500-fold higher fungicidal activity (Jeschke 2018).

The oxathiapiprolin (5R)-enantiomer is more effective than the corresponding (5S)-enantiomer. The decision to launch the racemic product into the market is probably based on its excellent activity, combined with the lower manufacturing costs associated with the use of non-chiral starting materials (Corsi and Lamberth 2015).

The production of tetrasubstituted biphenyl derivatives **52** illustrates the fact that crop protection chemistry is often at the forefront of organic chemistry. Derivatives **52** are key intermediates in the synthesis of bixafen (**53**), a broad-spectrum fungicide that blocks the fungal respiratory chain by inhibiting succinate dehydrogenase. As outlined in Fig. 8.42 (Lamberth et al. 2013), one way to obtain compound **52** is through Pd/Cu-catalyzed decarboxylative cross-coupling of the potassium benzoate **48** with 1-bromo-3,4-dichlorobenzene (**49**), which gives the nitrosubstituted trihalogenated biphenyl **52a** (Walter 2012). Alternatively, the phenylacetylene derivative **50**, which is readily prepared from the corresponding bromobenzene precursor in two steps, is converted to the anilide **52b** in a tandem Diels–Alder cycloaddition–cycloreversion sequence with 3,4-dichlorothiophene 1,1-dioxide (**51**) (McCann et al. 2012). Both key intermediates, **52a** and **52b**, are then transformed into the unprotected aniline, whose further acylation gives bixafen (**53**).

Structure-based design is an iterative and multidisciplinary process that is well established in the pharmaceutical industry (Klebe 2000; Anderson 2003). It has played an important role in the development of countless registered drugs and clinical candidates. In contrast, structure-based design is relatively recent in the agrochemical industry, and none of the currently available products is the direct result of this approach (Plant 2010). However, there are several examples of discovery programs on which structure-based design has had a strong impact.

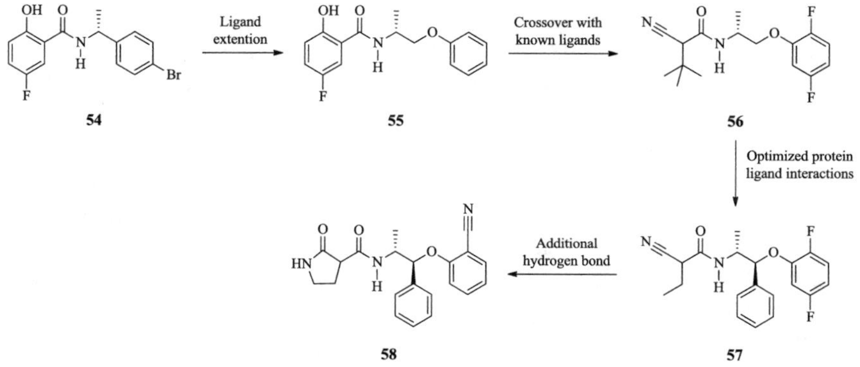

Fig. 8.43 Iterative structure-based design steps leading to fungicidal scytalone dehydratase inhibitors. Optimization of scytalone dehydratase inhibitors through multiple design–synthesis–test–analysis cycles (from Lamberth et al. 2013)

The development of scytalone dehydratase inhibitors as rice blast fungicides is one of the most detailed examples that have been reported in the field of crop protection so far (Walter 2002). Scytalone dehydratase is part of the fungal melanin biosynthetic pathway, and its disruption leads to fungal pathogenicity loss (Bechinger et al. 1999; Talbot 2003). The complex formed between *Magnaporthe grisea* scytalone dehydratase and the salicylamide ligand has served as the starting point for the design of novel inhibitors (Jordan et al. 1999). The binding pocket of the enzyme is large enough to accommodate elongated inhibitors. To investigate whether this is true, a phenoxypropyl salicylamide (compound **55** in Fig. 8.43) has been designed and synthesized. Although the activity of the enzyme drops ninefold in the presence of compound **55** as compared to the starting compound **54**, it remains in the picomolar range throughout the enzyme assay. Crossover compounds with diclocymet, a commercial rice blast fungicide, have been designed. Although the aryl ring in the acid moiety of compound **55** is replaced, the phenoxypropyl fragment is retained, to give compound **56**. The latter compound gives an inhibition constant of 20 pmol L^{-1} and provides excellent control of rice blast disease in field trials. Crystal structures of the enzyme in the complex with the new lead compounds have been solved, and a comparative analysis of these structures has indicated that conformationally flexible regions are present in the binding pocket. This information has been incorporated into the design, and inhibitor **57**, which carries an additional phenyl substituent, has been envisaged to optimize interactions with an adjacent phenylalanine (Basarab et al. 1999). On the basis of a new validation of the binding mode, confirmed by X-ray crystallography, scytalone dehydratase inhibitors are buried in a lipophilic binding site, so their potency can be improved by addition of lipophilic groups that fit in the binding pocket. However, activity on the whole organism depends not only on enzyme activity but also on bioavailability, which decreases when hydrophobicity becomes too high. Analysis of scytalone dehydratase crystal structures can generate original ideas to increase ligand hydrophilicity (Basarab et al. 2002). For example, the salicylamide **54** establishes a hydrogen bond with the Asp131 residue when the hydroxyphenyl group is replaced with a butyrolactam, as in the case of compound **58**; two hydrogen bonds can be formed with this same residue. As a result, polarity increases, leading to excellent in vivo activity, which is not possible in the case of the initial salicylamide **54** (Basarab et al. 1999).

Structure-based design is a growing area within crop protection research. One of the key guides has been the huge increase in the number of protein structures in the public domain, which has increased from about 13,000 to almost 100,000 in recent years (www.pdb.org) (Berman et al. 2000, 2002). This increase has been possible thanks to advances in gene sequencing, recombinant protein production and purification, protein crystallography, and organization of structural genomics centers (Berman 2008; Touw et al. 2015). Many crystal structures of agrochemicals bound to their target sites are now available to the public. One example is the structure of acetyl-coenzyme A carboxylase, which is an important enzyme in fatty acid biosynthesis. The herbicide (PDB code 3PGQ) and the insecticide (PDB code 2JGE) indications mention that this enzyme forms a complex with pinoxaden enolate (Yu et al. 2010) and with methamidophos (Hörnberg et al. 2007), respectively. This information can be used for structure-based design. Moreover, the structures of several ion channels—which are notoriously difficult to crystallize, but are of high interest to crop protection research—are public knowledge. For example, the crystal structure of a glutamate-gated chloride channel in a complex with ivermectin, a member of the mectin class of insecticides, was reported in 2011 (Hibbs and Gouaux 2011) and represents a solid starting point for the design of novel insecticidal glutamate-gated chloride channel activators.

On the basis of a crystal structure of a pentameric ion channel published in 2009 (Bocquet et al. 2009), researchers have generated a homology model for a related γ-aminobutyric acid (GABA)-gated chloride channel and proposed a binding mode for metadiamides (Nakao et al. 2013), a class of potent insecticides that are currently being investigated by many agrochemical companies. This last example demonstrates how valuable starting points for structure-based design can be obtained from the structure of a related protein when the actual target structure is unknown. Such a homology modeling approach can be very powerful if validated by experimental techniques like site-directed mutagenesis (Lamberth et al. 2013).

Advanced technologies offer additional opportunities to generate original leads in different and faster ways. Modern techniques like fragment-based design (Erlanson et al. 2004; Hao et al. 2016), virtual screening, and genome sequencing—developed primarily in the pharmaceutical or biotechnology industries—have been successfully applied to generate drug leads. These tools can be applied to the discovery of agrochemicals as long as the differences between pharmaceuticals and agrochemicals, such as their physicochemical properties, are taken into account. Thus far, however, published examples of fragment-based design in the agrochemical context have been comparatively rare.

8.7   Case Study

Mixture toxicity assisting the design of eco-friendlier plant protection products: A case study using a commercial herbicide that combines nicosulfuron and terbuthylazine (Queirós et al. 2018).

Agriculture relies on plant protection products to ensure improved crop quality and yield. However, the use of plant protection products may pose risks to the environment and the human health and give rise to significant costs. Regulatory agencies worldwide have already recognized this problem and have been developing tight screening protocols before plant protection products are given the green sign to enter the market. Regarding screening protocols, modeling tools have been developed to address plant protection product transport and accumulation in aquatic systems. In the European Union, for example, the FOCUS platform is used to assess **P**redicted **E**nvironmental **C**oncentrations (PEC values) in surface and groundwater, depending on application doses of plant protection products and on their physicochemical properties. PEC values can be compared with ecotoxicological benchmarks obtained after ecotoxicity tests, allowing conclusions about the environmental hazard potential of plant protection products to be reached. Moreover, plant protection products have been reaching surface water through different transport pathways, e.g., runoff and leaching. This often causes hazardous contamination scenarios where exposure to pesticide residues is often clearly linked to significant ecosystem risks.

To counteract these environmentally hazardous scenarios, the agrochemical industry has been impelled to innovate the formulation of its products. Common strategies to develop eco-friendlier plant protection product formulations include:

(i)   the use of natural products or greener equivalents in plant protection product formulations;

(ii) the improvement of plant protection product application techniques and target delivery;

(iii) the combination of already licensed active substances to obtain a synergistic behavior that aims to use lower quantities of active substances with the same levels of efficacy against the target.

The latter solution is frequently used to improve the control of a broader range of weeds, but its success is controversial. Whether some argue that a synergistic behavior can be reached, there is experimental evidence that pesticide mixtures rarely result in synergistic effects. Additionally, to predict joint action, the interactive effects between the formulation constituents should be considered and the toxic potential of each formulated plant protection product must be assessed; focus on each individual formulation component should be avoided. Well-developed and well-established tools can aid mixture toxicity assessment. This is the rationale that has motivated the authors of the case study described herein: They decided to focus on manipulating the ratio between two active substances of a commercial herbicide to evaluate the suitability of this strategy as an additional approach to formulate eco-friendlier products.

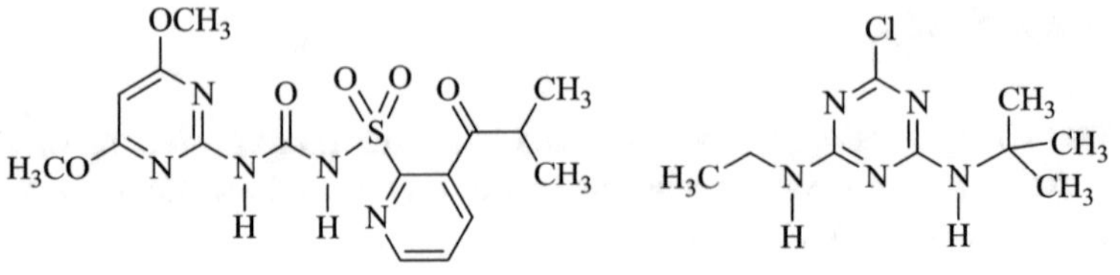

nicosulfuron   **terbuthylazine**

Fig. 8.44 Structures of the active substances nicosulfuron and terbuthylazine

For this case study, the authors selected a commercial herbicide combining nicosulfuron and terbuthylazine as active substances as a model of plant protection products (Queirós et al. 2018). Nicosulfuron and terbuthylazine (Fig. 8.44) belong to the chemical groups of sulfonylureas and 1,3,5-triazines, respectively. Nicosulfuron prevents the growth of susceptible plants by blocking the synthesis of branched-chain amino acids through acetohydroxyacid synthase inhibition, whereas terbuthylazine inhibits photosynthesis by blocking photosystem II.

The authors selected *P. oleracea*, commonly known as purslane, as a representative target weed within this case study because it is a major target of the commercial herbicide they decided to investigate. *P. oleracea* is a major worldwide weed affecting several crops, including maize, rice, wheat, cotton, and sugarcane, so it is a representative test species.

The testing involving the target weed *P. oleracea* and single as well as combined active substances showed that:

(i) the active substances behave antagonistically throughout the whole *P. oleracea* response surface;
(ii) no environmentally safe combination of active substances ensures target-efficacy;
(iii) nicosulfuron is apparently useless for the general activity of the herbicide;
(iv) at concentrations that are tenfold lower than the concentrations involved in commercially recommended application doses, terbuthylazine alone can effectively control *P. oleracea* with no potential environmental hazards.

Given that freshwater ecosystems are susceptible to contamination with plant protection product residues, the authors selected the macrophyte *Lemna minor*, commonly known as common duckweed, as the non-target test species to address the possibility of establishing eco-friendlier alternatives to the commercial combination of active substances. This option was ruled by (a) the established status of *Lemna* sp. as a standard ecotoxicological test species; (b) the herbicidal and systemic nature of the plant protection products, which a priori suggests that macrophytes should be more sensitive (thus more environmentally protective) than non-plant indicators and equivalent indicators lacking a vascular system such as microalgae; and (c) the fact that the available database on the ecotoxicity of each active substance confirms *Lemna* sp. as a very sensitive species. Single and mixture toxicity testing with *L.*

minor revealed an antagonistic joint action of the active substances, suggesting that the combination has an environmentally protective effect as compared to the active substances alone.

Therefore, this case study does not confirm the initial assumption that the active substances used in this commercial herbicide could act synergistically. In fact, the authors found that the substances display an antagonistic behavior toward the target weed and the non-target test species.

The results of this case study show that, due to environmental, toxicological, and regulatory demands, the reformulation of commercially available active substances can be an efficient alternative in terms of time and cost-effectiveness as compared to the huge expenses incurred during research and development of original active substances. However, the research and development phases, namely the phases involving laboratory and greenhouse tests and eventual field trials, should not be disregarded when the biological activity of the renewed product is confirmed. This applies to the formulations obtained after the best combination of active substances is established.

Overall, this case study suggests that modeling tools that are widely used in the field of environmental risk assessment of plant protection products might be applied in the stages of agrochemical design and may help to achieve a more efficient arrangement and to meet environmentally friendlier requirements.

8.8 Conclusions

This chapter has highlighted the important role agrochemicals play in agriculture. Agrochemicals have paved the way for all the improvements in food production and helped to tackle the issues related to feeding a growing world population over the last century.

The rising number of organisms that are resistant to commercially available agrochemicals makes the discovery of new molecules (natural or synthetic) that can replace the currently used active ingredients an urgent matter. These original compounds must act efficiently and selectively against the target organism, exhibit low toxicity against non-target organisms, and generate the lowest amount of residues as possible. Investigations into the extraction, isolation, structural identification, and synthesis or structural modification, as well as studies about the mechanism of action of these new agrochemicals are imperative.

The examples presented in this chapter show only a minimal fraction of the importance of organic synthesis (either total synthesis, partial synthesis, or the structural modification of an existing active ingredient) for the design and preparation of new compounds with potential biological activity against crop predators. We have disregarded the description of the theoretical and mechanistic aspects of the reactions because we aimed to emphasize the significance of obtaining the product of interest and the strategies related to its synthesis.

Furthermore, this chapter has not approached the viability of the industrial preparation of the molecules used in the examples. Our main purpose was to describe the use of classic processes coupled with the modern methods of organic synthesis with the objective of discovering original agrochemicals and producing the target molecules.

Finally, neither the synthesis of organic compounds using microorganisms or enzyme-catalyzed reactions nor crop protection by genetically modified organisms has been discussed because these subjects deserve a profound discussion that are far beyond the scope of this chapter.

References

Abad MJ, Ansuategui M, Bermejo P (2007) Active antifungal substances from natural sources. ARKIVOC vii:116–145

Alavanja MCR (2009) Pesticides use and exposure extensive worldwide. Rev Environ Health 24:303–309

Alavanja MCR, Bonner MR (2012) Occupational pesticide exposures and cancer risk: a review. J Toxicol Environ Health B 15:238–263

Alexandratos N, Bruinsma J (2012) World agriculture towards 2030/2050: the 2012 revision. ESA working paper no. 12-03, Food and Agriculture Organization of the United Nations, Rome, IT. Available at http://www.fao.org/docrep/016/ap106e/ap106e.pdf

Anastas PT, Warner JC (1998) Green chemistry: theory and practice. Oxford University Press, New York, US

Anderson AC (2003) The process of structure-based drug design. Chem Biol 10:787–797

Anderson JC, Ley SV (1990) Chemistry of insect antifeedants from *Azadirachta indica* (part 6): synthesis of an optically pure acetal intermediate for potential use in the synthesis of azadirachtin and novel antifeedants. Tetrahedron Lett 31:431–432

Baker DR, Umetsu NK (eds) (2001) Agrochemical discovery. American Chemical Society, Washington, US

Basarab GS, Jordan DB, Gehret TC, Schwartz RS, Wawrzak Z (1999) Design of scytalone dehydratase inhibitors as rice blast fungicides: derivatives of norephedrine. Bioorg Med Chem Lett 9:1613–1618

Basarab GS, Jordan DB, Gehret TC, Schwartz RS (2002) Design of inhibitors of scytalone dehydratase: probing interactions with an asparagine carboxamide. Bioorg Med Chem Lett 10:4143–4154

Batish DR, Setia N, Singh HP, Kohli RK (2004) Phytotoxicity of lemon-scented eucalypt oil and its potential use as a bioherbicide. Crop Prot 23:1209–1214

Batish DR, Singh HP, Setia N, Kohli RK, Kaur S, Yadav SS (2007) Alternative control of littleseed canary grass using eucalypt oil. Agron Sustain Dev 27:171–177

Batish DR, Singh HP, Kohli RK, Kaur S (2008) Eucalyptus essential oil as a natural pesticide. For Ecol Manage 256:2166–2174

Bechinger C, Giebel KF, Schnell M, Leiderer P, Deising HB, Bastmeyer M (1999) Optical measurements of invasive forces exerted by appressoria of a plant pathogenic fungus. Science 285:1896–1899

Begley MJ, Crombie L, Hadi HBA, Josephs JL (1989) Synthesis of novel labile rotenoids with unnatural trans-B/C ring systems. J Chem Soc Perkin Trans I:204–205

Berman HM (2008) The protein data bank: a historical perspective. Acta Crystallogr A 64:88–95

Berman HM, Westbrook J, Feng Z, Gilliland G, Bhat TN, Weissig H, Shindyalov IN, Bourne PE (2000) The protein data bank. Nucleic Acids Res 28:235–242

Berman HM, Battistuz T, Bhat TN, Bluhm WF, Bourne PE, Burkhardt K, Feng Z, Gilliland G, Iype L, Jain S, Fagan P, Marvin J, Padilla D, Ravichandran V, Schneider B, Thanki N, Weissig H, Westbrook JD, Zardecki C (2002) The protein data bank. Acta Crystallogr D 58:899–907

Bhan S, Mohan L, Srivastava CN (2018) Nanopesticides: a recent ecofriendly approach in insect pest management. J Entomol Res 42:263–270

Bilton JN, Broughton HB, Jones PS, Ley SV, Rzepa HS, Sheppard RN, Slawin AMZ, Williams DJ, Lidert Z, Morgan ED (1987) An X-ray crystallographic, mass spectroscopic, and NMR study of the limonoid insect antifeedant azadirachtin and related derivatives. Tetrahedron 43:2805–2815

Binetti R, Costamagna FM, Marcello I (2008) Exponential growth of new chemicals and evolution of information relevant to risk control. Ann Ist Super Sanità 44:13–15

Bocquet N, Nury H, Baaden M, Le Poupon C, Changeaux JP, Dalarue M, Corringer PJ (2009) X-ray structure of a pentameric ligand-gated ion channel in an apparently open conformation. Nature 457:111–114

Brocksom TJ, Vieira YW, Nakamura J, Finelli FG, Brocksom U (2007) A concise synthesis of the 1,6-disubstituted eudesmane sesquiterpene carbon skeleton. J Braz Chem Soc 18:448–452

Brocksom TJ, Donatoni MC, Uliana M, Vieira Y (2010) A reação de Diels-Alder no início do século vinte um. Quim Nova 33:2211–2218

Butterworth JH, Morgan ED (1968) Isolation of a substance that suppresses feeding in locust. Chem Commun 35:23–24

Carruthers W (1990) Cycloaddition reactions in organic synthesis. In: Tetrahedron organic chemistry series, vol 8. Pergamon Press, Oxford, GB

Carvalho FP (2006) Agriculture, pesticides, food security and food safety. Environ Sci Policy 9:685–692

Carvalho FP (2017) Pesticides, environment, and food safety. Food Energy Sec 6:48–60

Casida JE (2009) Pest toxicology: the primary mechanisms of pesticide action. Chem Res Toxicol 22:609–619

Chhipa H (2017) Nanopesticide: current status and future possibilities. Agric Res Technol 5:1–4

Cioc R, Ruijter E, Orru RVA (2014) Multicomponent reactions: advanced tools for sustainable organic synthesis. Green Chem 16:2958–2975

Corey EJ (2002) Catalytic enantioselective Diels-Alder reactions: methods, mechanistic fundamentals, pathways, and applications. Angew Chem Int Ed Engl 41:1650–1667

Corsi C, Lamberth C (2015) New paradigms in crop protection research: registrability and cost of goods. In: Maienfisch P, Stevenson TME (eds) Discovery and synthesis of crop protection products. American Chemical Society, Washington, US, pp 25–37

Crombie L, Freeman PW, Whiting DA (1973) A new synthesis of rotenoids. Application to 9-demethylmunduserone, mundeserone, rotenonic acid, dalpanol, and rotenone. J Chem Soc Perkin Trans I:1277–1285

Dagani R (1999) Presidential green chemistry challenge awards recognize five technologies for their environmental friendliness. Chem Eng News 77:30–32

Dejonghe W, Okamoto M, Cutler SR (2018) Small molecules probes of ABA biosynthesis and signaling. Plant Cell Physiol 59:1490–1499

Dostál J, Bochořáková H, Táborská E, Slavík J (1996) Structure of sanguinarine base. J Nat Prod 59:599–602

Durand-Reville T, Gobbi LB, Gray BL, Ley SV, Scott JS (2002) Highly selective entry to the azadirachtin skeleton via a Claisen rearrangement/radical cyclization sequence. Org Lett 4:3847–3850

Elango G, Rahuman A, Zahir AA, Marimuthu S (2010) Evaluation of repellent properties of botanical extracts against Culex tritaeniorhynchus Giles (Diptera: Culicidae). Parasitol Res 107:577–584

Erlanson DA, McDowell RS, O'Brien T (2004) Fragment-based drug discovery. J Med Chem 47:3463–3482

European Environment Agency (2013) Report no 1/2013. Late lessons from early warnings: science, precaution, innovation. Copenhagen, DK

Eurostat (2012) The REACH baseline study, 5 years update summary report. Available at http://ec.europa.eu/eurostat/en/web/products-statistical-working-papers/-/KS-RA-12-024

FAO (2006) Guidelines on efficacy evaluation for the registration of plant protection products. Food and Agricultural Organization of the United Nations. Available at http://www.fao.org/fleadmin/templates/agphome/documents/Pests_Pesticides/Code/Efficacy.pdf

FAO Statistics (2017) Food and Agriculture Organization of the United Nations. Available at http://www.fao.org/faostat/en/#home

FAO Statistical Yearbook (2013) World food and agriculture. Food and Agriculture Organization of the United Nations, Rome, IT

Forsyth SA, Gunaratne HQN, Hardacre C, McKeown A, Rooney DW (2006) One-pot multistep synthetic strategies for the production of fenpropimorph using an ionic liquid solvent. Org Process Res Dev 10:94–102

Fraceto LF, Grillo R, Medeiros GA, Scognamiglio V, Rea G, Bartolucci C (2016) Nanotechnology in agriculture: which innovation potential does it have? Front Environ Sci 4:20

Fukui K, Nakayama M, Harano T (1967) A new synthesis of dehydromunduserone. Experientia 23:613–614

Fukuzaki T, Kobayashi S, Hibi T, Ikuma Y, Ishihara J, Kanoh N, Murai A (2002) Studies aimed at the total synthesis of azadirachtin. A modeled connection of C-8 and C-14 in azadirachtin. Org Lett 4:2877–2880

Gilland B (2015) Nitrogen, phosphorus, carbon and population. Sci Progr 98:379–390

Gillij YG, Gleiser M, Zygadlo JA (2008) Mosquito repellent activity of essential oils of aromatic plants growing in Argentina. Bioresour Technol 99:2507–2515

Gnankiné O, Bassolé IHN (2017) Essential oils as an alternative to pyrethroids' resistance against Anopheles species complex Giles (Diptera: Culicidae). Molecules 22(10):1321

González JOW, Jesser EN, Yeguerman CA, Ferrero AA, Band BF (2017) Polymer nanoparticles containing essential oils: new options for mosquito control. Environ Sci Pollut Res 24:17006–17015

Govorushko S (2018) Human-insect interactions. CRC Press, Boca Raton, US

Guan AY, Liu C, Yang X, Dekeyser M (2014) Application of the intermediate derivatization approach in agrochemical discovery. Chem Rev 114:7079–7107

Guan AY, Liu CL, Sun XF, Xie Y, Wang MA (2016) Discovery of pyridine-based agrochemicals by using intermediate derivatization methods. Bioorg Med Chem 24:342–353

Hao GF, Jiang W, Ye YN, Wu FX, Zhu XL, Guo FB, Yang GF (2016) ACFIS: a web server for fragment-based drug discovery. Nucleic Acids Res 44:W550–W556

Hassankhani A (2015) A rapid, one-pot, multi-component route to 4,4′-(arylmethylene)bis(3-methyl-1-phenyl-1h-pyrazol-5-ols. J Mex Chem Soc 59:1–4

Helander JDM, Vaidya AS, Cutler SR (2016) Chemical manipulation of plant water use. Bioorg Med Chem 24:493–500

Henry KJJ, Fraser-Reid B (1994) Free radical methodology for carbohydrate to carbocycle transformations: an efficient synthesis of the tricyclic dihydrofuran portion of azadirachtin. J Org Chem 59:5128–5129

Hibbs RE, Gouaux E (2011) Principles of activation and permeation in an anion-selective Cys-loop receptor. Nature 474:54–60

Hörnberg A, Tunemalm AK, Ekström F (2007) Crystal structures of acetylcholinesterase in complex with organophosphorus compounds suggest that the acyl pocket modulates the aging reaction by precluding the formation of the trigonal bipyramidal transition state. Biochemistry 46:4815–4825

Hou QQ, Jing YF, Shao XS (2017) Synthesis and insecticidal activities of 1,8-naphthyridine derivatives. Chin Chem Lett 28:1723–1726

Huang L, Guan T, Qian Y, Huang M, Tang X, Li Y, Sun H (2011) Anti-inflammatory effects of maslinic acid, a natural triterpene, in cultured cortical astrocytes via suppression of nuclear factor-kappa B. Eur J Pharmacol 672:169–174

Huang Y, Liu Y, Liu Y, Song H, Wang Q (2016) C ring may be dispensable for β-carboline: design, synthesis, and bioactivities evaluation of tryptophan analog derivatives based on the biosynthesis of β-carboline alkaloids. Bioorg Med Chem 24:462–473

Huang JL, Min L, Xu H (2017) Semisynthesis of some matrine ether derivatives as insecticidal agents. RSC Adv 7:15997–16004

Huang J, Lv M, Thapa S, Xu H (2018) Synthesis of novel quinolinomatrine derivatives and their insecticidal/acaricidal activities. Bioorg Med Chem Lett 28:1753–1757

Huisgen R (1961) 1,3 dipolar cycloadditions. Proc Chem Soc 357–396

Hummelbrunner LA, Isman MB (2001) Acute, sublethal, antifeedant and synergistic effects of monoterpenoid essential oil compounds on the tobacco cutworm Spodoptera litura (Lep. Noctuidae). J Agric Food Chem 49:715–720

Hurst P, Hay A, Dudley N (1994) The pesticides handbook. Journeyman, London, GB

Ishihara J, Fukuzaki T, Murai A (1999a) Synthetic studies on azadirachtin (Part 3): asymmetric synthesis of the tricyclic dihydrofuran moiety of azadirachtin. Tetrahedron Lett 40:1907–1910

Ishihara J, Yamamoto Y, Kanoh N, Murai A (1999b) Synthetic studies on azadirachtin: construction of the highly functionalized decalin moiety of azadirachtin. Tetrahedron Lett 40:4387–4390

Ishihara J, Ikuma Y, Hatakeyama S, Suzuki T, Murai A (2003) Synthesis of the tricyclic dihydrofuran moiety of azadirachtin: efficient transformation of the Claisen rearrangement intermediate into a functionalized tricyclic dihydrofuran core. Tetrahedron 59:10287–10294

Isman MB (2000) Plant essential oils for pest and disease management. Crop Prot 19:603–608

Jeanmart S (ed) (2016) Recent developments in agrochemistry. Bioorg Med Chem 24:315–500

Jeanmart S, Edmunds AJF, Lamberth C, Pouliot M (2016) Synthetic approaches to the new agrochemicals. Bioorg Med Chem 24:317–341

Jeschke P (2012) The unique role of halogen substituents in the design of modern crop protection compounds. In: Jeschke P, Krämer W, Schirmer U, Witschel M (eds) Modern methods in crop protection research. Wiley-VCH, Weinheim, DE, pp 73–128

Jeschke P (2018) Current status of chirality in agrochemicals. Pest Manage Sci 74:2389–2404

Jia ZX, Luo YC, Xu PF (2011) Highly enantioselective synthesis of polysubstituted tetrahydroquinolines via organocatalytic Michael/Aza-Henry tandem reactions. Org Lett 13:832–835

Jordan DB, Lessen TA, Wawrzak Z, Bisaha JJ, Gehret TC, Hansen SL, Schwartz RS, Basarab GS (1999) Design of scytalone dehydratase inhibitors as rice blast fungicides: (N-phenoxypropyl)-carboxamides. Bioorg Med Chem Lett 9:1607–1612

Kah M, Kookana RS, Gogos A, Bucheli TD (2018) A critical evaluation of nanopesticides and nanofertilizers against their conventional analogues. Nat Nanotechnol 13:677–684

Khanna R, Gupta S (2018) Agrochemicals as a potential cause of ground water pollution: a review. Int J Chem Stud 6:985–990

Khater HF (2012) Prospects of botanical biopesticides in insect pest management. Pharmacologia 3:641–656

Kelly WL (2008) Intramolecular cyclizations of polyketide biosynthesis: mining for a "Diels-Alderase"? Org Biomol Chem 6:4483–4493

Klebe G (2000) Recent developments in structure-based drug design. J Mol Med 78:269–281

Koehler PG, Belmont RA (1998) Basic pesticide training manual (SM-59). Ohio State University, Columbus, US

Koprna R, De Diego N, Dundálková L, Spíchal L (2016) Use of cytokinins as agrochemicals. Bioorg Med Chem 24:484–492

Kraus W, Bokel M, Bruhn A, Cramer R, Klaiber I, Klenk A, Nagl G, Poehnl H, Sadlo H, Volger B (1987) Structure determination by NMR of azadirachtin and related compounds from Azadirachta indica A. Juss (Meliaceae). Tetrahedron 43:2817–2830

Kumar M, Shamsi T, Parveen R, Fatima S (2017) Application of nanotechnology in enhancement of crop productivity and integrated pest management. In: Prasad R, Kumar M, Kumar V (eds) Nanotechnology: an agricultural paradigm. Springer, Singapore, SG, pp 361–371

Kurihara N, Miyamoto J (eds) (1998) Chirality in agrochemicals. Wiley, Chichester, GB

Kusano M, Koshino H, Uzawa J, Fujioka S, Kawano T, Kimura Y (2000) Nematicidal alkaloids and related compounds produced by the fungus Penicillium cf. simplicissimum. Biosci Biotechnol Biochem 64:2559–2568

Lamberth C, Dinges J (2012) The significance of heterocycles for pharmaceuticals and agrochemicals. In: Lamberth C, Dinges J (eds) Bioactive heterocyclic compound classes: agrochemicals. Wiley-VCH, Weinheim, DE, pp 3–20

Lamberth C, Jeanmart S, Luksch T, Plant A (2013) Current challenges and trends in the discovery of agrochemicals. Science 341:742–746

Ley SV (2005) Development of methods suitable for natural product synthesis: the azadirachtin story. Pure Appl Chem 77:1115–1130

Li H, Guan A, Huang G, Liu CL, Li Z, Xie Y, Lan J (2016a) Design, synthesis and structure-activity relationship of novel diphenylamine derivatives. Bioorg Med Chem 24:453–461

Li L, Li Z, Wang K, Liu Y, Li Y, Wang Q (2016b) Synthesis and antiviral, insecticidal, and fungicidal activities of gossypol derivatives containing alkylimine, oxime or hydrazine moiety. Bioorg Med Chem 24:474–483

Lindell SD, Pattenden LC, Shannon J (2009) Combinatorial chemistry in the agrosciences. Bioorg Med Chem 17:4035–4046

Londershausen M (1996) Review: approaches to new parasiticides. Pest Manage Sci 48:269–292

Loso MR, Benko Z, Buysse A, Johnson TC, Nugent BM, Rogers RB, Sparks TC, Wang NX, Watson GB, Zhu Y (2016) SAR studies toward the pyridine moiety of the sap-feeding insecticide sulfoxaflor. Bioorg Med Chem 24:378–382

Lv P, Chen Y, Shi T, Wu X, Qing XL, Hua R (2018) Synthesis and fungicidal activities of san-guinarine derivatives. Pestic Biochem Physiol 147:3–10

McCann SF, Annis GD, Shapiro R, Piotrowski DW, Lahm GP, Long JK, Lee KC, Hughes MM, Myers BJ, Griswold SM, Reeves BM, March RW, Sharpe PL, Lowder P, Barnette WE, Wing KD (2001) The discovery of indoxacarb: oxadiazines as a new class of pyrazoline-type insecticides. Pest Manage Sci 57:153–164

McCann SF, Cordova D, Andaloro JT, Lahm GP (2012) Sodium channel-blocking insecticides: indoxacarb. In: Krämer W, Schirmer U, Jeschker P, Witschel M (eds) Modern crop protection compounds, 2nd edn. Wiley-VCH, Weinheim, DE, pp 1257–1273

McDougall P (2016) The cost of new agrochemical product discovery, development and registration in 1995, 2000, 2005–8 and 2010–2014. R&D expenditure in 2014 and expectations for 2019. Report, Croplife America, Washington, US. Available at https://croplife.org/wp-content/uploads/2016/04/Cost-of-CP-report-FINAL.pdf

Mergott DJ, Frank SA, Roush WR (2004) Total synthesis of (–)-spinosyn A. PNAS 101:11955–11959

Merzendorfer H (2013) Chitin synthesis inhibitors: old molecules and new developments. Insect Sci 20:121–138

Miyano M (1965) Rotenoids. XX. Total synthesis of rotenone. J Am Chem Soc 87:3958–3962

Miyano M, Kobayashi A, Matsui M (1960) Synthese and configurational elucidation of rotenoids. Part XVIII. The total synthesis of the natural rotenone. Bull Agric Chem Soc Jpn 24:540–542

Montilla MP, Agil A, Navarro MC, Jiménez MI, García-Granados A, Parra A, Cabo MM (2003) Antioxidant activity of maslinic acid, a triterpene derivative obtained from Olea europaea. Planta Med 69:472–474

Nakao T, Banba S (2016) Broflanilide: a meta-diamine insecticide with a novel mode of action. Bioorg Med Chem 24:372–377

Nakao T, Banba S, Nomura M, Hirase K (2013) Meta-diamide insecticides acting on distinct sites of RDL GABA receptor from those for conventional noncompetitive antagonists. Insect Biochem Mol Biol 43:366–375

Nejma AB, Znati M, Daich A, Othman M, Lawson AM, Jannet HB (2018) Design and semisynthesis of new herbicide as 1,2,3-triazole derivatives of the natural maslinic acid. Steroids 138:102–107

Nerio LS, Olivero-Verbel J, Stashenko E (2010) Repellent activity of essential oils: a review. Bioresour Technol 101:372–378

Nishikimi Y, Iimori T, Sodeoka M, Shibasaki M (1989) Synthetic studies of azadirachtin. Synthesis of the cyclic acetal intermediate in the naturally occurring form. J Org Chem 54:3354–3359

Nuruzzaman M, Rahman MM, Liu Y, Naidu R (2016) Nanoencapsulation, nano-guard for pesticides: a new window for safe application. J Agric Food Chem 64:1447–1483

Oberlander H, Silhacek DL (1998a) Mode of action of insect growth regulators in Lepidopteran tissue culture. Pest Manage Sci 54:300–302

Oberlander H, Silhacek DL (1998b) New perspectives on the mode of action of benzoylphenyl urea insecticides. In: Ishaaya I, Degheele D (eds) Insecticides with novel modes of action: mechanism and application. Springer, Berlin, DE, pp 92–105

Oerke EC (2006) Crop losses to pests. J Agric Sci 144:31–43

Oh MS, Yang JY, Lee HS (2012) Acaricidal toxicity of 2′-hydroxy-4′-methylacetophenone isolated from *Angelicae koreana* roots and structure-activity relationships of its derivatives. J Agric Food Chem 60:3606–3611

Oliveira JL, Campos EVR, Fraceto LF (2018) Recent developments and challenges for nanoscale formulation of botanical pesticides for use in sustainable agriculture. J Agric Food Chem 66:8898–8913

Omolo MO, Okinyo D, Ndiege IO, Lwande W, Hassanali A (2004) Repellency of essential oils of some Kenyan plants against *Anopheles gambiae*. Phytochemistry 65:2797–2802

PAN Pesticide Database (2016) Pesticide Action Network (PAN), North America (Oakland, CA). Available at www.pesticideinfo.org

Pasteris RJ, Hanagan MA, Bisaha JJ, Finkelstein BL, Hoffman LE, Gregory V, Andreassi JL, Sweigard JA, Klyashchitsky BA, Henry YT, Berger RA (2016) Discovery of oxathiapiprolin, a new oomycete fungicide that targets an oxysterol binding protein. Bioorg Med Chem 24:354–361

Pawar VC, Thaker VS (2006) In vitro efficacy of 75 essential oils against *Aspergillus niger*. Mycoses 49:316–323

Peshin R, Dhawan AK (eds) (2009) Integrated pest management. In: Innovation-development process, vol 1. Springer, Netherlands, NL

Plant A (2010) Crop protection chemistry: challenges and opportunities in the 21st century. Agrow Silver Jubilee Issue, XI–XV

Prasad R, Kumar V, Prasad KS (2014) Nanotechnology in sustainable agriculture: present concerns and future aspects. Afr J Biotech 13:705–713

Queirós L, Vidal T, Nogueira AJA, Gonçalves FJM, Pereira JL (2018) Mixture toxicity assisting the design of eco-friendlier plant protection products: a case study using a commercial herbicide combining nicosulfuron and terbuthylazine. Sci Rep 8:5547

Ray DE (1991) Pesticides derived from plants and other organisms. In: Hayes WJ Jr, Laws ER Jr (eds) Handbook of pesticide toxicology. Academic Press, San Diego, US, pp 585–636

Reyes-Zurita FJ, Pachon-Pena G, Lizárraga D, Rufino-Palomares EE, Cascante M, Lupianez JA (2011) The natural triterpene maslinic acid induces apoptosis in HT-29 colon cancer cells by a JNK-p 53-dependent mechanism. BMC Cancer 11:154

Rios MY, Córdova-Albores LC, Ramírez-Cisneros MA, King-Díaz B, Lotina-Hennsen B, Rivera IL, Miranda-Sánchez D (2018) Phytotoxic potential of *Zanthoxylum affine* and its major compound linarin as a possible natural herbicide. ACS Omega 3:14779–14787

Rodrigues SM, Demokritou P, Dokoozlian N, Hendren CO, Karn B, Mauter MS, Sadik OA, Safarpour M, Unrine JM, Viers J, Welle P, White JC, Wiesner MR, Lowry GV (2017) Nanotechnology for sustainable food production: promising opportunities and scientific challenges. Environ Sci Nano 4:767–781

Roser ME, Ortiz-Ospina E (2017) World population growth. Published online at OurWorldInData.org. Available at https://ourworldindata.org/world-population-growth/

Sakurai K, Tanino K (2015) Synthetic studies on azadirachtin: construction of the ABC ring system via the Diels-Alder reaction of a vinyl allenylsilane derivative. Tetrahedron Lett 56:496–499

Sasaki I, Yamashita K (1979) Synthesis of (–)-rotenone. Agric Biol Chem 43:137–139

Schleifer KJ (2007) Virtual screening in crop protection research. In: Ohkawa H, Miyagawa H, Lee PW (eds) Pesticide chemistry. Wiley-VCH, Weinheim, DE, pp 77–88

Sharpless KB, Kolb HC, Finn MG (2001) Click chemistry: diverse chemical function from a few good reactions. Angew Chem Int Ed Engl 40:2004–2021

Shi H, Tan C, Zhang W, Zhang Z, Long R, Gong J, Luo T, Yang Z (2016) Gold-catalyzed enantio- and diastereoselective syntheses of left fragments of Azadirachtin/Meliacarpin-type limonoids. J Org Chem 81:751–771

Singh D, Singh DJ (1991) Synthesis and antifungal activity of some 4-arylmethylene derivatives of substituted pyrazolones. J Indian Chem Soc 68:165–167

Singh RN, Saratchandra B (2005) The development of botanical products with special reference to seri-ecosystem. Casp J Environ Sci 3:1–8

Singh HP, Batish DR, Kohli RK (2003) Allelopathic interactions and allelochemicals: new possibilities for sustainable weed management. Crit Rev Plant Sci 22:239–311

Singh Z, Kaur J, Kaur R, Hundal SS (2016) Toxic effects of organochlorine pesticides: a review. Am J Biosci 4:11–18

Solomon KR, Stephenson GR, Corrêa CL, Zambrone FAD (2010) Praguicidas e o meio ambiente. ILSI Brasil, São Paulo, Brazil, pp 109–110

Sparks TC, Lorsbach BA (2017) Perspectives on the agrochemical industry and agrochemical discovery. Pest Manage Sci 73:672–677

Stanescu MD (2014) Pesticides: synthesis, activity and environmental aspects. Available at https://www.researchgate.net/publication/291688542_PESTICIDES_Synthesis_Activity_and_Environmental_Aspects

Stocking EM, Williams RM (2003) Chemistry and biology of biosynthetic Diels-Alder reactions. Angew Chem Int Ed Engl 42:3078–3115

Stoner K (2004) Bulletin #7144, Approaches to the biological control of insects. Available at https://extension.umaine.edu/publications/7144e/

Sun F, Zhu F, Shao X, Li Z (2015) One-pot, three-component synthesis of 1,8-naphthyridine derivatives from heterocyclic ketene aminals, malononitrile dimer, and aryl aldehydes. Synlett 26:2306–2312

Syngenta (2014) Introduction to agrochemicals and modern agronomy. Available at http://www.oxfordsynthesiscdt.ox.ac.uk/resources/SBM-CDT-Agrochemistry.pdf

Talbot NJ (2003) On the trail of a cereal killer: exploring the biology of *Magnaporthe grisea*. Annu Rev Microbiol 57:177–202

Tambo GMR, Bellus D (1991) Chirality and crop protection. Angew Chem Int Ed Engl 30:1193–1386

Tan C, Chen W, Mu X, Chen Q, Gong J, Luo T, Yang Z (2015) Synthetic progress toward azadirachtins. 2. Enantio- and diastereoselective synthesis of the right-wing fragment of 11-epi-azadirachtin I. Org Lett 17:2338–2341

Taylor MD, Klaine SJ, Carvalho FP, Barcelò D, Everaarts J (eds) (2003) Pesticide residues in coastal tropical ecosystems. Distribution, fate and effects. Taylor & Francis Publ., CRC Press, London, GB

Tebourbi O, Sakly M, Rhouma KB (2011) Molecular mechanisms of pesticide toxicity. In: Stoytcheva M (ed) Pesticides in the modern world: pests control and pesticides, exposure and toxicity assessment. InTech Publ. Available at http://www.intechopen.com/books/pesticides-in-the-modernworld-pests-control-and-pesticides-exposure-and-toxicity-assessment

Theodoridis G (2006) Fluorine-containing agrochemicals: an overview of recent development. In: Tressaud A (ed) Fluorine and the environment: agrochemicals, archaeology, green chemistry and water. Elsevier, Amsterdam, NL, pp 121–175

Tice CM (2001) Selecting the right compounds for screening: does Lipinski's rule of 5 for pharmaceuticals apply to agrochemicals? Pest Manage Sci 57:3–16

Tice CM (2002) Selecting the right compounds for screening: use of surface-area parameters. Pest Manage Sci 58:219–233

Tomizawa M, Casida JE (2005) Neonicotinoid insecticide toxicology: mechanisms of selective action. Annu Rev Pharmacol Toxicol 45:247–268

Touw WG, Baakman C, Black J, te Beek TAH, Krieger E, Joosten RP, Vriend G (2015) A series of PDB—related databanks for everyday needs. Nucleic Acids Res 43:D364–D368

Turner CJ, Tempesta MS, Taylor RB, Zagorski MG, Termini JS, Schroeder DR, Nakanishi K (1987) An NMR spectroscopic study of azadirachtin and its trimethyl ether. Tetrahedron 43:2789–2803

Ujvary I (2010) Pest control agents from natural products. In: Krieger R (ed) Hayes' handbook of pesticide toxicology, 3rd edn. Academic Press, New York, US, pp 119–229

United Nations (2015) Department of economic and social affairs, population division. World population prospects: the 2015 revision, key findings and advance tables. Working paper no. ESA/P/WP.241, United Nations, New York, US. Available at https://esa.un.org/unpd/wpp/publications/files/key_findings_wpp_2015.pdf

Veitch GE, Beckmann E, Burke BJ, Boyer A, Ayats C, Ley SV (2007) A relay route for the synthesis of azadirachtin. Angew Chem Int Ed Engl 46:7633–7635

Walter MW (2002) Structure-based drug design of agrochemicals. Nat Prod Rep 19:278–291

Walter H (2012) Pyrazole carboxamide fungicides inhibiting succinate dehydrogenase. In: Lamberth C, Dinges J (eds) Bioactive heterocyclic compound classes: agrochemicals. Wiley-VCH, Weinheim, DE, pp 175–193

War AR, Paulraj MG, Ahmad T, Buhroo AA, Hussain B, Ignacimuthu S, Sharma HC (2012) Mechanisms of plant defense against insect herbivores. Plant Signal Behav 7:1306–1320

Whitford F, Pike D, Burroughs F, Hanger G, Johnson B, Brassard D, Blessing A (2006) The pesticides marketplace. Discovery and developing new products. PPP-71. Pardue University, West Lafayette, US. Available at https://www.extension.purdue.edu/extmedia/PPP/PPP-71.pdf

World Health Organization (2012) The WHO recommended classification of pesticides by hazard and guidelines to classification. Geneva, IT

World Health Organization (2017) Agrochemicals, health and environment: directory of resources. Available at http://www.who.int/heli/risks/toxics/chemicalsdirectory/ en/index1.html

Xie Y, Chi HW, Guan AY, Liu CI, Ma HJ, Cui DI (2016) Synthesis and evaluation of substituted 3-(pyridine-2-yl)benzenesulfonamide derivatives as potent herbicidal agents. Bioorg Med Chem 24:428–434

Yang Y, Liu Y, Song H, Li S, Wang Q (2016) Additive effects on the improvement of insecticidal activity: design, synthesis and insecticidal activity of novel pymetrozine derivatives. Bioorg Med Chem 24:391–402

Yang R, Guo Y, Zhang Y, Xu H (2018) Semisynthesis of new ethers from furyl-ring-based acylation derivatives of fraxinellone as insecticidal agents against *Mythimna separata* Walker in vivo. Chin Chem Lett 6:995–997

Yin J, Wang Y, Gilbertson LM (2018) Opportunities to advance sustainable design of nano-enabled agriculture identified through a literature review. Environ Sci Nano 5:11–26

Yu LPC, Kim YS, Tong L (2010) Mechanisms for the inhibition of the carboxyltransferase domain of acetyl-coenzyme A carboxylase by pinoxaden. PNAS 107:22072–22077

Yu X, Liu Y, Li Y, Wang Q (2015) Design, synthesis, and acaricidal/insecticidal activities of oxazoline derivatives containing a sulfur ether moiety. J Agric Food Chem 63:9690–9695

Zhang YY, Xu H (2017) Recent progress in the chemistry and biology of limonoids. RSC Adv 7:35191–35220

Zhang L, Yan C, Guo Q, Zhang J, Ruiz-Menjivar J (2018) The impact of agricultural chemical inputs on environment: global evidence from informetrics analysis and visualization. Int J Low-Carbon Tech 13:338–352

Zhao X, Cui H, Wang Y, Sun C, Cui B, Zeng Z (2018) Development strategies and prospects of nano-based smart pesticide formulation. J Agric Food Chem 66:6504–6512

Chapter 9
Toxicological Aspects of Pesticides

Eloisa Dutra Caldas

Abstract Over 3 million tons of pesticide active ingredients were used worldwide in 2016, about 60% in China. Pesticides are primarily used in agriculture, but also to control disease insect vectors, domestic pests, and home gardening, among other uses. The general population is exposed to pesticides mainly through the consumption of treated food. Dietary intake assessment can be performed for specific population groups, such as toddlers, teenagers, and seniors, considering chronic and/or acute exposure, and may be combined with residential and water exposure. Most dietary exposure assessments conducted worldwide indicate low health risks to consumers. The main exposure route of pesticides for the occupational population is through skin contact, which occurs mainly from spilling pesticides and drift while mixing or applying pesticides. Epidemiological studies have shown association between pesticide exposure and various diseases, mainly cancer, although there is a lack of consensus among scientists on the link between pesticide exposure and cancer. The environment can also be affected by pesticide use, as the products may move off-site via spray drift, leaching, and runoff and affect non-target aquatic and terrestrial organisms. Pesticides with high water solubility and mobility have the potential to reach groundwater, and those with low mobility have a tendency to remain in soil/sediments, but they can reach water systems through superficial runoff. Exposure of the biota to pesticides at sublethal levels during the embryonic period may interfere with the development and survival of embryos and hatchling, with a potential impact on population survival, biodiversity, and ecological equilibrium.

Keywords Toxicity · Dietary exposure · Occupational exposure · Environment · Biota

E. D. Caldas (✉)
Department of Pharmacy, Faculty of Health Sciences, University of Brasilia,
Campus Darci Ribeiro, 70910-900 Brasilia, DF, Brazil
e-mail: eloisa@unb.br

© Springer Nature Switzerland AG 2019
S. Vaz Jr. (ed.), *Sustainable Agrochemistry*,
https://doi.org/10.1007/978-3-030-17891-8_9

9.1   Definition of Main Terms in Toxicology

Definitions were taken from different sources: IPCS (2004a), Duffus et al. (2007) and US EPA (2012).

Acceptable daily intake (ADI)—Estimated maximum amount of an agent, expressed on a body mass basis, to which individuals in a (sub)population may be exposed daily over their lifetimes without appreciable health risk.

Active ingredient—Component responsible for the biological effects of the pesticide.

Acute/short-term exposure—Exposure of short duration, normally a single event or during the period of 24 h.

Acute reference dose (ARfD)—Estimate of the amount of a substance in food and/or drinking—water, normally expressed on a body—weight basis, that can be ingested in a period of 24 h or less, without appreciable health risk to the consumer, on the basis of all the known facts at the time of the evaluation.

Adverse effect—Change in the morphology, physiology, growth, development, reproduction, or life span of an organism, system, or (sub)population that results in an impairment of functional capacity, an impairment of the capacity to compensate for additional stress, or an increase in susceptibility to other influences.

Benchmark dose (BMDL)—Statistically calculated lower 95% confidence limit on the dose that produces a defined response (usually 5 or 10%) of an adverse effect compared to background.

Bioaccumulation—Progressive increase in the amount of a substance in an organism or part of an organism that occurs because the rate of intake exceeds the organism's ability to remove the substance from the body.

Biodegradation—Breakdown of a substance catalyzed by enzymes in vitro or in vivo.

Codex Alimentarius—Collection of internationally adopted food standards drawn up by the Codex Alimentarius Commission, the principal body implementing the joint FAO/WHO Food Standards Program.

Chronic/long-term exposure—Continued exposure or exposures occurring over an extended period of time, or a significant fraction of the test species' or of the group of individuals', or of the population's lifetime.

Effective concentration (EC)—Concentration of a substance that causes a defined magnitude of *response* in a given system. EC50 is the median concentration that causes 50% of maximal response.

Ecotoxicologically relevant concentration (ERC)—Concentration of a pesticide (active ingredient, formulations, and relevant metabolites) that is likely to affect a determinable ecological characteristic of an exposed system.

Estimated/expected environmental concentration (EEC)—Predicted concentration of a substance, typically a pesticide, within an environmental compartment based on estimates of quantities released, discharge patterns, and inherent disposition of the substance (fate and distribution) as well as the nature of the specific receiving ecosystems.

Good agricultural practice (GAP) in the use of pesticides—Nationally authorized safe uses of pesticides under actual conditions necessary for effective and reliable pest control.

Half-life, $t_{1/2}$—Time required for the concentration of a substance in a given reaction to reach a value that is the arithmetic mean of its initial and final (equilibrium) values.

Hazard—Set of inherent properties of a substance, mixture of substances, or a process involving substances that, under production, usage, or disposal conditions, make it capable of causing *adverse effects* to organisms or the environment, depending on the degree of *exposure*.

Hazard quotient (HQ)—Ratio of toxicant exposure to a reference value regarded as corresponding to a threshold of toxicity. A hazard quotient of less than 1 indicates that no adverse effects are likely over a lifetime of exposure.

Lethal concentration (LC_{50})—Statistically derived median *concentration* of a substance in an environmental medium expected to kill 50% of organisms in a given population under a defined set of conditions.

Lethal dose (LD_{50})—Statistically derived median *dose* of a chemical or physical agent (radiation) expected to kill 50% of organisms in a given population under a defined set of conditions.

Maximum residue limit (MRL)—Maximum contents of a pesticide residue ($mg\,kg^{-1}$) to be legally permitted in or on food commodities and animal feeds. MRLs are based on data obtained following good agricultural practice. Foods derived from commodities that comply with the respective MRLs are intended to be safe.

Non-target organism—Organism affected by a pesticide although not the intended object of its use.

No-observed-adverse-effect concentration or level (NOAEC/NOAEL)—Greatest concentration or dose of a substance which causes no detectable adverse alteration of morphology, functional capacity, growth, development, or life span of an organism under defined conditions of exposure.

Octan-1-ol–water partition coefficient, Kow—Ratio of the solubility of a chemical in octan-1-ol divided by its solubility in water.

Pesticide residue—Any substance or mixture of substances found in humans or animals or in food and water following use of a pesticide: The term includes any specified derivatives, such as degradation and conversion products, metabolites, reaction products, and impurities considered to be of toxicological significance.

Risk—Probability of adverse effects caused under specified circumstances by an agent in an organism, a population, or an ecological system.

Safety/uncertainty factor—Value used in extrapolation from experimental animals to man (assuming that man may be more sensitive) or from selected individuals to the general population. For example, a value applied to the *no-observed-adverse-effect level* (NOAEL) to derive an *acceptable daily intake* (ADI) or *tolerable daily intake* (TDI).

9.2   Human Exposure to Pesticides

In 2016, over 3 million tons of pesticide active ingredients were used worldwide, mainly in China (1.8 million tons), USA (408,000 tons), and Brazil (377,000 tons) (FAO 2018a). Pesticide use occurs in rural and urban areas, during the production, storage, and transport of food commodities, in wood production, for the control of human disease insect vectors, in domestic settings for killing nuisance pests, in home gardening, and in veterinary medicine. Some pesticide products are also used directly in human for lice and mite control (WHO 2006a).

Humans are exposed to pesticides through three main routes—oral, dermal, and respiratory (inhalation). Although all populations are exposed to some extent to all routes, oral exposure is mostly relevant to the general population, mainly through the consumption of treated food and contaminated water, and dermal and inhalation routes are more relevant to the occupational population. About 45% of the world's population live in rural areas from which ~80% are farmers (FAO 2018b). In addition to the use of pesticide during farming, occupational activities that are source for pesticide exposure include public health campaigns to control insect-borne diseases, wood preservation and park maintenance, and pesticide manufacture (Gangemi et al. 2016). Additionally, the general population can be exposed to domestic and residential pesticides (e.g., treatment of insects or weeds in the garden), and non-occupational exposure can occur to those living in agricultural areas or in residences nearby.

The level of pesticide exposure can be estimated by measuring concentrations in food, water, residential and occupational environments to which a population or an individual is in contact (external exposure approach), or to monitor the levels of pesticides, and/or their metabolites, in biological samples (internal exposure approach).

9.2.1   Oral Exposure to Pesticides

Human exposure to pesticides starts in the uterus, because of the mother exposure to these compounds from all sources (dietary, environmental, residential and, in some cases, occupational) and continues during lactation at early infancy and during the whole life, mainly from the consumption of treated food.

The dietary risks from the exposure to pesticides can be estimated within the context of the risk (or safety) assessment process that also includes the hazard identification and characterization, the exposure (intake) assessment, and the risk characterization steps (IPCS 2009). Pesticide dietary risk assessment is conducted at international level by the FAO/WHO Joint Meeting on Pesticide Residues (JMPR) (Ambrus 2016) and by national or regional authorities within the regulatory process (Humphrey et al. 2017). Pesticide dietary risk assessment conducted by the JMPR supports the establishment of maximum residue limits (MRLs) by the Codex Alimentarius, which have the objective of guarantee a fair international trade of food commodities and safety food to consumers (Codex Alimentarius 2018).

In the identification/hazard characterization steps, the no-observed-adverse-effect level (NOAEL) found for the identified critical effect, normally in animal model (mostly rodents), is divided by a safety factor, generally 100, which accounts for the extrapolation between the rodent species to human ($10\times$) and the variability within the human species ($10\times$) to derive the accepted daily intake (ADI) or reference dose (RfD) for chronic exposure and the acute reference dose (ARfD) for acute exposure (WHO 2015).

The exposure assessment (dietary intake) is estimated using information on food consumption per body weight (bw) and residue of the pesticide in the food, according to Eq. 9.1 (IPCS 2009).

$$\text{Intake} = \frac{\text{food consumption} \times \text{pesticide residue concentration}}{\text{body weight}} \tag{9.1}$$

The estimation can be performed for the total population, or for specific population groups, such as toddlers, teenagers, and seniors, which have different dietary patterns. The estimation can consider long-term (chronic, lifetime) and/or short-term (acute, within 24 h) exposure, be performed for a single pesticide or for a group of compounds with the same mechanism of action or same effect (cumulative dietary exposure), and may be combined with other routes (aggregate exposure, which include residential and water) (IPCS 2009; US EPA 2001; Caldas 2017). More recently, a less-than-lifetime exposure has also been considered (FAO/WHO JMPR 2018).

Food consumption information can be obtained at a population level from food production statistics, which represent foods available for consumption by the whole population, typically in the form as the food is produced, and include the 17 GEMS/Food Cluster Diets used by the JMPR to estimate pesticide dietary exposure at international level (Ambrus 2016; Caldas 2017). However, a more refined data can be obtained through household budget and individual food consumption surveys conducted at national level (Valsta et al. 2017).

Pesticide residue data in food to estimate dietary intake can be obtained from various sources. For chronic exposure, MRL can be used in a tier 1 approach, although it greatly overestimates the exposure. The MRL reflects the maximum level of a pesticide expected to be found in food when treated according to most critical good agricultural practice (cGAP), including a maximum application rate, and the shortest application and post-harvest intervals. Using MRL as a proxy for concentration of the residue in Eq. 9.1 assumes a daily consumption of all foods for which the pesticide is registered, that all foods contain residue levels at the MRL and that there is no degradation of the pesticide during transport, storage, or food processing (e.g., peeling, milling, cooking) (IPCS 2009). The JMPR uses the supervised trial median residue (STMR) from trials conducted according to the cGAP to estimate the International Estimated Daily Intake (IEDI) for chronic exposure to pesticides at international level (Ambrus 2016; FAO/WHO JMPR 2018). At national or regional level, the best approach is to use pesticide residue monitoring data (Boon et al. 2015; Sieke et al. 2018; Jardim et al. 2018a, b).

While in most cases, dietary exposure is estimated for individual pesticides, and those with the same mechanism of toxicity can be grouped to form a cumulative assessment group (CAG), for which the total intake is estimated. In cumulating residues of pesticides belonging to one CAG, the residue of each pesticide is multiplied by its relative potency factor related to the index compound of the group (US EPA 2017a; EFSA 2009). The Environmental Protection Agency of the United States of America (US EPA) has established CAG for organophosphorus, N-methyl carbamates, pyrethrins/pyrethroids, triazines, and chloroacetanilides (US EPA 2016), and a CAG for triazole compounds was proposed by the European Food Safety Authority (EFSA 2009). Some authors grouped nicotinoid pesticides (Chang et al. 2018a) and dithiocarbamates (Jardim et al. 2018a; Sieke et al. 2018) pesticides to estimate the cumulative exposure. Larsson et al. (2018) grouped pesticides from different chemical group and mechanisms of action to estimate total exposure.

The dietary intake can be estimated using deterministic (point estimate) or probabilistic approaches. In the deterministic approach, a single value for each parameter in Eq. 9.1 is used for the estimation. For chronic exposure, the mean food consumption and body weight and a mean or median pesticide residue values are used, and the intakes from each food are added to calculate the total dietary exposure. The JMPR uses the deterministic approach to estimate chronic (IEDI) and acute exposures (International Estimate of Short-Term Intake, IESTI). The IEDI is the summation of the intakes from the consumption of all foods for which a MRL was recommended, using the 17 cluster diets and the STMR values. The IESTI is calculated for each food crop using the 97.5th of a consumption distribution (provided by national authorities) and a highest residue (HR) or a STMR value from supervised trials (Ambrus 2016). In the IESTI, a variability factor (equal to 3) of the residues in individual units of a food (e.g., apple) is also considered in Eq. 9.1. The European Union uses the MRL as residue concentration and higher variability factors for the IESTI calculation (EFSA 2015a).

In the probabilistic approach to estimate dietary exposure to pesticides, at least one variable is represented by a distribution of values, instead of a single value, and a total dietary intake distribution (chronic or acute) is generated after several thousand iterations. This approach represents a more refined and realistic assessment as it considers individual members of the population that experience different levels of exposure, mainly due to personal dietary preferences (IPCS 2009). There are a number of models used for probabilistic dietary exposure to pesticides including the Monte Carlo Risk Assessment (MCRA) software, developed by Biometris, Wageningen University and Research Centre and the National Institute for Public Health and the Environment (RIVM) in the Netherlands (de Boer et al. 2016) and the DEEM (Dietary Exposure Evaluation Model)/CALENDEX^{TM}—FCID developed by the US EPA (2018a).

The last step of the dietary risk assessment of pesticides is the risk characterization, when the estimated chronic intake is compared with the ADI and the acute intake with the ARfD. Risk may exist when the intake exceeds the ADI and/or the ARfD. When the probabilistic approach is used in the exposure assessment, a given percentile of the intake distribution (normally higher than 95th) is used for the risk characterization.

Table 9.1 Summary results of dietary risk assessment of pesticides conducted worldwide[a]

Country	Method; model	Pesticide or CAG	Results
Argentina, Maggioni et al. (2018)	Deterministic; IESTI (MRL)	2–5 years: 28 pesticides/75 foods	39.1% pesticide/food combinations > ARfD
		10–49 years: 9 pesticides/59 foods	31.3% pesticide/food combinations > ARfD
Brazil, Jardim et al. (2018a, b)	Probabilistic; MCRA[b]	Triazoles	0.9% ADI; 0.5% ARfD[c]
		Dithiocarbamates	5.3% ADI
		Organophosphorus	36% ARfD
		Carbamates	9% ARfD
		Pyrethroids	2% ARfD
China, Li et al. (2017)	Probabilistic; MCRA[b] (F&V)	Organophosphorus and carbamates	36.6% ARfD (<7 years) 31.4% ARfD (7–18 years) 10.2% ARfD (>18 years)
Denmark, Larsson et al. (2018)	Deterministic; cumulative	42 pesticides	43.6% ADI (4–6 years) 15.9% ADI (adults)
Germany, Sieke et al. (2018)	Probabilistic; MCRA[b]	Dithiocarbamates Imazalil Chlorpyrifos	32% ADI 46% ADI; 23% ARfD 21% ADI; 99% ARfD
USA, Chang et al. (2018a)	Probabilistic; MIXTRAN[b] (F&V)	Neocotinoids	35% RfD (chronic)
Europe, Stephenson et al. (2018)	Deterministic; IESTI (STMR, HR or MRL) Probabilistic; MCRA[b]	Glyphosate	126% ARfD for wild fungi (German children, MRL) Up to 1.1% ARfD
International (FAO/WHO JMPR 2017)	Deterministic; IEDI (17 Cluster diets) IESTI (STMR or HR)	27 pesticides 19 pesticides	from 0 to 80% ADI from 110 to 310% ARfD for fenpyroximate (5 food crops)

[a]Unless specified, individual consumption and monitoring residue data were considered to estimate the intake, for the total population, and include all foods for which data is available
[b]Risk characterization performed at 99.9th percentile of the intake distribution
[c]For women of child-bearing age; ADI: accepted daily intake; ARfD: acute reference dose; CAG: cumulative assessment group; IEDI: International Estimated Daily Intake Estimation; IESTI: International Estimate of Short-Term Intake

Table 9.1 summarizes some dietary risk assessment studies conducted worldwide and by the JMPR. The deterministic IESTI approach, mainly using MRL as residue level, gives the worst-risk scenario, and in all cases, probabilistic assessments to estimate the exposure indicate no health risks at 99.9th percentile.

Another approach to characterize the risk from the exposure to pesticides is to divide the toxicological parameter, normally the benchmark dose (BMDL), by the total intake to estimate the margin of exposure (MOE). Risk may exist when the MOE is lower than a target value (Colnot and Dekant 2017). This approach is used by the US EPA to estimate the cumulative/aggregated exposure to pesticides using probabilistic models (US EPA 2017a). For the CAG of pyrethroids/pyrethrins (deltamethrin as index compound), the minimum MOE at the 99.9th percentile (dietary and residential acute exposures) was 420 for children (1–2 years) and 810 for adults, higher than the target levels of 300 and 100, respectively. For the N-methyl carbamate CAG (oxamyl as index compound), the MOE at the 99.9th percentile (dietary, residential, and water acute exposures) was approximately 8 for all populations, lower than the target value of 10. Food was the main source of exposure in both cases, followed by residential and water (from surface water sources) (US EPA 2007).

Exposure to pesticides from different sources can also be estimated by analyzing biological samples for pesticides and their metabolites (internal exposure), mainly urine. With this purpose, Sinha and Banda (2018) analyzed dialkyl phosphate metabolites (DAP) of organophosphorus pesticides (OP) in urine samples from 377 Indian children aged 6–15 years. Girls showed higher DAP detection frequency than boys, and those aged 11–15 years had significantly higher mean levels than boys (4.98 and 2.48 μmol L^{-1}, respectively), probably due to their higher fruit consumption. DAP levels found in this study were much higher than what was found in Canada, USA, and Italy (<0.4 μmol L^{-1}; Sinha and Banda 2018). Using literature data on the urinary DAP excretion in various countries (occupational and environmental exposure), Katsikantami et al. (2019) estimated the highest intake of OP for farmers, followed by children and pregnant women; in all cases, the hazard index was below 1, indicating no health concerns.

In a study conducted with 6–11 years old children from Spain, 3,5,6-trichloro-2-pyridinol (TCPy, metabolite of chlorpyrifos), diethyl phosphate (DEP, generic metabolite of organophosphorus), and 2-isopropyl-4-methyl-6-hydroxypyrimidine (IMPY, metabolite of diazinon) were present in most of the 125 urine samples (Roca et al. 2014). TCPy and IMPY had the highest mean levels, mainly associated with age, vegetable consumption, and pesticide residential use; home location and vegetable consumption were associated with DEP levels.

9.2.2 Dermal and Inhalation Exposures to Pesticides

The main exposure route for occupational pesticide users is through skin contact and occurs from spilling pesticides on unprotected skin, by drift while mixing or applying pesticides, by wearing pesticide-contaminated clothing, and during cleaning of pesticide application equipment (MacFarlane et al. 2013). Furthermore, agricultural workers are also exposed to pesticides during reentry activities, such as removing branches and wires, and during crop harvesting. The amount of pesticide absorbed depends on the pesticide formulation (emulsifiable concentrates are mostly readily absorbed) and on the part of the body affected (the genital area and the scalp are highly absorptive, while the hands are more resistant to absorption) (WHO 2006b). Although absorption by the respiratory tract appears to be more limited due to the low vapor pressures of most pesticides, protecting the lungs is especially important since pesticide powders, dusts, gases, vapors, or very small spray droplets can be inhaled during pesticide handling, especially in confined areas.

During occupational agricultural activities, the use of automated equipment can minimize pesticide exposure, including closed mixing and loading systems and spray tractors with enclosed cabs. Closed transfer systems allow pesticides to be transferred directly from the container into the sprayer via a closed route, and air-filtering systems in enclosed tractor cabs also provide respiratory protection for the pesticide applicator. Other engineering control systems, such as low-drift nozzles, handwash water supply, and tank rinse systems, are also important to reduce exposure during pesticide handling (Coffman et al. 2009).

Personal protective equipment (PPE), such as respirators with appropriated filters, gloves, coverall and chemically resistant aprons, boots, and eyewear, can provide protection in situations where closed systems are not available (IPCS 2004b). Figure 9.1 shows farmers applying pesticides using a backpack sprayer (top) or with the help of a tractor (below). The farmer applying the pesticide using a backpack sprayer uses a mask, but inappropriate cap, clothes, and boots. The farmer behind the truck uses a mask (not shown), but the truck driver is totally exposed to the pesticide spray.

Tsakirakis et al. (2014) estimated the dermal and inhalation exposure of operators during fungicide application to Greek vineyards using a handheld single-nozzle spray guns connected to a tractor tank. The dermal exposure was measured using the whole-body dosimetry method, and the inhalation exposure was measured with the personal air sample devices with XAD tubes on the operator's breathing zone. Pesticide dermal exposure levels were mostly derived from the exposure of the head (89%), with hand exposure accounting from less than 10%. The mean inhalation exposure accounted for about 0.2% of the total exposure. The use of protective coveralls provided about 98.4% protection for the operators. A study conducted with 702 certified pesticide applicators in USA showed that applicators working on large farms, users of boom and hydraulic sprayers, and growers of field crops were more likely to use engineering devices. Respondents reported a high level of PPE use, with chemical-resistant gloves showing the highest level of compliance (Coffman et al. 2009).

However, in many countries, the compliance of PPE use is rather low. In the Central-West Region of Brazil, less than 20% of the farmers used masks, impermeable clothes, or gloves during pesticide application, with most complaining that these devices are too uncomfortable to be used in the warm climate (Recena et al. 2006). In a study conducted in the Federal District of Brazil, 44.6% of the farmers use complete PPE, and 16.7% of them showed blood acetylcholinesterase depletion over 30% (a biomarker of organophosphorus and carbamate exposure) during the exposure period

Fig. 9.1 Application of pesticides using backpack sprayer in a Brazilian papaya orchard in the state of Bahia (top), and a farmer applying the pesticide carried by a tractor in a mango orchard in the Federal District of Brazil (below). Photographs taken by the author

compared with the baseline level (non-exposure period), which indicates a risk to health (Pasiani et al. 2012). In this situation, the worker should be removed from the activities involving pesticide until the blood levels return to normal (OEHHA 2015). Ribeiro et al. (2012) found that greenhouse workers in the São Paulo State of Brazil may be at higher risk of pesticide exposure due to inadequate welfare facilities, poor pesticide storage, use and disposal conditions, use of highly toxic pesticides and incorrect use and maintenance of PPE and lack of control on reentry intervals after pesticide application. In Iran, about one-third of the farmers showed unwillingness or were unsure about using PPE, mainly due to the low availability and high price of these devices (Sharifzadeh et al. 2017). In Greece, about half of the farmers showed potentially unsafe behavior with respect to PPE, and few used respirator; those who had experienced an episode of intoxication used PPE more frequently, and elderly farmers tended not to use them (Damalas and Abdollahzadeh 2016).

Dermal contact and inhalation of dust in the residence can be an important source of exposure to pesticides for non-occupational population and occupational population out of working hours, mainly due to spray drifting from the field (Pasiani et al. 2012; Butler-Dawson et al. 2018; Deziel et al. 2017, 2018). Using data from studies conducted in the USA from 1995 to 2015, Deziel et al. (2017) found that homes near-treated fields, homes of farmers who applied pesticides more frequently or recently, and of those who applied pesticides around the house had higher pesticide concentrations in the dust compared to their reference groups. In the study conducted in the Federal District of Brazil (Pasiani et al. 2012), one-third of the residents living on the farms also had significant blood acetylcholinesterase depletion compared with their control group and to their baseline level measured when the farmers were not exposed. Figure 9.2 shows a Brazilian farmer using no PPE applying the pesticide in a field very close to the residence.

Pesticides used in and around homes, schools, offices, golf courts, and other urban areas may result in potential human exposure via the oral (hand-to-mouth activity by children), dermal, and inhalation routes, during or after the application. Examples of post-application activities include weeding and harvesting gardens, mowing and playing on lawns, and playing golf. Furthermore, some pesticides (such as the N-methyl carbamates carbaryl and propoxur) are formulated as impregnated pet collars, and adults and children can be exposed when in contact with pets (US EPA 2007).

Hung et al. (2018) found cypermethrin (a pyrethroid) and chlorpyrifos (a organophosphorus) in indoor and outdoor dust collected from homes in a rural county of Taiwan (80% of the indoor samples and 40–48% of outdoor samples, respectively). Permethrin, prallethrin, and tetramethrin, other pyrethroid residential insecticides, were also frequently found in indoor dust. Permethrin and cypermethrin were the main pyrethroids found in residential material, including carpet dust, hard floor surface wipes, vacuum dust bags, food and air from various USA locations, at levels that may represent a risk to the children (Morgan 2012; Lu et al. 2013). The pyrethroid metabolite 3-phenoxybenzoic acid was detected in over 67% of the urine samples from children 2–11 years of age (Morgan 2012).

Fig. 9.2 Application of pesticides using backpack sprayer in a familiar system in the Central-West Region of Brazil (Caldas 2016)

9.3 Toxicological Effects of Pesticides to Humans

All pesticides possess an inherent toxicity to some living organism to be of practical use. However, their selectivity to the target species is not always totally developed, and they can present a health risk to humans and other organisms. Pesticides can cause a wide effect on humans, including cancer, neurotoxicity, pulmonotoxicity, reproductive and development toxicity, and metabolic toxicity (Gangemi et al. 2016). Common mechanisms include oxidative stress, mitochondrial dysfunction, inflammatory responses, immune dysregulation, and endocrine disruption. Most of the evidences come from epidemiological studies with the occupational population, but studies with the general population are also reported, including children (Mostafalou and Abdollahi 2017).

9.3.1 Cancer

The largest set of epidemiological studies that investigated the health impact of pesticides on human health is related to cancer, accounting for 54% of the studies reviewed by Mostafalou and Abdollahi (2017), and included brain cancer (adult and children), leukemia (adult and children), esophageal, stomach and colorectal cancers, liver cancer, non-hodgkin lymphoma, and bladder cancer.

Bonner et al. (2017) evaluated the association of 43 pesticides and 654 lung cancer cases after 10 years of additional follow-up in the Agricultural Health Study, a prospective cohort study comprising 57,310 pesticide applicators from Iowa and North Carolina, USA. The authors found additional evidence for an association between pendimethalin, dieldrin, and parathion use and lung cancer risk, in addition to chlorimuron ethyl, an herbicide for which lung cancer had not been previously reported. Another cohort study (AGRIculture and CANcer) followed for about 7 years 181,842 French farmers and found 381 incident cases of brain tumors, with hazard ratios for gliomas ranging from 1.18 for thiofanox to 4.60 for formetanate, and for meningiomas from 1.51 for carbaryl to 3.67 for thiofanox (Piel et al. 2018). van Maele-Fabry et al. (2017, 2019) found associations between residential exposure to pesticides and childhood leukemia and brain tumors, and although causality could not be established, the results indicated the need for limiting the use of household pesticides during pregnancy and childhood.

DDT (p,p'-dichloro-diphenyl-trichloroethane) is a very efficient insecticide introduced in the early 1940s to control agricultural pests and vectors for diseases, such as malaria. The compound was widely used until it was banned for agricultural use in most countries in the 1970s and 1980s due to its impact on the environment (Jarman and Ballschmiter 2012). DDT is highly persistent in the environment and in biological system, with DDE (p,p'-dichloro-diphenyl-dichloroethylene) the main degradation product/metabolite. Due to its estrogenic properties, since the early 1990s, various studies have investigated the association between and DDT and/or DDE and the risk of cancer, particularly breast cancer, although the subject still lacks a consensus among researchers. With the objective to elucidate the contradicting results, probably due to methodological differences, López-Cervantes et al. (2004) performed a meta-analysis of 22 studies conducted in various countries up to 2001 regarding DDT and breast cancer. The summary odds ratio was 0.97, and the authors concluded that there was a strong evidence to discard the relationship between DDE and breast cancer risk. Ingber et al. (2013) updated this meta-analysis by including studies published through June 2012 and confirmed the previous conclusion that the existing information does not support the hypothesis that exposure to DDT/DDE increases the risk of breast cancer in humans.

However, in the study conducted by Cohn et al. (2007), which was included in the meta-analysis conducted by Ingber et al., blood samples collected from women during the peak DDT use period in USA (1959–1967) showed high levels of serum DDT, and the authors reported a statistically significant fivefold increased risk of breast cancer among women who were born after 1931. Furthermore, Cohn et al. (2018) revisited the same study and suggested that vulnerability to breast cancer before the age of 50 may be associated with an inverse correlation between body mass index and serum DDTs. Chang et al. (2018b) found that women born between 1951 and 1959 exposed to DDT (for malaria control) in Taiwan had an increased risk of breast cancer in adulthood. Cohn et al. (2015) linked the o,p'-DDT (a minor constituent of technical DDT) exposure in utero to the risk of breast cancer, a result that needs further investigation (Paumgartten 2015).

Another controversial issue is glyphosate, the herbicide most used worldwide. In 2015, the International Agency for Research on Cancer (IARC) classified glyphosate as "probably carcinogenic to humans" (Group 2A), due to sufficient evidence of carcinogenicity in animals, limited evidence of carcinogenicity in humans (increased risks for non-Hodgkin lymphoma), and strong evidence for two carcinogenic mechanisms (Guyton et al. 2015). Although this conclusion had support from many researchers (Portier et al. 2016; Myers et al. 2016), others contested the IARC classification (Acquavella et al. 2016; Tarone 2018) and national authorities, including the US EPA (2017b), the Brazilian Health Regulatory Agency (ANVISA 2019) and the Australian Pesticides and Veterinary Medicines Authority (APVMA 2016), the EFSA (2015b), the European Chemical Agency (ECHA 2017), and the FAO/WHO JMPR (2016), have concluded that current data support the conclusion that glyphosate is unlikely to cause cancer in human. The IARC conclusion, however, has raised public

Table 9.2 Classification of pesticides regarding their carcinogenicity to humans according to the International Agency for Research on Cancer (IARC 2018)

Pesticide (year)	Classification	Criteria
Lindane (2018)	I (carcinogenic)	Sufficient evidence of carcinogenicity in humans
DDT (2018)	2A (probably carcinogenic)	Limited evidence of carcinogenicity in humans and sufficient evidence of carcinogenicity in experimental animals
Diazinon (2015/2017)		
Dieldrin and aldrin metabolized do dieldrin (in preparation)		
Glyphosate (2015/2017)		
Malathion (2015/2017)		
Parathion (2015/2017)		
2,4-D (2015)	2B (possibly carcinogenic)	Limited evidence of carcinogenicity in humans and less than sufficient evidence of carcinogenicity in experimental animals
Hexachlorocyclohexanes (1987)		
Tetrachlorvinfos (2015/2017)		
Aldicarb (1991)	3 (not classifiable as to its carcinogenicity)	Evidence of carcinogenicity is inadequate in humans and inadequate or limited in experimental animals
Atrazine (1999)		
Captan (1987)		
Carbaryl (1987)		
Chlorpyrifos (1987)		
Dicofol (1987)		
Maneb (1987)		
Methyl parathion (1987)		

concerns around the world and the glyphosate registration status has been reconsidered by the European Commission (EC 2017). Table 9.2 summarizes the current IARC classification of pesticides. Lindane, a persistent organochlorine insecticide banned in most countries, is the only pesticide classified as carcinogenic to humans.

9.3.2 Neurotoxicity

The organophosphorus (OP), *N*-methyl carbamates (NMC), and pyrethroids (PY) insecticides are neurotoxic compounds that act in mammals, including humans, through the same mechanism by which they exert their acute toxic effects to the target insects (Casida and Durkin 2013; Soderlund 2012). OP and NMC inhibit the acetylcholinesterase (AChE) in the central and peripheral (humans only) nervous systems, an enzyme that hydrolyzes the neurotransmitter acetylcholine, and PY interacts with the voltage-gated sodium channels leading to delayed repolarization of the nervous signal. High acute exposure to AChE inhibitors, mainly OP, results in accumulation of acetylcholine at the synaptic cleft leading to excessive stimulation and impairment of the physiological functions controlled by the cholinergic, muscarinic, and central nervous systems and can be fatal (Peter et al. 2014). Symptoms of intoxications include salivation, gastric cramps, emesis, neuromuscular weakness, respiratory, and cardiovascular effects. Fatal poisonings with AChE inhibitors have been reported worldwide, mainly suicide, but also accidental cases involving children and under occupational circumstances (Dawson et al. 2010; Yimaer et al. 2017; Magalhães and Caldas 2018).

Various studies have shown an association between pesticide exposure, mainly OP, and neurodegenerative diseases, including Alzheimer, Parkinson, and Amyotrophic Lateral Sclerosis (Mostafalou and Abdollahi 2017, 2018). Koh et al. (2017) found a significant positive association between >20-year period of pesticide use and depression among Korean adults. High exposure to pesticides of rural workers and residents may increase the risk of depression or other psychiatric disorders and suicidal behavior; chlorpyrifos was related to increased suicide mortality (Freire and Koifman 2013).

9.3.3 Reproductive and Development Toxicity

Many studies have investigated the association between exposure to pesticides and increase the risk of developing reproductive and developmental disorders, mainly endocrine disruptor pesticides. A population-based case-control study conducted by García et al. (2017) on Spain found significantly higher prevalence rates and risk of miscarriage, low birth weight, hypospadias, cryptorchidism, and micropenis in areas of higher pesticide use in relation to those with lower use. Various studies have investigated the association between pesticides and semen quality (Zamkowska et al.

2018). Melgarejo et al. (2015) found a significant positive association between DAP concentration in urine of men from a Spanish infertility clinic and sperm concentration and total sperm count. Similar results were found in China with healthy men (Perry et al. 2011).

Epidemiological studies reviewed by Burke et al. (2017) reported statistically significant correlations between prenatal exposures to chlorpyrifos and postnatal neurological deficit, including impaired cognition and motor function, attention deficit hyperactive disorder, deficits in working memory and reduced full-scale intelligence quotient. Kongtip et al. (2017) found higher DEP and/or total DAP organophosphorus metabolite levels from the mother at 28 weeks' gestation in Taiwan, which were significantly associated with reduced motor and cognitive composite scores of the infants.

9.3.4 Toxicity to the Respiratory Tract

The link between occupational exposures to chemicals, including pesticides, with asthma, bronchitis and other respiratory diseases, has long been reported (Ye et al. 2017; Mostafalou and Abdollahi 2017). In USA, from an estimated 2.1 million farm operators, 40.0% used pesticides, 30.8% had lifetime allergic rhinitis, and 5.1% had current asthma (Patel et al. 2018). Insecticide and herbicide use were significantly associated with lifetime allergic rhinitis and current asthma, with 2,4-D and carbaryl significantly associated with lifetime allergic rhinitis. A cohort study conducted with Canadian grain farmers also showed that lifetime exposure to phenoxy herbicides, which include 2,4-D, is associated with an increased risk of asthma (Cherry et al. 2018). In a study conducted with children (3–10 years old) living in a French vineyard rural area, there was an association between ethylene thiourea urinary concentration, a metabolite of dithiocarbamate fungicides, asthma, and rhinitis symptoms (Raherison et al. 2018).

Paraquat is a highly toxic herbicide to humans, which use has been restricted or withdrawn in many countries. Paraquat is not well absorbed by inhalation in agricultural/occupational setting, but the inhalation of paraquat droplets may produce nasal and tracheobronchial irritation. Most of the poisoning cases occur by self-poisoning, when it accumulates in the lungs and death occurs by respiratory failure, particularly due to pulmonary fibrosis (Dinis-Oliveira et al. 2008). A retrospective study of 62 paraquat poisoning cases in French Guiana from 2008 to 2015 includes 44 adults and 18 children younger than 16 years of age; 48% of them died (Elenga et al. 2018). In Korea, the mean mortality rate of 1,056 cases that occurred from 2010 and 2014 was 73.0%, with a significant decreased after the compound was banned in the country in 2012 (Ko et al. 2018).

9.4 Environmental Fate of Pesticides and Impact to the Biota

The fate of the pesticide in the various environmental compartments is illustrated in Fig. 9.3. The chemical may move off-site via spray drift, volatilization, leaching, and runoff, be degraded in the air by photolysis, water and in the soil to less or more toxic products, and can affect aquatic and terrestrial organisms, and non-target insects, such as bees. Additionally, predatory invertebrates may become contaminated by consuming pests such as leafhoppers or aphids that feed on treated crops, and other beneficial plant-feeding invertebrates may be exposed directly through the diet. Other routes of exposure include contact with treated surfaces, exposure to sprays, or consumption of guttation droplets (Khani et al. 2012; Pisa et al. 2015).

The fate of pesticides in the environment depends primarily on their physical—chemical and other properties, and soil characteristics (Table 9.2), which give information on the persistence, potential transport pathways, and bioavailability of chemicals in the environment (US EPA 2009). Table 9.3 shows the main properties of some pesticides, in addition to toxicological benchmark to non-target organisms.

Compounds with a high water solubility and mobility (low K_{OC}), such as 2,4-D, have the potential to reach groundwater, and those with a higher K_{OC} (very low mobility), such as flumetralin and glyphosate, have a tendency to remain in soil/sediments (Table 9.2), but they can reach water systems through superficial runoff (Fig. 9.1).

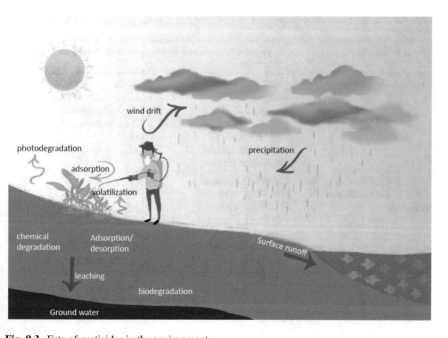

Fig. 9.3 Fate of pesticides in the environment

Pesticide with high vapor pressure is likely to be transported over long distances, which is not the case of any pesticide shown in Table 9.3. Furthermore, physical—chemical properties are closely related to the toxicity of some pesticides of ecotoxicological relevance (Table 9.4).

Table 9.3 Summary of information needed to predict the fate of a pesticide in the environment (US EPA 2009)

Physical–chemical properties	Solubility in water, octan-1-ol-water partition coefficient (K_{ow}/log K_{ow}), vapor pressure/volatility, Henry's Law Constant (air-water partition coefficient), octan-1-ol-air partition coefficient (K_{OA}), dissociation constant in water for week acids and bases (pK_a or pK_b), UV/visible light absorption
Other properties	Degradation (water and/or soil): hydrolysis, photolysis, aerobic, and anaerobic Sorption: soil/sediment-water distribution coefficients (K_d and K_f) and organic-carbon-normalized distribution coefficient (K_{OC}). The pesticide can be classified as highly mobile in the soil ($K_{oc} <$ 10 mL g^{-1}) to immobile ($K_{oc} > 100000$ mL g^{-1}) Field dissipation—half lives ($t_{1/2}$) in terrestrial and aquatic media bioconcentration in fish
Soil characteristics	pH, hydroxyl radical concentration, microbial community, organic carbon content

Table 9.4 Physical–chemical properties and acute toxicity of some pesticides of ecotoxicological relevance[a]

Pesticide (class)	Physical–chemical and other properties	Toxicological benchmark
Ametryn (triazine herbicide)	MM: 227.3 g mol^{-1} Sol$_w$: 170–200 mg L^{-1} Log k_{ow}: 2.63 Vapor pressure 2.74 $\times$ 10^{-6} mm Hg at 25 °C K_{OC}: 170–390 mL g^{-1} $t_{1/2}$ (soil): 60 days	Algae EC$_{50}$ = 3.6–4.06 μg L^{-1} Microcrustaceans EC$_{50}$ = 28–45.29 mg L^{-1} Fish LC$_{50}$ = 4.24–14.1 mg L^{-1}
2,4-Dichloro-phenoxy acetic acid (2,4-D)(Phenoxy acetic herbicide)	MM: 221.0 g mol^{-1} Sol$_w$: 23,180 mg L^{-1} Log k_{ow}: −0.83 Vapor pressure: 1.4 $\times$ 10^{-7} mm Hg at 20 °C K_{OC}: 70–117 mL g^{-1} $t_{1/2}$ (soil): 1–14 days $t_{1/2}$ (water): 15 days	Algae EC$_{50}$ = 24.2 mg L^{-1} Microcrustaceans: EC$_{50}$ = 100 mg L^{-1} Fish LC$_{50}$ = 63.4 mg L^{-1} Earthworms (*Lumbricus rubellus*) LC$_{50}$: 61.6 μg cm^{-1} Bees (*Apis melífera*) LD$_{50}$: 10 μg per bee (oral)

(continued)

Table 9.4 (continued)

Pesticide (class)	Physical–chemical and other properties	Toxicological benchmark
Glyphosate (glycine herbicide)	MM: 169.1 g mol^{-1} Sol$_w$: 12 g L^{-1} Log k_{ow}: -3.4 Vapor pression: 9.8×10^{-8} mm Hg at 25 °C K_{OC}: 300–20100 mL g^{-1} $t_{1/2}$ (soil): 77 days $t_{1/2}$ (water): up to 91 days	Crustraceous (*Palaemonetas vulgaris*) LC$_{50}$ = 281 mg L^{-1} Fish (rainbow trout) LC$_{50}$ = 140 mg L^{-1} Earthworm (biomass) EC$_{50}$: >5,000 μg g^{-1} soil Bee (*Apis melífera*) LD$_{50}$: >100 μg per bee (oral)
Carbendazim (Benzimidazole fungicide)	MM: 191.2 g mol^{-1} Sol$_w$: 7.0 mg L^{-1} Log k_{ow}: 1.51 Vapor pressure: 7.5×10^{-10} mm Hg at 20 °C K_{OC}: 200 mL g^{-1} $t_{1/2}$ (soil): 120 days	Algae EC$_{50}$ = 2.57–7.7 mg L^{-1} Microcustaceans EC$_{50}$ = 0.11–0.19 mg L^{-1} Fish LC$_{50}$ = 0.36–5.5 mg L^{-1}
Flumetralin (2,6-dinitroaniline herbicide)	MM: 421.7 g mol^{-1} Sol$_w$: 5.85 μg L^{-1} Log k_{ow}:5.45 Vapor pressure: 1.3×10^{-8} mm Hg at 20 °C K_{OC}: 100,000 mL g^{-1} $t_{1/2}$ (soil): 20 days	Algae EC$_{50}$ = 0.85–8.58 mg L^{-1} Crocrustaceans EC$_{50}$ = 0.0028–0.42 mg L^{-1} Fish LC$_{50}$ = 0.014–0.024 mg L^{-1}
Imidacloprid (neonicotinoid insecticide)	MM: 255.7 g mol^{-1} Sol$_w$: 610 mg L^{-1} Log k_{ow}: 0.57 Vapor pressure: 3×10^{-12} mm Hg at 20 °C K_{OC}: 249–336 mL g^{-1} $t_{1/2}$ (field dissipation): 40–124 days $t_{1/2}$ (water/sediment): 30–162 days	Algae EC$_{50}$ = 10–389 mg L^{-1} Crustaceans (*Americamysis bahia*) EC$_{50}$ = 34.6 μg L^{-1} Fish (rainbow trout) LC$_{50}$ = 21 mg L^{-1} Earthworm (biomass) EC$_{50}$: 0.8 mg kg^{-1} dry soil Bees (*Apis melífera*) LD$_{50}$: 3.7–40.9 ng per bee (oral) LD$_{50}$: 59.7–242.6 ng per bee (contact)

MM molar mass; *Sol$_w$* water solubility, at 20 °C; *K$_{ow}$* n-octanol–water partition coefficient; *pK$_a$* dissociation constant; *K$_{OC}$* organic-carbon-normalized distribution coefficients; *CE$_{50}$* effective concentration; *LC$_{50}$* lethal concentration; *LD$_{50}$* lethal dose

[a]Information obtained from: https://pubchem.ncbi.nlm.nih.gov/; http://npic.orst.edu/factsheets/archive/; and/or http://www.efsa.europa.eu/en/publications/?f%5B0%5D=im_field_subject%3A62081&f%5B1%5D=sm_field_so_type%3Aconclusion_on_pesticides&f%5B2%5D=sm_field_so_type%3Atechnical_report_post_11; Rebelo and Caldas 2014

9.4.1　Impact of Pesticides on Aquatic Organisms

Pesticides reach surface waters mainly through atmospheric deposition (spray drift) and by surface runoff (Fig. 9.1). The uptake of pesticides by aquatic invertebrates occurs through respiration (gills and trachea), feeding, and through the epidermis (Pisa et al. 2015). Toxic pesticide residues in aquatic systems may eliminate aquatic species, reduce biodiversity, and compromise the functioning of ecosystems (Carvalho 2017). Toxicological benchmarks for aquatic organisms are shown in Table 9.2 for some pesticides. Ametryn is the most toxic to algae ($LC_{50} \sim 4\ \mu g\ L^{-1}$), flumetralin, and carbendazim the most toxic to crustaceans ($EC_{50} < 1\ mg\ L^{-1}$) and flumetralin the most toxic to fish ($LC_{50} < 0.05\ mg\ L^{-1}$).

To estimate the exposure of aquatic organisms, computer simulation models (e.g., GENeric Estimated Environmental Concentration, GENEEC) can be used to estimate the environmental concentrations (EECs) in surface water using degradation data (half-life) and organic-carbon-normalized distribution coefficient (K_{OC}), considering the pesticide is applied at its maximum application rates (US EPA 2018b). A more refined assessment can be performed based on actual use site conditions. To characterize the risk of a given pesticide to the organisms, the EEC is compared to the toxicological benchmark values (e.g., LC50) to estimate the risk quotient (RQ). This RQ is then compared with the levels of concern (LOCs) for direct effects, which can be defined by the environmental agency to analyze potential risk to non-target organisms under the pesticide regulatory process. Table 9.5 shows the LOCs defined by the US EPA (2018b). A regulatory action (e.g., use restriction) can be taken when the RQ is higher than the LOC.

Various studies have been conducted to evaluate the pesticide environmental concentrations and the potential impact on aquatic organisms worldwide. In a study conducted in the Great Barrier Reef, Australia, the runoff of pesticides, mainly herbicides from sugar cane cultivation, resulted in the presence of several pesticides (mainly herbicides) in both freshwater and coastal marine waters with a potential to reduce the productivity of marine plants and corals (Lewis et al. 2009). Chen et al. (2018) investigated the levels of 31 pesticides in water of Dongjiang River (China), including persistent organochlorines, organophosphorus, N-methyl carbamates and

Table 9.5 Levels of concern for the risk characterization of pesticides for aquatic organisms (US EPA 2018b)	Risk presumption	Risk quotient	Level of concern
	Acute risk	EEC/LC_{50} or EC_{50}	0.5
	Acute restricted use	EEC/LC_{50} or EC_{50}	0.1
	Acute endangered species	EEC/LC_{50} or EC_{50}	0.05
	Chronic risk	EEC/NOAEC	1

EEC Estimated environmental concentrations and lowest tested EC_{50}, LC_{50}, or NOAEC for freshwater fish and invertebrates and estuarine/marine fish and invertebrates from acute toxicity tests

pyrethroids insecticides, and found residues at levels ranging up to 1,198 ng L^{-1}. Ecological risk assessments indicated that most of the pesticides posed a high level of risk to the aquatic organisms.

Ernst et al. (2018) reported multiple pesticide residues in muscle tissue of wild fish species from two large rivers in South America (Uruguay and Negro Rivers) at level-up to 194 μg kg^{-1}, with the incidence directly related to the properties of the chemical (K_{ow}, environmental persistence and mobility), pesticide use and cultivated land area. Trifloxystrobin, metolachlor, and pyraclostrobin showed the highest rates of occurrence, and the results suggest a regular exposure of aquatic wild biota to sublethal concentrations of multiple pesticides.

Using a dynamic multimedia model for the Caño Azul River drainage area (Costa Rica), which is heavily influenced by banana and pineapple plantations, Mendez et al. (2018) estimated the levels of diuron, ethoprofos, and epoxiconazole in water, air, soil, and sediments based on pesticide properties, emission patterns, and environmental conditions. Concentration in the environment was highly variable, reaching peak concentrations in water that can exceed thresholds for ecosystem health. Another study conducted in Costa Rica (Tempisque river basin) during 2007 and 2012 found the pesticides carbendazim, diuron, endosulfan, epoxiconazole, propanil, triazophos, and terbutryn showing non-acceptable risk for the ecosystem (Carazo-Rojas et al. 2018).

A monitoring study conducted in 2010 and 2011 in Guadalquivir River Basin (Spain) showed that pesticides are widespread in surface waters and sediments, with organophosphorus, triazines, and carbamates the pesticides most detected, indicating a high-risk scenario (Masiá et al. 2013). Atrazine and terbuthylazine degradation products were found at higher concentrations than parent pesticides.

9.4.2 Impact on Non-target Insects

Wild bee communities are important agricultural pollinators and bee abundance and diversity are of great benefit to farms (Eilers et al. 2011). Many studies have shown that chronic exposure to multiple stressors is driving honeybee colony losses and declines of wild pollinators, including habitat loss owing to agricultural intensification and exposure to pesticides, in addition to parasites and pathogens (Goulson and Nicholls 2016; Park et al. 2015). Bees are exposed to pesticides through direct contact with dust during drilling, guttation drops from seed-treated crops, and consumption of pollen and nectar from wildflowers and trees growing near-treated crops (Bonmatin et al. 2014). Pesticide exposure can impair detoxification mechanisms and immune responses, rendering bees more susceptible to parasites, and can affect bee behavior, including mobility, learning and orientation (Pisa et al. 2015; Goulson and Nicholls 2016; Thorbek et al. 2017).

Among the pesticides, the neonicotinoid insecticides, including thiamethoxam and imidacloprid, and fipronil are considered the most toxic to bees (Bonmatin et al. 2014; Pisa et al. 2015). These pesticides are generally used as seed coating/dressing treatment, generating a "toxic" dust around the planting machine at concentration enough to kill bees passing through the cloud (Girolami et al. 2012). In response to the concern over honeybee colony declines and of wild bee distribution and abundance, the European Commission (EC 2013) prohibited the use and sale of seeds treated with imidacloprid, thiamethoxam, and clothianidin. However, the impact of this regulatory action on bee colonies in Europe is still unclear and probably can only be assessed in the future (Blacquière and van der Steen 2017).

Thorbek et al. (2017) found that the contact with oilseed rape or sunflower treated with neonicotinoids affected bee behavior, with poor brood care impacting the most the colony; good forage mitigated the effects substantially. Queenless bumblebee (*Bombus terrestris*) micro-colonies exposed to thiamethoxam pollen paste and sugar for a 28-day period up to 10 ng g^{-1} (which correspond to the maximum field application) consumed significantly less sugar solution than control colonies and colonies fed at the highest concentration had reduced nest-building activity and produced significantly fewer eggs and larvae (Elston et al. 2013).

An extensive review on the impact of neonicotinoid use on bee colonies and other organisms conducted by Wood and Goulson (2017) showed that proximity to treated flowering crops increases bee exposure to neonicotinoids, mainly through pollen and nectars. In general, wild bees have similar sensitivity to neonicotinoids compared to honeybees when considering direct mortality, although there is a wide variability between bee species, genera, and families. In addition to bees, correlational studies conducted in various countries have suggested a link between neonicotinoid use and population of butterflies and insectivorous birds. The authors suggested that bee declines are the indicators of environmental damage that is likely to impact broadly on biodiversity and the ecosystem services it provides.

9.4.3 Impact on Non-target Terrestrial Organisms

Earthworms constitute an important part of agricultural soil animal biomass, playing a critical role in the development and maintenance of soil properties and fertility (Sharma et al. 2017). They are one of the most important bioindicators in the terrestrial environment and can be exposed to pesticides by direct contact with the applied granules or treated seeds, or contaminated soil and water (Katagi and Ose 2015; Pisa et al. 2015).

Populations with short generation times and/or high dispersal capacity are likely to recover from pesticide-induced toxicity, although soil-persistent pesticides, such as neonicotinoids may impact subsequent generations (Pisa et al. 2015). Wang et al. (2012) investigated the toxicity of 45 pesticides to *Eisenia fetida* and found clothianidin (a neonicotinoid), fenpyroximate, and pyridaben the most toxic after dermal exposure (LC_{50} lower than 1 μg cm^{-2}), and clothianidin and picoxystrobin showing the highest toxicity after oral exposure (LC_{50} lower than 10 μg kg^{-1}).

Arbuscular mycorrhizal (AM) fungi are ubiquitous soil microorganisms that facilitate plant uptake of water and nutrients and receive carbohydrates from the plant in return to complete their life cycle. Helander et al. (2018) showed that glyphosate reduced the mycorrhizal colonization and growth of both target and non-target grasses, and the magnitude of reduction depended on tillage and soil properties due to cultivation history of endophyte symbiotic grass. Li et al. (2013) demonstrated that the herbicide prometryn and acetochlor exerted negative effects on the AM fungus and symbiosis at increasing concentrations, with prometryn apparently being more toxic.

Some argue that pesticides of natural origin are an environment-friendly alternative to synthetic pesticides. Romdhane et al. (2019) applied leptospermone, a natural β-triketone herbicide, and sulcotrione, a synthetic one, to soil microcosms at recommended field doses. Both compounds fully dissipated over the incubation period of 45 days, but CMBA, the major metabolite of sulcotrione, remained in soil microcosms. For both herbicides, the diversity of the soil bacterial community was still not completely recovered by the end of the experiment.

Figure 9.4 shows a generic conceptual model used by the US EPA for pesticide effects on terrestrial organisms, considering drinking water and inhalation exposure pathways for terrestrial vertebrates, and ingestion in dew by terrestrial invertebrates (US EPA 2018c). Additionally, dermal exposure can occur in the treated field during pesticide application or in adjacent areas, during contact to contaminated soils or to contaminated puddles water on the treated field or in areas impacted by drift and runoff. In addition to death, exposure to pesticides can affect the food chain

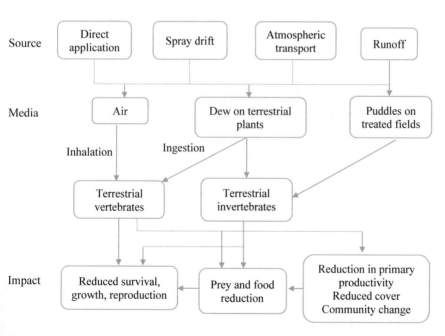

Fig. 9.4 Major routes of potential exposure for terrestrial organisms. Adapted from US EPA (2018c)

and food availability, growth and reproduction. Schaumburg et al. (2016) showed that glyphosate-based formulations have genotoxic effects on the tegu lizard (*Salvator merianae*) at sublethal concentrations during the embryonic period, which may interfere with the development and survival of embryos and hatchling, with a potential impact on population survival, affecting the local biodiversity and ecological equilibrium.

References

Acquavella J, Garabrant D, Marsh G, Sorahan T, Weed DL (2016) Glyphosate epidemiology expert panel review: a weight of evidence systematic review of the relationship between glyphosate exposure and non-Hodgkin's lymphoma or multiple myeloma. Crit Rev Toxicol 46(sup1):28–43

Ambrus Á (2016) FAO manual on the submission and evaluation of pesticide residues data for the estimation of maximum residue levels in food and feed, 3rd edn. Fod and Agriculture Organization, Rome

ANVISA (Agência Nacional de Vigilância Sanitária) (2019) *Glifosato: aberto prazo da consulta pública*. http://portal.anvisa.gov.br/noticias/-/asset_publisher/FXrpx9qY7FbU/content/glifosato-aberto-prazo-da-consulta-publica/219201?p_p_auth=LE14tVRt&inheritRedirect=false&redirect=http%3A%2F%2Fportal.anvisa.gov.br%2Fnoticias%3Fp_p_auth%3DLE14tVRt%26p_p_id%3D101_INSTANCE_FXrpx9qY7FbU%26p_p_lifecycle%3D0%26p_p_state%3Dnormal%26p_p_mode%3Dview%26p_p_col_id%3Dcolumn-3%26p_p_col_count%3D5

APVMA (2016) Regulatory position: consideration of the evidence for a formal reconsideration of glyphosate. Australian Pesticides and Veterinary Medicines Authority. https://apvma.gov.au/sites/default/files/publication/20701-glyphosate-regulatory-position-report-final.pdf

Blacquière T, van der Steen JJ (2017) Three years of banning neonicotinoid insecticides based on sub-lethal effects: can we expect to see effects on bees? Pest Manag Sci 73(7):1299–1304

Bonmatin J-M, Giorio C, Girolami V, Goulson D, Kreutzweiser D, Krupke C, Liess M, Long E, Marzaro M, Mitchell E, Noome D, Simon-Delso N, Tapparo A (2014) Environmental fate and exposure; neonicotinoids and fipronil. Environ Sci Pollut Res 22:35–67

Bonner MR, Freeman LE, Hoppin JA, Koutros S, Sandler DP, Lynch CF, Hines CJ, Thomas K, Blair A, Alavanja MC (2017) Occupational exposure to pesticides and the incidence of lung cancer in the agricultural health study. Environ Health Perspect 125(4):544–551

Boon PE, van Donkersgoed G, Christodoulou D, Crépet A, D'Addezio L, Desvignes V, Ericsson BG, Galimberti F, Ioannou-Kakouri E, Jensen BH, Rehurkova I, Rety J, Ruprich J, Sand S, Stephenson C, Strömberg A, Turrini A, van der Voet H, Ziegler P, Hamey P, van Klaveren JD (2015) Cumulative dietary exposure to a selected group of pesticides of the triazole group in different European countries according to the EFSA guidance on probabilistic modelling. Food Chem Toxicol 79:13–31

Burke RD, Todd SW, Lumsden E, Mullins RJ, Mamczarz J, Fawcett WP, Gullapalli RP, Randall WR, Pereira EFR, Albuquerque EX (2017) Developmental neurotoxicity of the organophosphorus insecticide chlorpyrifos: from clinical findings to preclinical models and potential mechanisms. J Neurochem 142(Suppl 2):162–177

Butler-Dawson J, Galvin K, Thorne PS, Rohlman DS (2018) Organophosphorus pesticide residue levels in homes located near orchards. J Occup Environ Hyg 23:1–24

Caldas ED (2016) Pesticide poisoning in Brazil. In: Reference module earth systems and environmental sciences. https://doi.org/10.1016/B978-0-12-409548-9.10282-9

Caldas ED (2017) Principles of safety assessment of pesticides at national levels. In: Food safety assessment of pesticide residues. Chapter 3. World Scientific Publishing Europe Ltd., New Jersey, USA

Carazo-Rojas E, Pérez-Rojas G, Pérez-Villanueva M, Chinchilla-Soto C, Chin-Pampillo JS, Aguilar-Mora P, Alpízar-Marín M, Masís-Mora M, Rodríguez-Rodríguez CE, Vryzas Z (2018) Pesticide monitoring and ecotoxicological risk assessment in surface water bodies and sediments of a tropical agro-ecosystem. Environ Pollut 241:800–809

Carvalho FP (2017) Pesticides, environment, and food safety. Food Energy Secur 6(2):48–60. https://doi.org/10.1002/fes3.108

Casida JE, Durkin KA (2013) Anticholinesterase insecticide retrospective. Chem Biol Interact 203(1):21–225

Chang CH, MacIntosh D, Lemos B, Zhang Q, Lu C (2018a) Characterization of daily dietary intake and the health risk of neonicotinoid insecticides for the U.S. Population. J Agric Food Chem 66(38):10097–10105

Chang S, El-Zaemey S, Heyworth J, Tang MC (2018b) DDT exposure in early childhood and female breast cancer: evidence from an ecological study in Taiwan. Environ Int 121(Pt 2):1106–1112

Chen Y, Yu K, Hassan M, Xu C, Zhang B, Gin KY, He Y (2018) Occurrence, distribution and risk assessment of pesticides in a river-reservoir system. Ecotoxicol Environ Saf 166:320–327

Cherry N, Beach J, Senthilselvan A, Burstyn I (2018) Pesticide use and asthma in Alberta grain farmers. Int J Environ Res Public Health 15(3)

Codex Alimentarius (2018) Purpose of the Codex Alimentarius. http://www.fao.org/fao-who-codexalimentarius/about-codex/en/

Coffman CW, Stone J, Slocum AC, Landers AJ, Schwab CV, Olsen LG, Lee S (2009) Use of engineering controls and personal protective equipment by certified pesticide applicators. J Agric Saf Health 15:311–326

Cohn BA, Wolff MS, Cirillo PM, Sholtz RI (2007) DDT and breast cancer in young women: new data on the significance of age at exposure. Environ Health Perspect 115(10):1406–1414

Cohn BA, La Merrill M, Krigbaum NY, Yeh G, Park JS, Zimmermann L, Cirillo PM (2015) DDT Exposure in utero and breast cancer. J Clin Endocrinol Metab 100(8):2865–2872

Cohn BA, Cirillo PM, La Merrill MA (2018) Correlation of body mass index with serum DDTs predicts lower risk of breast cancer before the age of 50: prospective evidence in the Child Health and Development Studies. J Expo Sci Environ Epidemiol. https://doi.org/10.1038/s41370-018-0072-7

Colnot T, Dekant W (2017) Approaches for grouping of pesticides into cumulative assessment groups for risk assessment of pesticide residues in food. Regul Toxicol Pharmacol 83:89–99

Damalas CA, Abdollahzadeh G (2016) Farmers' use of personal protective equipment during handling of plant protection products: determinants of implementation. Sci Total Environ 571:730–736

Dawson AH, Eddleston M, Senarathna L, Mohamed F, Gawarammana I, Bowe SJ, Manuweera G, Buckley NA (2010) Acute Human Lethal Toxicity of Agricultural Pesticides: A Prospective Cohort Study. PLoS Med. 7(10):e1000357

de Boer WJ, Goedhart PW, Hart A, Kennedy MC, Kruisselbrink J, Owen H, Roelofs W, van der Voet H (2016) MCRA 8.2 a web-based program for Monte Carlo risk assessment. Reference Manual. December 2016. Biometris, Wageningen UR, Food and Environmental Research Agency (Fera) and National Institute for Public Health and the Environment (RIVM), Wageningen, Bilthoven, the Netherlands and York, UK

Deziel NC, Freeman LE, Graubard BI, Jones RR, Hoppin JA, Thomas K, Hines CJ, Blair A, Sandler DP, Chen H, Lubin JH, Andreotti G, Alavanja MC, Friesen MC (2017) Relative contributions of agricultural drift, para-occupational, and residential use exposure pathways to house dust pesticide concentrations: meta-regression of published data. Environ Health Perspect 125:296–305

Deziel NC, Beane Freeman LE, Hoppin JA, Thomas K, Lerro CC, Jones RR, Hines CJ, Blair A, Graubard BI, Lubin JH, Sandler DP, Chen H, Andreotti G, Alavanja MC, Friesen MC (2018) An algorithm for quantitatively estimating non-occupational pesticide exposure intensity for spouses in the Agricultural Health Study. J Expo Sci Environ Epidemiol. https://doi.org/10.1038/s41370-018-0088-z

Dinis-Oliveira RJ, Duarte JA, Sánchez-Navarro A, Remião F, Bastos ML, Carvalho F (2008) Paraquat poisonings: mechanisms of lung toxicity, clinical features, and treatment. Crit Rev Toxicol 38(1):13–71

Duffus JH, Nordberg M, Templeton DM (2007) Glossary of terms used in toxicology, 2nd edn (IUPAC Recommendations 2007) Pure Appl Chem 79:1153–1344

EC (European Commission) (2013) Commission implementing regulation (EU) No 485/2013 of 24 May 2013 amending Implementing Regulation (EU) No 540/2011, as regards the conditions of approval of the active substances clothianidin, thiamethoxam and imidacloprid, and prohibiting the use and sale of seeds treated with plant protection products containing those active substances. Off J Eur Union 139:12–26

EC (European Commission) (2017) Glyphosate. https://ec.europa.eu/food/plant/pesticides/glyphosate_en

ECHA (European Chemical Agency) (2017) Glyphosate not classified as a carcinogen by ECHA. https://echa.europa.eu/-/glyphosate-not-classified-as-a-carcinogen-by-echa

EFSA (European Food Safety Authority) (2009) Scientific Opinion on risk assessment for a selected group of pesticides from the triazole group to test possible methodologies to assess cumulative effects from exposure through food from these pesticides on human health. EFSA J 7(9):1167 (187 pp)

EFSA (European Food Safety Authority) (2015a) Revisiting the international estimate of short-term intake (IESTI equations) used to estimate the acute exposure to pesticide residues via food. EFSA Scientific Workshop, co-sponsored by FAO and WHO. 8/9 Sept 2015, Geneva, Switzerland. https://efsa.onlinelibrary.wiley.com/doi/epdf/10.2903/sp.efsa.2015.EN-907

EFSA (European Food Safety Authority) (2015b) Conclusion on the peer review of the pesticide risk assessment of the active substance glyphosate. EFSA J 13:4302

Eilers EJ, Kremen C, Greenleaf SS, Garber AK, Klein A-M (2011) Contribution of pollinator-mediated crops to nutrients in the human food supply. PLoS ONE 6:e21363

Elenga N, Merlin C, Le Guern R, Kom-Tchameni R, Ducrot YM, Pradier M, Ntab B, Dinh-Van KA, Sobesky M, Mathieu D, Dueymes JM, Egmann G, Kallel H, Mathieu-Nolf M (2018) Clinical features and prognosis of paraquat poisoning in French Guiana: a review of 62 cases. Medicine (Baltimore) 97(15):e9621

Elston C, Thompson HM, Walters KM (2013) Sublethal effects of thiamethoxam, a neonicoti-noid pesticide, and propiconazole, a DMI fungicide, on colony initiation in bumblebee (Bombus terrestris) micro-colonies. Apidologie 44:563

Ernst F, Alonso B, Colazzo M, Pareja L, Cesio V, Pereira A, Márquez A, Errico E, Segura AM, Heinzen H, Pérez-Parada A (2018) Occurrence of pesticide residues in fish from South American rainfed agroecosystems. Sci Total Environ 631–632:169–179

FAO (Food and Agricultural Organization) (2018a) Statistical pocketbook. http://www.fao.org/3/CA1796EN/ca1796en.pdf

FAO (Food and Agricultural Organization) (2018b) FAOSTAT pesticide use http://www.fao.org/faostat/en/#data/RP

FAO/WHO JMPR (2016) Pesticide residues in food (2016) Report of the special session of the joint meeting of the FAO panel of experts on pesticide residues in food and the environment and the WHO core assessment group on pesticide residues Geneva, Switzerland, 9–13 May 2016

FAO/WHO JMPR (2017) Pesticide residues in food (2017) Report, joint FAO/WHO meeting on pesticide residues. FAO Plant Production and Protection Paper 232. Rome. http://www.fao.org/fileadmin/templates/agphome/documents/Pests_Pesticides/JMPR/Report2017/web_2017_JMPR_Report_Final.pdf

FAO/WHO JMPR (2018) Pesticide residues in food 2018. Report, Joint FAO/WHO Meeting on Pesticide Residues. FAO plant production and protection paper, Rome

Freire C, Koifman S (2013) Pesticides, depression and suicide: a systematic review of the epidemi-ological evidence. Int J Hyg Environ Health 216(4):445–460

Gangemi S, Miozzi E, Teodoro M, Briguglio G, De Luca A, Alibrando C, Polito I, Libra M (2016) Occupational exposure to pesticides as a possible risk factor for the development of chronic diseases in humans (Review). Mol Med Rep 14:4475–4488

García J, Ventura MI, Requena M, Hernández AF, Parrón T, Alarcón R (2017) Association of reproductive disorders and male congenital anomalies with environmental exposure to endocrine active pesticides. Reprod Toxicol 71:95–100

Girolami V, Marzaro M, Vivan L, Mazzon L, GreattiM Giorio C, Marton D, Tapparo A (2012) Fatal powdering of bees in flight with particulates of neonicotinoids seed coating and humidity implication. J Appl Entomol 136:17–26

Goulson D, Nicholls E (2016) The canary in the coalmine; bee declines as an indicator of environmental health. Sci Prog 99(3):312–326

Guyton KZ, Loomis D, Grosse Y, El Ghissassi F, Benbrahim-Tallaa L, Guha N, Scoccianti C, Mattock H, Straif K (2015) International agency for research on cancer monograph working group, IARC, Lyon, France. Carcinogenicity of tetrachlorvinphos, parathion, malathion, diazinon, and glyphosate. Lancet Oncol 16(5):490–491

Helander M, Saloniemi I, Omacini M, Druille M, Salminen JP, Saikkonen K (2018) Glyphosate decreases mycorrhizal colonization and affects plant-soil feedback. Sci Total Environ 15(642):285–291

Humphrey P, Margerison S, van der Velde-Koerts T, Doherty MA Rowland J (2017) Principles of safety assessment of pesticides at national levels. In Food safety assessment of pesticide residues. Chapter 3. World Scientific Publishing Europe Ltd. New Jersey, USA

Hung CC, Huang FJ, Yang YQ, Hsieh CJ, Tseng CC, Yiin LM (2018) Pesticides in indoor and outdoor residential dust: a pilot study in a rural county of Taiwan. Environ Sci Pollut Res Int. https://doi.org/10.1007/s11356-018-2413-4

IARC (International Agency for Cancer Research) (2018) IARC monographs on the classification of cancer risks to human. World Health Organization. https://monographs.iarc.fr/wp-content/uploads/2018/09/List_of_Classifications.pdf

Ingber SZ, Buser MC, Pohl HR, Abadin HG, Murray HE, Scinicariello F (2013) DDT/DDE and breast cancer: a meta-analysis. Regul Toxicol Pharmacol 67(3):421–433

IPCS (International Programme on Chemical Safety) (2004a) Risk Assessment Terminology Part 1: IPCS/OECD key generic terms used in chemical hazard/risk assessment. International Programme on Chemical Safety and Inter-Organization Programme for the Sound Management of Chemicals. World Health Organization, Geneva. http://www.inchem.org/documents/harmproj/harmproj/harmproj1.pdf

IPCS (International Programme on Chemical Safety) (2004b) Guidelines on the prevention of toxic exposures. Education and Public Awareness Activities; World Health Organization: Geneva. https://www.who.int/ipcs/features/prevention_guidelines.pdf

IPCS (International Programme on Chemical Safety) (2009) Dietary exposure assessment of chemicals in food (Chapter 6). Principles and methods for the risk assessment of chemicals in food. Environmental Health Criteria 240. International Programme on Chemical Safety World Health Organization, Geneva

Jardim ANO, Mello DC, Brito AP, van der Voet H, Boon PE, Caldas ED (2018a) Probabilistic dietary risk assessment of triazole and dithiocarbamate fungicides for the Brazilian population. Food Chem Toxicol 118:317–327

Jardim ANO, Brito AP, van Donkersgoed G, Boon PE, Caldas ED (2018b) Dietary cumulative acute risk assessment of organophosphorus, carbamates and pyrethroids insecticides for the Brazilian population. Food Chem Toxicol 112:108–117

Jarman WM, Ballschmiter K (2012) From coal to DDT: the history of the development of the pesticide DDT from synthetic dyes till Silent Spring. Endeavour 36(4):131–142

Katagi T, Ose K (2015) Toxicity, bioaccumulation and metabolism of pesticides in the earthworm. Pestic Sci 40:69–81

Katsikantami I, Colosio C, Alegakis A, Tzatzarakis MN, Vakonaki E, Rizos AK, Sarigiannis DA, Tsatsakis AM (2019) Estimation of daily intake and risk assessment of organophosphorus pesticides based on biomonitoring data—the internal exposure approach. Food Chem Toxicol 23:57–71

Khani A, Ahmadi F, Ghadamyari M (2012) Side effects of imidacloprid and abamectin on the Mealybug destroyer, Cryptolaemus montrouzieri. Trakia J Sci 10:30–35

Ko DR, Chung SP, You JS, Cho S, Park Y, Chun B, Moon J, Kim H, Kim YH, Kim HJ, Lee KW, Choi S, Park J, Park JS, Kim SW, Seo JY, Park HY, Kim SJ, Kang H, Hong DY, Hong JH (2018) Effects of paraquat ban on herbicide poisoning-related mortality. Yonsei Med J 58(4):859–866

Koh SB, Kim TH, Min S, Lee K, Kang DR, Choi JR (2017) Exposure to pesticide as a risk factor for depression: a population-based longitudinal study in Korea. Neurotoxicology 62:181–185

Kongtip P, Techasaensiri B, Nankongnab N, Adams J, Phamonphon A, Surach A, Sangprasert S, Thongsuksai A, Srikumpol P, Woskie S (2017) The impact of prenatal organophosphate pesticide exposures on Thai infant neurodevelopment. Int J Environ Res Public Health 14(6)

Larsson MO, Sloth Nielsen V, Bjerre N, Laporte F, Cedergreen N (2018) Refined assessment and perspectives on the cumulative risk resulting from the dietary exposure to pesticide residues in the Danish population. Food Chem Toxicol 111:207–267

Lewis SE, Brodie JE, Bainbridge ZT, Rohde KW, Davis AM, Masters BL, Maughan M, Devlin MJ, Mueller JF, Schaffelke B (2009) Herbicides: a new threat to the Great Barrier Reef. Environ Pollut 157(8–9):2470–2484

Li X, Miao W, Gong C, Jiang H, Ma W, Zhu S (2013) Effects of prometryn and acetochlor on arbuscular mycorrhizal fungi and symbiotic system. Lett Appl Microbiol 57(2):122–128

Li F, Yuan Y, Meng P, Wu M, Li S, Chen B (2017) Probabilistic acute risk assessment of cumulative exposure to organophosphorus and carbamate pesticides from dietary vegetables and fruits in Shanghai populations. Food Addit Contam Part A Chem Anal Control Expo Risk Assess 34(5):819–831

López-Cervantes M, Torres-Sanchez L, Tobias A, Lopez-Carrillo L (2004) Dichlorodiphenyldichloroethane burden and breast cancer risk: a meta-analysis of the epidemiologic evidence. Env Health Perspect 112:207–214

Lu C, Adamkiewicz G, Attfield KR, Kapp M, Spengler JD, Tao L, Xie SH (2013) Household pesticide contamination from indoor pest control applications in urban low-income public housing dwellings: a community-based participatory research. Environ Sci Technol 47(4):2018–2025

MacFarlane E, Carey R, Keegel T, El-Zaemay S, Fritschi L (2013) Dermal exposure associated with occupational end use of pesticides and the role of protective measures. Saf Health Work 4(3):136–141

Magalhães AFA, Caldas ED (2018) Underreporting of fatal poisonings in Brazil—a descriptive study using data from four information systems. Forensic Sci Int 287:136–141

Maggioni DA, Signorini ML, Michlig N, Repetti MR, Sigrist ME, Beldomenico HR (2018) National short-term dietary exposure assessment of a selected group of pesticides in Argentina. J Environ Sci Health B 19:1–13

Masiá A, Campo J, Vázquez-Roig P, Blasco C, Picó Y (2013) Screening of currently used pesticides in water, sediments and biota of the Guadalquivir River Basin (Spain). J Hazard Mater 263(Pt 1):95–104

Melgarejo M, Mendiola J, Koch HM, Moñino-García M, Noguera-Velasco JA, Torres-Cantero AM (2015) Associations between urinary organophosphate pesticide metabolite levels and reproductive parameters in men from an infertility clinic. Environ Res 137:292–298

Mendez A, Castillo LE, Ruepert C, Hungerbuehler K, Ng CA (2018) Tracking pesticide fate in conventional banana cultivation in Costa Rica: a disconnect between protecting ecosystems and consumer health. Sci Total Env 613–614:1250–1262

Morgan MK (2012) Children's exposures to pyrethroid insecticides at home: a review of data collected in published exposure measurement studies conducted in the United States. Int J Environ Res Public Health 9(8):2964–2985

Mostafalou S, Abdollahi M (2017) Pesticides: an update of human exposure and toxicity. Arch Toxicol 91(2):549–599

Mostafalou S, Abdollahi M (2018) The link of organophosphorus pesticides with neurodegenerative and neuro developmental diseases based on evidence and mechanisms. Toxicology 409:44–52

Myers JP, Antoniou MN, Blumberg B, Carroll L, Colborn T, Everett LG, Hansen M, Landrigan PJ, Lanphear BP, Mesnage R, Vandenberg LN, Vom Saal FS, Welshons WV, Benbrook CM (2016) Concerns over use of glyphosate-based herbicides and risks associated with exposures: a consensus statement. Env Health 15:19

OEHHA (Office of Environmental Health Hazard Assessment) (2015) California environmental protection agency. https://oehha.ca.gov/pesticides/program/cholinesterase-monitoring-agricultural-pesticide-applicators

Park MG, Blitzer EJ, Gibbs J, Losey JE, Danforth BN (2015) Negative effects of pesticides on wild bee communities can be buffered by landscape context. Proc Biol Sci 282(1809):20150299

Pasiani JO, Torres P, Silva JR, Diniz BZ, Caldas ED (2012) Knowledge, attitudes, practices and biomonitoring of farmers and residents exposed to pesticides in Brazil. Int J Environ Res Public Health 9:3051–3068

Patel O, Syamlal G, Henneberger PK, Alarcon WA, Mazurek JM (2018) Pesticide use, allergic rhinitis, and asthma among US farm operators. J Agromed 23(4):327–335

Paumgartten FJ (2015) Letter to the editor: DDT exposure in utero and breast cancer. J Clin Endocrinol Metab 100(11):L104–L105

Perry MJ, Venners SA, Chen X, Liu X, Tang G, Xing H, Barr DB, Xu X (2011) Organophosphorous pesticide exposures and sperm quality. Reprod Toxicol 31(1):75–79

Peter JV, Sudarsan TI, Moran JL (2014) Clinical features of organophosphate poisoning: a review of different classification systems and approaches. Indian J Crit Care Med 18(11):735–745

Piel C, Pouchieu C, Migault L, Béziat B, Boulanger M, Bureau M, Carles C, Grüber A, Lecluse Y, Rondeau V, Schwall X, Tual S, Lebailly P, Baldi I; AGRICAN group (2018) Increased risk of central nervous system tumours with carbamate insecticide use in the prospective cohort AGRICAN. Int J Epidemiol. https://doi.org/10.1093/ije/dyy246

Pisa LW, Amaral-Rogers V, Belzunces LP, Bonmatin JM, Downs CA, Goulson D, Kreutzweiser DP, Krupke C, Liess M, McField M, Morrissey CA, Noome DA, Settele J, Simon-Delso N, Stark JD, Van der Sluijs JP, Van Dyck H, Wiemers M (2015) Effects of neonicotinoids and fipronil on non-target invertebrates. Environ Sci Pollut Res Int 22(1):68–102

Portier CJ, Armstrong BK, Baguley BC, Baur X, Belyaev I, Bellé R, Belpoggi F, et al (2016) Differences in the carcinogenic evaluation of glyphosate between the International. Agency for Research on Cancer (IARC) and the European Food Safety Authority (EFSA). J Epidemiol Commun Health 70(8):741–745

Raherison C, Baldi I, Pouquet M, Berteaud E, Moesch C, Bouvier G, Canal-Raffin M (2018) Pesticides exposure by air in vineyard rural area and respiratory health in children: a pilot study. Environ Res 169:189–195

Rebelo RM, Caldas ED (2014) Avaliação de risco ambiental de ambientes aquáticos afetados pelo uso de agrotóxicos. Quím Nova 37:1199–1208

Recena M, Pires D, Pontes E, Caldas ED (2006) Pesticides exposure in Culturama, Brazil, knowledge, attitudes, and practices. Environ Res 102:230–236

Ribeiro MG, Colasso CG, Monteiro PP, Pedreira Filho WR, Yonamine M (2012) Occupational safety and health practices among flower greenhouses workers from Alto Tietê region (Brazil). Sci Total Environ 416:121–126

Roca M, Miralles-Marco A, Ferré J, Pérez R, Yusà V (2014) Biomonitoring exposure assessment to contemporary pesticides in a school children population of Spain. Environ Res 131:77–85

Romdhane S, Devers-Lamrani M, Beguet J, Bertrand C, Calvayrac C, Salvia MV, Jrad AB, Dayan FE, Spor A, Barthelmebs L, Martin-Laurent F (2019) Assessment of the ecotoxicological impact of natural and synthetic β-triketone herbicides on the diversity and activity of the soil bacterial community using omic approaches. Sci Total Environ 15, 651(Pt 1):241–249

Schaumburg LG, Siroski PA, Poletta GL, Mudry MD (2016) Genotoxicity induced by Roundup® (Glyphosate) in tegu lizard (Salvator merianae) embryos. Pestic Biochem Physiol 130:71–78

Sharifzadeh MS, Damalas CA, Abdollahzadeh G (2017) Perceived usefulness of personal protective equipment in pesticide use predicts farmers' willingness to use it. Sci Total Environ 609:517–523

Sharma DK, Tomar S, Chakraborty D (2017) Role of earthworm in improving soil structure and functioning. Curr Sci 113:1064–1071

Sieke C, Michalski B, Kuhl T (2018) Probabilistic dietary risk assessment of pesticide residues in foods for the German population based on food monitoring data from 2009 to 2014. J Expo Sci Environ Epidemiol 28(1):46–54

Sinha SN, Banda VR (2018) Correlation of pesticide exposure from dietary intake and bio-monitoring: the different sex and socio-economic study of children. Ecotoxicol Environ Saf 162:170–177

Soderlund DM (2012) Molecular mechanisms of pyrethroid insecticide neurotoxicity: recent advances. Arch Toxicol 86(2):165–181

Stephenson CL, Harris CA, Clarke R (2018) An assessment of the acute dietary exposure to glyphosate using deterministic and probabilistic methods. Food Addit Contam Part A Chem Anal Control Expo Risk Assess 35(2):258–272

Tarone RE (2018) On the international agency for research on cancer classification of glyphosate as a probable human carcinogen. Eur J Cancer Prev 27(1):82–87

Thorbek P, Campbell PJ, Thompson HM (2017) Colony impact of pesticide-induced sublethal effects on honeybee workers: a simulation study using BEEHAVE. Environ Toxicol Chem 36(3):831–840

Tsakirakis AN, Kasiotis KM, Charistou AN, Arapaki N, Tsatsakis A, Tsakalof A, Machera K (2014) Dermal and inhalation exposure of operators during fungicide application in vineyards. Evaluation of coverall performance. Sci Total Environ 470–471:282–289

US EPA (Environmental Protection Agency) (2001) General principles for performing aggregate exposure and risk assessments for pesticides. https://www.epa.gov/sites/production/files/2015-07/documents/aggregate.pdf

US EPA (Environmental Protection Agency) (2007). Revised N-methyl carbamate cumulative risk assessment. Office of Pesticide Programs. https://archive.epa.gov/pesticides/reregistration/web/pdf/nmc_revised_cra.pdf

US EPA (Environmental Protection Agency) (2009) Guidance for reporting on the environmental fate and transport of the stressors of concern in problem formulations for registration review, registration review risk assessments, listed species litigation assessments, new chemical risk assessments, and other relevant risk assessments. https://www.epa.gov/pesticide-science-and-assessing-pesticide-risks/guidance-reporting-environmental-fate-and-transport#I

US EPA (Environmental Protection Agency) (2012) Glossary of pesticides terms. Office of Chemical Safety and Pollution Prevention/Office of Pesticides Programs/Information Technology and Resources Management Division. USA Environmental Protection Agency

US EPA (Environmental Protection Agency) (2016) Pesticide cumulative risk assessment: framework for screening analysis purpose. Office of pesticide programs. https://www.regulations.gov/document?D=EPA-HQ-OPP-2015-0422-0019

US EPA (Environmental Protection Agency) (2017a) Cumulative assessment of risk from pesticides. https://www.epa.gov/pesticide-science-and-assessing-pesticide-risks/cumulative-assessment-risk-pesticides

USA EPA (Environmental Protection Agency) (2017b) Revised Glyphosate issue paper: evaluation of carcinogenic potential EPA's office of pesticide programs. 12 Dec 2017. https://www.regulations.gov/document?D=EPA-HQ-OPP-2009-0361-0073

USA EPA (Environmental Protection Agency) (2018a). Models for pesticide risk assessment. https://www.epa.gov/pesticide-science-and-assessing-pesticide-risks/models-pesticide-risk-assessment#health

USA EPA (Environmental Protection Agency) (2018b). Technical overview of ecological risk assessment: risk characterization. https://www.epa.gov/pesticide-science-and-assessing-pesticide-risks/technical-overview-ecological-risk-assessment-risk

USA EPA (Environmental Protection Agency) (2018c) Guidance for the development of conceptual models for a problem formulation developed for registration review. https://www.epa.gov/pesticide-science-and-assessing-pesticide-risks/guidance-development-conceptual-models-problem#section2

Valsta L, Ocké M, Lindtner O (2017) Towards a harmonized food consumption survey methodology for exposure assessment. In Food safety assessment of pesticide residues. Chapter 3. World Scientific Publishing Europe Ltd., New Jersey, USA

van Maele-Fabry G, Gamet-Payrastre L, Lison D (2017) Residential exposure to pesticides as risk factor for childhood and young adult brain tumors: a systematic review and meta-analysis. Environ Int 106:69–90

van Maele-Fabry G, Gamet-Payrastre L, Lison D (2019) Household exposure to pesticides and risk of leukemia in children and adolescents: updated systematic review and meta-analysis. Int J Hyg Environ Health 222(1):49–67

Wang Y, Wu S, Chen L, Wu C, Yu R, Wang Q, Zhao X (2012) Toxicity assessment of 45 pesticides to the epigeic earthworm Eisenia fetida. Chemosphere 88:484–491

WHO (World Health Organization) (2006a) Pesticides and their application for control of vectors and pests of public health importance, 6th edn. WHO/CDS/NTD/WHOPES/GCDPP/2006.1

WHO (World Health Organization) (2006b) Dermal absorption. Env Health Criteria 235. World Health Organization, Geneva. http://www.who.int/iris/handle/10665/43542

WHO (World Health Organization) (2015) Pesticide residues in food. Guidance document for WHO monographers and reviewers WHO Core Assessment Group on Pesticide Residues. World Health Organization, Geneva. WHO/HSE/FOS/2015.1

Wood TJ, Goulson D (2017) The environmental risks of neonicotinoid pesticides: a review of the evidence post 2013. Environ Sci Pollut Res Int 24(21):17285–17325

Ye M, Beach J, Martin JW, Senthilselvan A (2017) Pesticide exposures and respiratory health in general populations. J Environ Sci (China) 51:361–370

Yimaer A, Chen G, Zhang M, Zhou L, Fang X, Jiang W (2017) Childhood pesticide poisoning in Zhejiang, China: a retrospective analysis from 2006 to 2015. BMC Public Health 17:602

Zamkowska D, Karwacka A, Jurewicz J, Radwan M (2018) Environmental exposure to non-persistent endocrine disrupting chemicals and semen quality: an overview of the current epidemiological evidence. Int J Occup Med Env Health 31(4):377–414

Chapter 10
Green Chemistry and Agrochemistry

Sílvio Vaz Jr.

Abstract This chapter presents and discusses the green chemistry principles, highlighting their application in agrochemistry. From this, the understanding of methods and practices can help to change the conventional agrochemistry.

Keywords Agrochemicals · 12 green principles · Green strategy

10.1 Introduction

Green chemistry emerges in the 1990s in countries such as the USA and England, spreading very fast to the world, as a new philosophy in academia and industry to break old paradigms of chemistry such as large waste generation and intensive use of petrochemicals through a holistic view of processes in laboratories and industries (Anastas and Kirchhoff 2002; Anastas and Warner 1998). This approach described in 12 principles—seen later—proposes to consider, among other aspects, the reduction of waste generation, the atomic and energy economy, and the use of renewable raw materials.

These principles, which promote the clean production and green innovations, are already relatively widespread for industrial applications—as the production of agrochemicals—particularly in countries with a well-developed chemical industry and with strict control over the emission of pollutants. They are based on the assumption that chemical processes with potential to impact negatively on the environment will be replaced by less polluting or non-polluting processes. Clean technology, reduction of pollutants at source, environmental chemistry, and green chemistry are denominations that have emerged and were minted over the last two decades to translate concerns for chemical sustainability (Sheldon 2014).

S. Vaz Jr. (✉)
National Research Center for Agroenergy (Embrapa Agroenergy), Embrapa Agroenergia, Parque Estação Biológica, Brazilian Agricultural Research Corporation, s/n, Av. W3 Norte (final), Brasília, DF 70770-901, Brazil
e-mail: silvio.vaz@embrapa.br

© Springer Nature Switzerland AG 2019
S. Vaz Jr. (ed.), *Sustainable Agrochemistry*,
https://doi.org/10.1007/978-3-030-17891-8_10

Then, it is expected that the application of green chemistry principles will reduce deleterious effects from the production and use of agrochemicals.

10.2   The 12 Green Chemistry Principles

The 12 fundamental principles of green chemistry are as follows (ACS Green Chemistry Institute 2018):

I. **Prevention**
   It is better to prevent waste than to treat or clean up waste after it has been created.

II. **Atom Economy**
   Synthetic methods should be designed to maximize the incorporation of all materials used in the process into the final product.

III. **Less Hazardous Chemical Syntheses**
   Wherever practicable, synthetic methods should be designed to use and generate substances that possess little or no toxicity to human health and the environment.

IV. **Designing Safer Chemicals**
   Chemical products should be designed to affect their desired function while minimizing their toxicity.

V. **Safer Solvents and Auxiliaries**
   The use of auxiliary substances (e.g., solvents, separation agents, etc.) should be made unnecessary wherever possible and innocuous when used.

VI. **Design for Energy Efficiency**
   Energy requirements of chemical processes should be recognized for their environmental and economic impacts and should be minimized. If possible, synthetic methods should be conducted at ambient temperature and pressure.

VII. **Use of Renewable Feedstocks**
   A raw material or feedstock should be renewable rather than depleting whenever technically and economically practicable.

VIII. **Reduce Derivatives**
   Unnecessary derivatization (use of blocking groups, protection/deprotection, temporary modification of physical/chemical processes) should be minimized or avoided if possible, because such steps require additional reagents and can generate waste.

IX. **Catalysis**
   Catalytic reagents (as selective as possible) are superior to stoichiometric reagents.

X. **Design for Degradation**
   Chemical products should be designed so that at the end of their function they break down into innocuous degradation products and do not persist in the environment.

XI. **Real-time analysis for Pollution Prevention**
Analytical methodologies need to be further developed to allow for real-time, in-process monitoring and control prior to the formation of hazardous substances.

XII. **Inherently Safer Chemistry for Accident Prevention**
Substances and the form of a substance used in a chemical process should be chosen to minimize the potential for chemical accidents, including releases, explosions, and fires.

The 12 principles are applicable, on a larger or smaller scale, to the production and use of agrochemicals, as we will see ahead. For instance:

- Principles II, III, IV, VI, VII, VIII, IX, and X are directly useful to the development and production of agrochemical molecules;
- Principles I, V, XI, and XII are directly useful to the agrochemical use.

These applications are detailed in the next item.

10.3 Green Chemistry Principles Applied to Agrochemistry

Nowadays, we can observe some relevant challenges to agrochemistry:

- Development of environmentally friendly agrochemicals, with a reduction in the toxicological activity of these molecules and their degradation without the generation of hazardous by-products;
- Reduction or elimination of risk to the environment and public health, as a consequence of the reduction of toxicological activity;
- Greater effectiveness *versus* less application, to generate the best dose–response relation to reduce environmental and human risks due to the exposition to pesticides;
- Monitoring and control of market and application (e.g., combating counterfeiters), to avoid the indiscriminate and uncontrolled use;
- Design and synthesis of green molecules, according to the 12 principles of green chemistry—it is not easy to attend all of them but consider the maximum of principles always when it is possible;
- Release standards in the environment (e.g., coefficient of partition) and procedures based on reliable scientific data, taking into account the specificities of each country, according to requirements of Food and Agriculture Organization of the United Nations (FAO) and World Health Organization (WHO) (World Health Organization 2018);
- Presence of emerging pollutants (e.g., antibiotics) that represents a potential risk for environment and health which are not yet well understood;
- Nanotechnology *versus* nanotoxicology, without enough data to support the understanding the toxicological properties of the agrochemicals in the nanoscale.

Table 10.1 Contribution of green chemistry to overcome challenges related to the production and use of agrochemicals

Challenges to agrochemicals	Contribution of green chemistry
Development of environmentally friendly agrochemicals	Replacement of conventional synthetic agrochemicals by means of innovative routes applying the 12 principles, highlighting the use of catalysts (principle 9)
Reduction or elimination of risk to the environment and public health	Replacement of conventional synthetic agrochemicals by means of innovative routes using the 12 principles, highlighting the principles 7 (renewable feedstock) and 10 (design for degradation)
Greater effectiveness *versus* less application	Development of carriers for controlled release from renewable feedstock (principle 7)
Monitoring and control of market and application	Development of molecular markers for agrochemical formulation using renewable feedstock (principle 7) as lignin derivatives
Design and synthesis of green molecules	Replacement of conventional synthetic agrochemicals by means innovative routes using the 12 principles, highlighting the design of safer (agro)chemicals (principle 4)
Release standards and procedures based on reliable scientific data	Development and validation of green analytical methods to study, for instance, the sorption of agrochemicals to soil components
Presence of emerging pollutants	Development and validation of green analytical methods to study the presence of these pollutants in different environmental matrixes
Nanotechnology *versus* nanotoxicology	Development and validation of green analytical methods to study the presence and influence of nano-agrochemicals in different environmental matrixes and biological samples

Table 10.1 depicts the contribution of green chemistry to overcome those challenges.

A general outlook about green chemistry can be seen in the book *Green Chemistry—Theory and Practice* (Anastas and Werner 2000). Regarding to the application of green analytical methods, a good reference material is the book *Challenges in Green Analytical Chemistry* (De la Guardia and Garrigues 2011).

Both books highlight the practical use of the 12 principles presented in the item 10.2.

10.4   Green Strategy to Develop Green Agrochemicals

We can apply the green chemistry principles in several steps related to the development of a new agrochemical, as we can see in Fig. 10.1.

More detailed information about each step involved in the development of new molecules of pesticides can be obtained in the text *The Pesticide Market Place* (Purdue University 2018).

According to the content of Fig. 10.1, we can propose principles for the most relevant steps highlighted in the green boxes:

- Design of molecules: application of principles II, III, VI, V, VI, VII, VIII, IX, X, and XII;
- Development of molecular analogs: application of principles II, III, IV, V, VI, VII, VIII, IX, X, and XII;
- Exploratory toxicity tests: application of principles I, XI, and XII;
- Initial field tests: application of principles I, XI, and XII;
- Testing on other pests: application of principles I, XI, and XII;
- Identifying soil type response: application of principles I, XI, and XII.

Molecular design has a fundamental role in the development of agrochemicals with a minimized hazard potential. According to Anastas (2009), this step can take into account a hierarchy of knowledge based on:

- Bioavailability (*Tier IV*)—the extent of absorption of a substance by a living organism compared to a standard system.
- Kinetics and dynamics (*Tier III*)—the rate of a reaction or a related quantity is measured and utilized to determine concentrations, and the simulation procedure consisting of the computation of the motion of atoms in a molecule or of individual atoms or molecules in solids, liquids, and gases, according to Newton's laws of motion, respectively.
- Quantitative structure–activity relationship (QSAR) (*Tier II*)—the mathematical relationships linking chemical structure and pharmacological activity in a quantitative manner for a series of compounds.
- Mechanism of action (*Tier I*)—the description of the molecular sequence of events (covalent or non-covalent) that lead to the manifestation of a response in a biological system.

The number of design options increases from the bioavailability (*Tier IV*) to the mechanisms of action (*Tier I*).

This knowledge, when associated with that from synthetic chemistry—mainly the mechanism of reaction—will promote more safe synthetic agrochemicals.

To conclude this approach, aspects of industrial ecology are desirable to be considered, as we can see in Fig. 10.2. We can observe disposal pathways and recycling and reuse pathways. The disposal will take into account the raw material and the process of fabrication, producing residues to be treated by the most adequate technology.

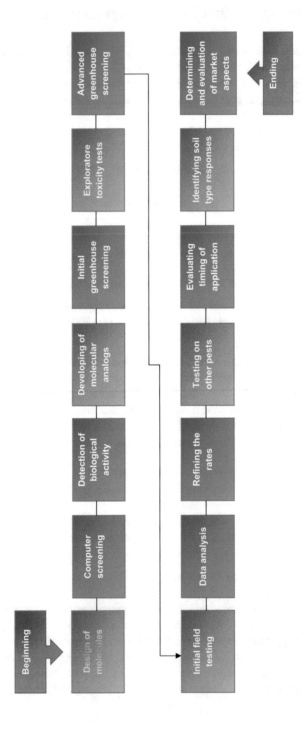

Fig. 10.1 A flowchart of the process of development of a new agrochemical molecule. Green boxes highlight those steps where green chemistry can be directly applied

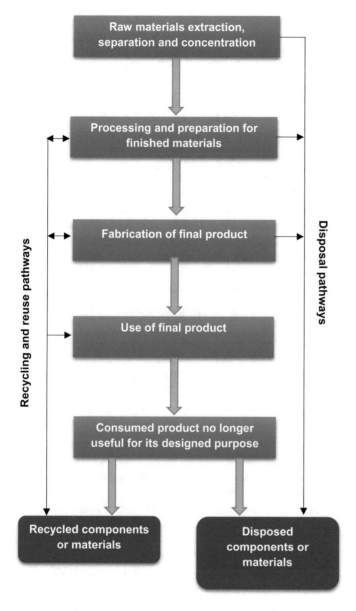

Fig. 10.2 Flowchart of a complete industrial ecosystem. *Source* Adapted from Manahan (2000)

Unfortunately, it can be transformed into environmental liability. On the other hand, recycling and reuse will take into account the process of fabrication and the use of the final product. For instance, based on the U.S. Environmental Protection Agency (2018a), recycling promotes:

- Reduction in the amount of waste sent to landfills and incinerators;
- Conservation of the natural resources such as timber, water and minerals;
- Increasing in the economic security by tapping a domestic source of materials;
- Prevention of the pollution by reducing the need to collect new raw materials;
- Energy economy;
- Support to manufacturing and conserving of valuable resources;
- Help the creation of jobs in the recycling and manufacturing industries.

Furthermore, according to also the U.S. Environmental Protection Agency (2018b) reuse promotes:

- Prevention of pollution caused by reducing the need to harvest new raw materials;
- Energy economy;
- Reduction of greenhouse gas emissions that contribute to global climate change;
- Help the sustention of the environment for future generations;
- Money economy;
- Reduction in the amount of waste that will need to be recycled or sent to landfills and incinerators;
- Allowing products to be used to their fullest extent.

Several points presented above are, naturally, closely related to the green chemistry principles in their essence.

To complete the considerations about industrial ecology, we can introduce the life-cycle analysis (LCA) as a tool to measure the impact on the environment from the production and use of a certain agrochemical. It can be measured by means softwares as GaBi® (Thinkstep 2018) and SimaPro® (Simapro 2018). In a practical way, the best scores in the LCA are achieved, according to Miller (2009), if:

- Reduce the hazard of the materials used—avoid the use of materials that are persistent, bio-accumulative or are named on any list of high-risk materials published by a safety regulator;
- Use renewable raw materials—for example, carbohydrates, lipids or biopolymers;
- Use recycled materials for the product;
- Use raw materials obtained from a waste stream of another process;
- Reduce the number of different raw materials used;
- Minimize the transportation of raw materials—can locally produced raw materials be used, or can manufacturing be shifted to where the raw materials are available?
- Avoid raw materials whose production generates a lot of waste;
- Avoid raw materials whose production is particularly energy-intensive;
- Avoid raw materials produced in an environmentally damaging way.

Lapkin and Constable (2009) address in detail different metrics of green chemistry for products and processes, providing basis for the product design, or more specifically for "green product design," that stimulate R&D and implementation of new concepts.

From this general strategy, we could obtain a green new agrochemical or greening a certain conventional agrochemical. Some examples of green pesticides using some aspects of these strategies are:

- **Intelligent pesticides** with precise controlled release modes that can respond to micro-ecological environment changes such as light sensitivity, thermosensitivity, humidity sensitivity, soil pH, and enzyme activity. Moreover, the establishment of intelligent and controlled pesticide release technologies using nanomaterials could increase pesticide-loading, improve the dispersibility and stability of active ingredients, and promote target ability (Huang et al. 2018).
- **Plant secondary metabolites** in the discovery of new potential **biopesticides**. Among the major challenges to be faced by the candidate products to reach the market are the sustainable use of raw materials, the standardization of chemically complex extracts, and the regulatory requirements and approval. The unique set of secondary metabolites produced by plants may play an important role in a sustainable pest management as new products directly, as novel chemical frameworks for synthesis and/or for identifying original modes of action. The potential of plants and their secondary metabolites for plant health could be used in different strategies: employing the whole plant, crop residues, and part of plants, and using plant chemicals and extracts in integrated or ecological pest management acting directly on the target pest or inducing resistance (Pino et al. 2013).
- **Greening of pesticide–environment interactions** involving improved pest specificity, less nontarget toxicity, lower persistence, and reduced use rates. These successes were sometimes accompanied by unexpected problems, unanticipated hazards, and even major environmental accidents, most of which were solved or placed in risk perspective by fundamental investigations. Safety has been substantially increased by integrating information related to pharmacokinetic and pharmacodynamic behaviors and operational factors (targeting and use rates), and as our knowledge continues to improve, we can look forward to even greener pesticide–environment interactions (Casida 2012).
- **Plant essential oils** have a broad spectrum of activity against pest insects and plant pathogenic fungi—some oils have a long tradition of use in the protection of stored products. Investigations indicate that some chemical constituents of these oils interfere with the octopaminergic nervous system in insects. As this target site is not shared with mammals, most essential oil chemicals are relatively non-toxic to laboratory animals and fish in ecotoxicological tests and meet the criteria for "reduced risk" pesticides. Some of these oils and their constituent chemicals are widely used as flavoring agents in foods and beverages and are even exempt from pesticide registration in the USA. This special regulatory status combined with the wide availability of essential oils from the flavor and fragrance industries has made it possible to fast-track commercialization of essential oil-based pesticides in the USA. Though well received by consumers for use against home and garden pests, these "green pesticides" can also prove effective in agricultural situations, particularly for organic food production (Isman 2004).

As highlighted by Pino et al. (2013), for a successful research and development process leading to a commercial product—the green agrochemical—a wide range of criteria (biological, environmental, toxicological, regulatory, and commercial) must be satisfied from the beginning. In addition, these are the compliance requirements to be satisfied to achieve the market.

10.5 Case Study

10.5.1 Application of Kraft Lignin in Controlled Release Formulation for Pest Management

Agrochemicals move a huge global market, and it is expected that this market achieves 250.5 Billion USD by 2020 (Statista 2018). However, they are one of the main classes of chemical pollutants, with serious negative impacts on public health and the environment. The search for alternatives to conventional agrochemicals presents itself as an excellent opportunity for the development of sustainable agricultural technologies and for opening new businesses. Agricultural chemistry is, undoubtedly, one of the fields of research whose impact is felt throughout the world, since we all need to eat to survive. Added to this is the fact that, increasingly, technology is intertwining with modern agriculture, both with regard to new production strategies and the reduction of negative environmental impacts (Herman 2015).

Lignin (Fig. 10.3), an abundant residue in the pulp and paper industry, was used as a support to promote the controlled release of the semiochemical *cis*-jasmone (Fig. 10.4) for pest management in crops. A simple adsorption process was devel-

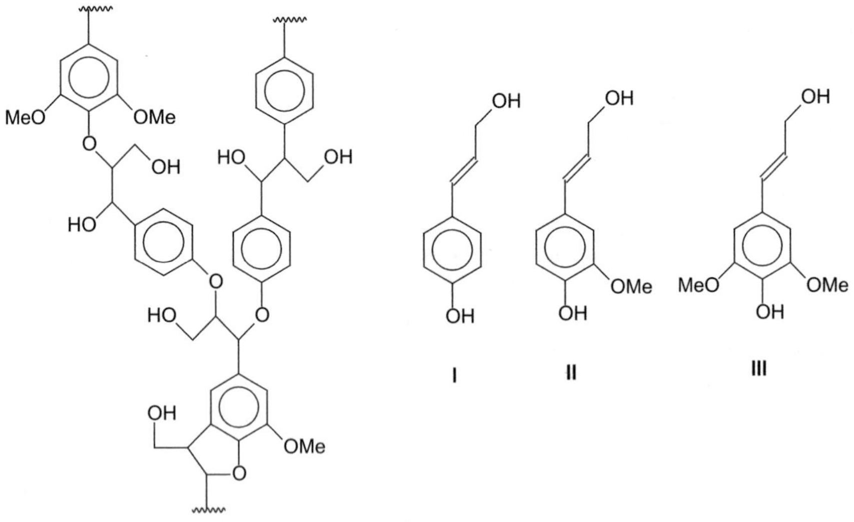

Fig. 10.3 Chemical structure of lignin (left) and its primary precursors (right), being: (I) *p*-coumaric alcohol, (II) coniferyl alcohol, and (III) sinapyl alcohol

Fig. 10.4 Chemical structure of the *cis*-jasmone molecule (left). It can be extracted from the bergamot oranges (*Citrus bergamia*, right) Courtesy of Brazilian Agricultural Research Corporation, Brasília

Fig. 10.5 Adhesive trap containing the lignin-semiochemical formulation randomly disposed in cotton cropping area. *Source* Author

oped by solubilizing the lignin in organic solvent, followed by addition of the semio-chemical. The lignin used in this study was obtained by the precipitation of black liquor from the kraft wood pulping process—a waste with polluting potential—of the eucalyptus species *E. grandis*.

Laboratory experiments using olfactometry were carried out to attract the parasitoid *T. podisi*, with a positive result using the lignin-semiochemical formulation. Furthermore, field tests in cotton cropping (Fig. 10.5) confirmed the efficiency of the formulation. Thus, it was possible to observe that lignin presents a high potential to be used as support for the promotion of the controlled release of the semiochemical *cis*-jasmone, which may be extended to other molecules of agrochemicals.

The substitution of inorganic supports (e.g., silica) for controlled release by lignin, a degradable and non-toxic polymer from renewable source, meets principle 7 of green chemistry—the use of renewable feedstock.

Advantages: In a practical way, the use of this formulation can reduce the use of pesticides for plague control for several crops, decreasing damages to the environment and to the health and reducing costs.

References

ACS Green Chemistry Institute (2018) 12 principles of green chemistry. https://www.acs.org/content/acs/en/greenchemistry/what-is-green-chemistry/principles/12-principles-of-green-chemistry.html. Accessed in Aug 2018

Anastas PT, Kirchhoff MM (2002) Origins, current status and future challenges of green chemistry. Acc Chem Res 35:686–694

Anastas PT, Warner JC (1998) Green chemistry: theory and practice. Oxford University Press, New York, p 30

Anastas PT, Warner JC (2000) Green chemistry—theory and practice. Oxford University Press, New York

Anastas ND (2009) Incentives for using green chemistry and the presentation of an approach for green chemical design. In: Lapkin A, Constable D (eds) Green chemistry metrics—measuring and monitoring sustainable processes. Wiley, Chichester, pp 27–40

Casida JE (2012) The greening of pesticide-environment interactions: some personal observations. Environ Health Perspect 120:487–493

De la Guardia M, Garrigues S (2011) Challenges in green analytical chemistry. RSC Publishing, Cambridge

Huang B, Chen F, Shen Y, Qian K, Wang Y, Sun C, Zhao X, Cui B, Gao F, Zeng Z, Cui H (2018) Advances in targeted pesticides with environmentally responsive controlled release by nanotechnology. Nanomaterials. https://doi.org/10.3390/nano8020102

Herman C (2015) Agricultural chemistry: new strategies and environmental perspectives to feed a growing global population. American Chemical Society, Washington

Isman MB (2004) Plant essential oils as green pesticides for pest and disease management. In: Agricultural applications in green chemistry. ACS symposium series, vol 887, pp 41–51

Lapkin A, Constable D (ed) (2009) Green chemistry metrics—measuring and monitoring sustainable processes. Wiley, Chichester

Manahan S (2000) Environmental chemistry. CRC Press LLC, Boca Raton

Miller R (2009) Green product design. In: Lapkin A, Constable D (eds) Green chemistry metrics—measuring and monitoring sustainable processes. Wiley, Chichester, pp 41–68

Pino O, Sánchez Y, Rojas MM (2013) Plant secondary metabolites as an alternative in pest management. I: background, research approaches and trends. Revista de Protección Vegetal 28:81–94

Purdue University (2018) The pesticide market place. https://www.extension.purdue.edu/extmedia/
PPP/PPP-71.pdf. Accessed in August 2018

Sheldon RA (2014) Green and analytical manufacture of chemicals from biomass: state of the art.
Green Chem 16:950–963

Simapro (2018) Enabling fact-based sustainability. https://simapro.com/. Accessed in Aug 2018

Statista (2018) Worldwide agrochemical market value in 2014 and 2020 (in billion
U.S. dollars). https://www.statista.com/statistics/311957/global-agrochemical-market-revenue-
projection/. Accessed in Aug 2018

Thinkstep (2018) Life cycle assessment. http://www.gabi-software.com/international/solutions/
life-cycle-assessment. Accessed in Aug 2018

U.S. Environmental Protection Agency (2018a) Recycling basics. https://www.epa.gov/recycle/
recycling-basics. Accessed in Aug 2018

U.S. Environmental Protection Agency (2018b) Reducing and reusing basics. https://www.epa.gov/
recycle/reducing-and-reusing-basics. Accessed in Aug 2018

World Health Organization (2018) Joint FAO/WHO meeting on pesticide residues (JMPR). http://
www.who.int/foodsafety/areas_work/chemical-risks/jmpr/en/. Accessed in Aug 2018

Chapter 11
Exposure Characterization Tools for Ecological Risk Assessment of Pesticides in Water

Claudio A. Spadotto and Rafael Mingoti

Abstract Risk assessment and management of pesticides are directly related to sustainable agriculture concept because, besides playing an important role in intensified agriculture by protecting crops from pests and diseases and reducing competition from weeds, the use of pesticides can cause human health and ecological problems. Several pesticides have been shown to reduce water quality and result in adverse effects to sensitive organisms, aquatic ecosystems, and human health. Pesticides enter water systems through different pathways, and therefore, it is important to understand the environmental behavior and fate of pesticides and assess their potential exposure and associated risks to the environment. Ecological risk assessment—ERA—has been adopted in many countries for regulatory purpose and as basis for management of pesticides. Models can be used during different stages of the ERA process and include fate-exposure models, exposure-effect models, and integrated models. In this chapter, definitions of ERA are stated. Pesticide environmental behavior processes and modeling approaches are briefly discussed. Tools for ecological exposure characterization in the regulatory context of agricultural pesticides concerning surface water and groundwater bodies are presented.

Keywords Environmental fate · Behavior · Model · Regulation · Management

11.1 Introduction

The use of pesticides in agriculture is important to ensure plant health and crop yield. Intensive agriculture relies on the use of pesticides for different purposes (Lorenz et al. 2017), because pesticides play a key role in enabling agricultural intensification

C. A. Spadotto (✉)
Brazilian Agricultural Research Corporation, Embrapa, Parque Estação Biológica, S/N, Av. W3 Norte (Final), Brasilia 70770-901, DF, Brazil
e-mail: claudio.spadotto@embrapa.br

R. Mingoti
Brazilian Agricultural Research Corporation, Embrapa Territorial, Av. Soldado Passarinho, 303, Campinas 13070-115, SP, Brazil

© Springer Nature Switzerland AG 2019
S. Vaz Jr. (ed.), *Sustainable Agrochemistry*,
https://doi.org/10.1007/978-3-030-17891-8_11

by protecting crops from damage by insect pests and pathogenic diseases and by reducing competition from weed plants (Carriquiriborde et al. 2014).

On the other hand, the use of pesticides can cause some human health and ecological problems. Thus, risk assessment and management of pesticides have a direct relation to sustainable agriculture, which involves practices that have to fulfill requirements on environmental and human issues.

Pesticides are biologically active compounds designed to interfere with metabolic processes (Matsumura 1985; Manahan 1992; Rice et al. 2007); thus, they may also be hazardous to nontarget organisms. Consequently, it is necessary to assess whether the use of pesticides might pose potential risks to nontarget organisms, including those in off-target habitats (Carriquiriborde et al. 2014).

The application of pesticides to targeted areas inevitably results in the transport of a portion of these chemicals and their degradation products to surrounding nontarget areas, and the intensive application of pesticides in modern agriculture to ensure production quantity and quality has resulted in pollution in soil, surface water, and groundwater (Rice et al. 2007).

Several monitoring studies mention the presence of pesticides in surface water and groundwater (e.g., Scott et al. 1987; Hallberg 1989; Belluck et al. 1991; Seiler et al. 1992; Hoffman et al. 2000; Papastergiou and Papadopoulou-Mourkidou 2001; Harman-Fetcho et al. 2005; Meyer et al. 2006; USGS 2006; Gilliom 2007; Gonçalves et al. 2007; Malaj et al. 2014; Hull et al. 2015). Thus, off-site movement of pesticides to surface water and ground water is a major environmental concern.

A number of these compounds have been shown to reduce water quality and result in adverse effects to sensitive organisms, aquatic ecosystems, and human health, depending on their concentration and duration of exposure (Scott et al. 1987; Clark et al. 1993; Margni et al. 2002; Schulz 2004; Munns Jr 2006). The detection of pesticides and their degradation products in water and reported adverse effects to nontarget organisms and ecosystems at environmentally relevant levels have invoked public concern (Rice et al. 2007), and the evaluation of the risk to the environment of pesticides is considered an important step in the registration process of pesticides in several countries (UNEP 2009).

Pesticides enter water systems through different pathways (Wauchope 1978; Kreuger 1998; Van de Zande et al. 2000; Leu et al. 2004a; Schulz 2004; Hagemann et al. 2006; Rabiet et al. 2010; Taghavi et al. 2010; Bereswill et al. 2012; Stehle and Schulz 2015; Lorenz et al. 2017). Therefore, it is important to understand the environmental behavior and fate of pesticides and assess their potential exposure and associated risks to the environment.

Understanding the physical, chemical, and biological processes that control the behavior of pesticides in the environment, and their effects on nontarget species, is needed to minimize adverse impacts. Pesticide environmental fate and exposure assessments are necessary to better understand associated risks, to develop mitigation and remediation strategies, and to establish sound science-based regulations (Rice et al. 2007).

As noted by Ghirardello et al. (2014), the problem of nonpoint source pollution together with the effects of pesticides on nontarget organisms has been one of the most studied environmental issues for almost half-a-century, and the need to reduce the risk for nontarget terrestrial and aquatic ecosystems is acknowledged in legislation of many countries.

The negative consequences of pesticides in deterioration of ecosystems and human health hazards have led governments to develop pesticide regulation procedures. Much work has taken place over the past years with the aim of developing procedures that are appropriate for incorporation into risk assessment legislation (Girling et al. 2000).

Ecological risk assessment—ERA—of regulated products such as agricultural pesticides is an important process to safeguard the desired level of protection of the environment and biodiversity. ERA evaluates the potential adverse effects on the environment of certain actions and is an important analytical scientific procedure to support regulatory decision-making. Significant advances have been made in the field in recent years (Devos et al. 2016).

ERA has been adopted in many countries for regulatory purpose, as well as basis for management of pesticides. The terms environmental risk, ecotoxicological risk, and ecological risk are many times used interchangeably. In this chapter, 'ecological risk' is adopted.

Models differ in relation to their purpose, scope and level of analysis, and they can be used in the regulatory procedure, as well as to base management decisions. Thus, models have been used during different stages of the ERA process and include fate-exposure models, exposure-effect models, and integrated models.

Fate-exposure models (in short, exposure models) represent the first half of the source-to-outcome continuum—Fig. 11.1—and, as noted by Williams et al. (2010), they assess the transport and transformation of pesticides in the environment and predict the concentrations in different environmental media (e.g., atmosphere, water, soil). The outputs of these models therefore represent concentrations to which receptors have the potential for exposure. These estimates are often used instead of actual exposure and can serve as an input to exposure-effect models (in short, effect models), which are used in the final part of the source-to-outcome continuum of the ERA.

This chapter focuses on tools for ecological exposure characterization, in the stressor domain, during the regulatory (registration and use control) context of agricultural pesticides, concerning surface water and groundwater bodies and examples of models are presented. Definitions of ERA in general and specifically for pesticides, processes involved in the behavior of pesticides in the environment, and modeling approaches are discussed.

11.2 Ecological Risk Assessment of Pesticides

Risk assessment is taken to organize, structure, and compile scientific information in order to identify existing hazardous situations, anticipate potential problems, establish priorities, and provide a basis for regulatory controls and corrective actions (WHO 2004).

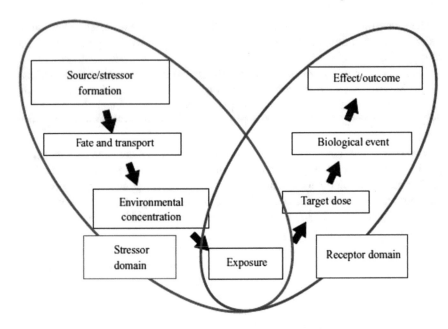

Fig. 11.1 Source-to-outcome continuum in ecological risk assessment—ERA. *Source* Williams et al. (2010)

The risk assessment can be taken as the process of defining the components of a risk in quantitative terms (NRC 1996). The definition given by Westman (1985) used the term 'assessment' as the broadest one, and in this case, the term 'analysis' is used more narrowly to refer to the quantitative techniques considering risk components: hazard and exposure. As noted by Forbes et al. (2008), it would not be possible experimentally to study all the combinations of factors that are thought to influence risk.

In the early 1990s, the United States Environmental Protection Agency—USEPA—generated a framework and guidance documents for conducting ecological risk assessment—ERA—(USEPA 1992), and similar documents were developed in Europe, Canada, and Australia. The publication in 1998 of the Guideline on Ecological Risk Assessment by the USEPA (1998) was considered a milestone for a new paradigm of ERA.

The ERA process is used to systematically assess and organize data, information, estimates, and uncertainties in order to understand and predict the relationships between agents and ecological effects, and it has several features that contribute to effective environmental decision-making (USEPA 1998). It evaluates the likelihood that adverse ecological effects may occur or are occurring as a result of exposure to one or more agents or stressors (USEPA 1998; Posthuma et al. 2002; Strassemeyer et al. 2017). According to USEPA (1998); adverse ecological effects are undesirable changes that alter valued structural or functional characteristics of ecosystems or their components.

The conceptual meanings of the words 'hazard' and 'risk' are different. The hazard of a chemical is a function of two broad considerations: the potential of the chemical to harm biological systems (or damage other systems) and its potential for exposure such that harm or damage can occur (OECD 1982). As noted in USEPA (1998), the term 'hazard assessment' has been used to mean either (1) evaluating the intrinsic effects of a stressor (USEPA 1979) or (2) defining a margin of safety or quotient by comparing a toxicologic effects concentration with an exposure estimate (SETAC 1987). The term risk is conceptually more complex because it also includes the quantitative estimate of the probability of an adverse effect on a biological target exposed to an agent, as a chemical substance (Solomon 1996; Finizio and Villa 2002).

ERA of pesticides is a special case of risk assessment applied to the use of upwards of 700 pesticide active ingredients, with a wide range of physical, chemical, and biological properties (Solomon 2010). To obtain registration of pesticides, extensive ERAs are carried out to show that the use will cause no unacceptable effects on non-target organisms (Schmolke et al. 2010). The goal of prospective ERA of pesticides is to quantify the risk that a concentration of a given chemical would impair on the structure and function of natural ecosystems (De Laender et al. 2014).

ERA is an important component of pesticide registration procedures in Europe, the USA, and other industrialized countries. This risk assessment procedure consists of two parts:

i. Assessment of effects to these organisms derived from ecotoxicological experiments (effect assessment);
ii. Assessment of concentration levels to which organisms will be exposed in the field after pesticide application (exposure assessment).

Part (i) is the domain of ecotoxicology, and part (ii) is the domain of environmental chemistry (Boesten et al. 2007).

In 1989, the United Nations Food and Agriculture Organization—FAO (1989)—published a guide on environmental criteria for registration of plant protection products (agricultural pesticides), outlining the principles of how ERA should be conducted.

In the USA, the National Research Council (NRC 1983) developed a general risk assessment scheme, which was later modified and adopted by the USEPA. The USEPA scheme added to the principles presented by FAO (FAO 1989) the concept of three formal environmental risk assessment steps. When the USEPA assesses the risk of a pesticide to the environment, it considers the ecotoxicity of the pesticide and the amount of pesticide to which the environment compartments may be exposed.

ERAs of pesticides are usually conducted in series of tiers (WWF 1992; Suter et al. 1993; SETAC 1994; Solomon 1996; USEPA 1998; ACP/ECP 1999; ECOFRAM 1999), and in the tiered approach, the initial use of conservative criteria allows substances or situations that truly do not present a risk to be eliminated from the ERA process, thus allowing a shift of resources to situations with potentially greater risk. As one progresses through the tiers, the estimates of exposure and effects become more realistic as uncertainty is reduced through the acquisition of more or better-quality data. Thus, tiers are normally designed such that the earlier tiers are more conservative, while the later tiers are more realistic. Because the earlier tiers are designed to be protective, failing to meet the criteria for these tiers is merely an indication that an assessment based on more realistic data is needed before a regulatory or risk management decision can be reached (Solomon and Sibley 2002).

Therefore, the first tier of a pesticide ERA is designed to allow rapid identification of those that pose no significant risk to the environment. At this level, the calculations of environmental concentrations invariably overestimate the exposure, resulting in a conservative assessment. As assessment is refined in higher tiers, with more likely estimates of environmental concentrations, less conservative and more realistic criteria are used, culminating, when necessary, with a monitoring work. In monitoring, rather than concentration estimates, the exposure is characterized by the determined concentration.

In summary, in ERA of pesticides, the initial tiers are characterized by being conservative and not very close to reality, keeping a large 'margin of safety.' In advanced tiers of ERA, less conservative and closer to reality procedures are adopted, which requires more extensive calculations and larger amounts of input data that are often difficult to obtain.

ERA of pesticides is typically based on a framework as depicted in Fig. 11.2 (USEPA 1998). Despite the many challenges faced in the implementation and acceptance in risk assessment, adoption of this framework is essential to make a scientific informed decision on the admittance of a pesticide on the market (Brock et al. 2006; Van den Brink et al. 2013; Teklu et al. 2015).

Thus, ERA process includes three primary phases: problem formulation, analysis, and risk characterization, and it is based on two major elements: characterization of effects and characterization of exposure (USEPA 1998). As described by USEPA (1998), in problem formulation, the purpose for the assessment is articulated, the problem is defined, and a plan for analyzing and characterizing risk is determined.

Analysis of risk is directed by the products of problem formulation. During the analysis phase, data are evaluated to determine how exposure to stressors is likely to occur (characterization of exposure) and the potential and type of ecological effects that can be expected (characterization of effects). The first step in analysis is to determine the strengths and limitations of data on exposure, effects, and ecosystem and receptor characteristics. The products from these analyses are two profiles, one for exposure and one for stressor response. These products provide the basis for risk characterization (USEPA 1998). In the analysis phase, characterization of ecological effects involves consideration of the results of laboratory tests, in which the concentration and duration of exposure are varied, taking into account the response of organisms.

Characterization of exposure is the complementary portion of the analysis phase of ERA that evaluates the interaction of the stressor with one or more ecological entities. Exposure can be expressed as co-occurrence, contact, or dose in diet, depending on the stressor and ecological component involved. Exposure characterization could

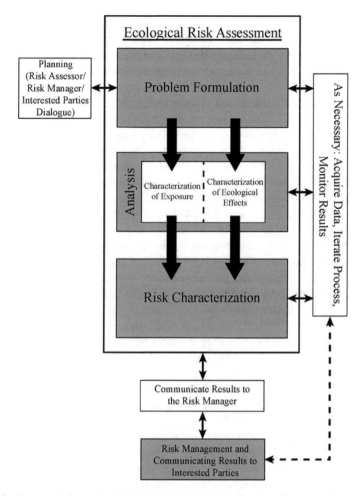

Fig. 11.2 Framework for ecological risk assessment—ERA. *Source* USEPA (1998)

apparently be regarded as a simple task but becomes much more complicated considering the variety of environments and ecosystems (Di Guardo and Hermens 2013).

In general, ERA procedures compare single-point estimates of exposure with ecosystem-level thresholds inferred from toxicity data for individual-level endpoints (De Laender et al. 2014). The majority of regulatory ERA is carried out by means of ecotoxicity-exposure ratios in which exposure is compared with ecotoxicological endpoints measured in laboratory or semi-field studies. The protection goal is populations rather than individuals (Streissl 2010; Wogram 2010). These conventional methods are likely to remain the primary tool for lower tiers of risk assessment because they are simple and rapid, and are appropriate for use as screening tool, provided they are sufficiently conservative (Hart 2001).

The most widely used method in ERA is quotient methods. Thus, the environmental concentration of a stressor is compared to an effect concentration (Urban and Cook 1986; Calabrese and Baldwin 1993; Solomon and Sibley 2002). To allow for unquantified uncertainty in the effect and exposure estimates, the quotient method may be made more conservative by the use of an uncertainty factor (CWQG 1999) or by comparison with predefined criteria, as Levels of Concern—LOCs (Urban and Cook 1986). LOC values may vary, depending on the type of effect (acute or chronic) or whether endangered species are likely to be exposed. Because they frequently make use of worst-case or extreme data, hazard and risk quotient methods are designed to be protective of almost all possible situations that may occur.

Risk characterization is a phase of ERA that integrates exposure and stressor response profiles to evaluate the likelihood of adverse ecological effects associated with exposure to a stressor. During the risk characterization, the exposure and stressor response profiles are integrated through the risk estimation process. Risk characterization includes a summary of assumptions, scientific uncertainties, and strengths and limitations of the analyses. The final product is a risk description in which the results of the integration are presented, including an interpretation of ecological adversity and descriptions of uncertainty and lines of evidence (USEPA 1998). In other words, data on effects on different organisms (to a large extent average toxicity data or limit values) are compared with estimated (expected) concentrations in relevant environmental compartments or in the dietary elements of organisms.

In Europe, the European and Mediterranean Plant Protection Organization—EPPO—has developed ERA guides for pesticides with flexible procedures that can be adapted to be used according to the different priorities of the European Community member countries (UNEP 2009). European pesticide authorization procedures follow a tiered approach (FOCUS 2000; FOCUS 2001a, b). The concentration estimation methodology was developed as a different tiered approach with four levels of assessment: (1) relatively simple calculation based on a maximal loading and a fixed scenario; (2) multiple applications and regional variation across Europe; (3) scenarios developed; and (4) detailed site-specific approach. Thus, the predicted environmental concentration—PEC—is evaluated in a number of sequential steps, taking into account more detailed or more specific information in each sequential step (Tiktak et al. 2013). Other countries such as Canada (Stephenson and Solomon 2007), Australia (UNEP 2009), and Japan (Miyamoto et al. 2008) have ERA procedures for pesticides established.

The drawbacks of the conventional, deterministic ERA have led to the development of more probabilistic techniques in this area (Hart 2001). The use of probabilistic approach for characterizing effects and exposure data has been suggested as a means of considering the range of sensitivities to many substances and, in particular, pesticides, and it has been recommended for later tiers in the ERA process (e.g., SETAC 1994; Solomon 1996; ECOFRAM 1999; USEPA 2001). The major advantage of probabilistic ERA is that it uses all relevant single-species toxicity data and, when combined with exposure distributions, allows quantitative estimations of risks. However, the method does have some disadvantages; more effects and exposure data are usually needed, all sources of uncertainty are not addressed, and results have not been widely calibrated against field observations (Solomon and Sibley 2002).

Even with the importance of probabilistic approach, it may be unnecessary to perform a probabilistic analysis when well-conducted screening tiers result in risks clearly below levels of concern. It is argued here the importance of carrying out ERA, even if it is not entirely probabilistic (despite the stochastic nature of the risk) at the initial stage. This is particularly important in developing countries, where lack of data and resources can limit the adoption of ERA, which represents a methodological step forward in considering possible environmental problems associated with pesticides and other stressors. In Brazil, for example, the Brazilian Institute of Environment and Renewable Natural Resources—IBAMA—has adopted the conceptual framework of the ERA process defined by the USEPA in the evaluation of pesticides (IBAMA 2012).

Many types of data can be used for risk assessment, and they may come from laboratory or field studies or may be produced as output from a model (USEPA 1998). Mathematical models are important tools for estimating exposure and effects of different pesticides in various environmental and use conditions. Effect models focus on the results or consequences on organisms and ecological systems. Ecological effect models may be able to bridge the gap between laboratory studies and population-level endpoints or extrapolate the results of semi-field or field studies to the ecosystem level (Pastorok et al. 2002; Van den Brink et al. 2007; Forbes et al. 2008, 2009; Schmolke et al. 2010). A broad range of effect models have been applied to chemicals in the scientific literature, and the use of them in risk assessments is compatible with current regulations of pesticides in Europe (Wogram 2010) and the USA (Schmolke et al. 2010).

Exposure models, in turn, focus on the behavior and fate of chemicals and estimate their concentrations in the environment. Predicting the environmental concentrations of pesticides by means of exposure models is an essential part of analysis phase (Tiktak et al. 2013).

11.3 Behavior and Fate of Pesticides in the Environment

Along with the active ingredient properties as well as the type and composition of the pesticide, application technology used and weather conditions during and soon after the application directly affect its initial distribution. Therefore, a key factor for agronomic efficiency and consequently for environmental safety of pesticides is the suitability of application technology for the characteristics of the pesticide (formulated product) and meteorological conditions.

Environmental pesticide behavior and fate are governed by the processes of transport, retention, and transformation. Differences in pesticide properties and environmental conditions affect these processes. Land slope, soil type, composition, and activity of microorganisms in the soil, volume and intensity of rain, and management practices adopted, for example, affect the behavior of pesticides in the environment.

Fig. 11.3 Schematic
representation of the main
processes that determine the
behavior and fate of
pesticides in the environment

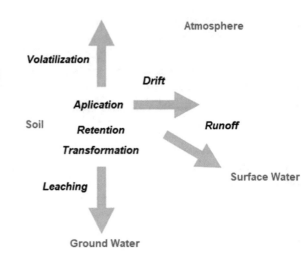

Transport of pesticides between environmental compartments occurs mainly by volatilization into the atmosphere, offside drift, surface runoff, and leaching in soil—- Fig. 11.3. Drift, which occurs to some degree during application, is one of the most common problems related to the use of pesticides.

The transport of pesticides into the atmosphere occurs by direct volatilization and covaporation with water. Volatilization can take place during and after application, from surface of plants, surface and matrix of soils, as well as from surface and column of water bodies. The volatilized pesticide is transported into the atmosphere and sometimes is deposited long distances from the application site. Pesticide may volatilize again and may be transported further via the atmosphere (Van Jaarsveld and Van Pul 1999). Several factors are cited in the literature as important in the emission and transport of pesticides into the atmosphere and deposition in adjacent areas of application sites. Laabs et al. (2002b) pointed out the importance of air transport of pesticides in tropical conditions due to high temperatures.

Runoff, when water flowing on land surface when soil infiltration capacity has been overcome, carries dissolved substances and suspended particles. Depending on soil conservation practices, the fate of runoff material is a lake, pond, stream, or river, what can lead to the pollution of water and sediment. The primary transport routes for pesticides, particularly to small surface water bodies in non-irrigation agriculture, are surface runoff and tile drainage induced by heavy precipitation events (Leu et al. 2004; Rabiet et al. 2010; Taghavi et al. 2010; Bereswill et al. 2012; Stehle and Schulz 2015; Lorenz et al. 2017). Runoff is a route that rapidly changes the mass balance of the pesticide in the soil in cases where the application has been carried out just before a rain of medium to high intensity.

Pesticide leaching in soil, in turn, tends to result in pollution of groundwater, and in this case, chemicals are mainly taken in solution together with the water that percolates in soil and feeds aquifers. Leaching can be classified as matrix or preferential transport. Matrix transport of pesticides occurs in soils with uniform flow conditions by convection-dispersion. In soils with large pores and well drained, it can be assumed that pesticide transport occurs mainly by convection, where the mass of solute (pesticide) is concentrated at a specific point in the water column and not dispersed in the soil profile (piston flow). This transport process is predominantly vertical, with some retardation in relation to water that percolates in soil. In cases where soil does not have good porosity and it is excessively compacted or even unstructured, transport of the solute by diffusion and dispersion becomes more important. The nonuniform flow condition is referred to as preferential transport of solute. The preferential (or rapid) transport of pesticides in soil has been observed, and its importance has been highlighted in several works (Levanon et al. 1993; Flury et al. 1994; Lennartz 1999; Reichenberger et al. 2002; Scorza 2002).

The most important process affecting the transport of pesticides through soil is sorption-desorption behavior because it controls the amount of pesticide available for transport (Koskinen et al. 2002; Rice et al. 2007). Partition between solid and liquid phases determines the proportion of pesticide in solution and sorbed on soil particles in surface runoff and, together with the degradation, conditions the persistence and leaching of the pesticide in soil profile. Sorption is, in fact, very important in the overall behavior of pesticides in the environment. Retention in solid phase tends to limit bioavailability and biodegradation of pesticides and volatilization can also be influenced by sorption. Thus, interactions of pesticides with soil components determine their mobility and persistence, with agronomic and environmental implications.

Pesticides belong to different classes of chemicals, and there are various types of interactions with soil colloids. Thus, the interactions of pesticides with soil components are complex, and several soil properties affect the mechanism and degree of sorption. The predominance of sorption of organic nonionic compounds in soil organic matter has been extensively documented. However, some pesticides are ionizable; thus, a factor that has an important influence on sorption of these compounds is pH (Hornsby et al. 1996). This aspect is especially important in tropical soils without liming, which present acid conditions (Spadotto and Hornsby 2003; Spadotto et al. 2005). Studies have also highlighted the influence of dissolved organic matter on sorption and leaching of pesticides (Cornejo et al. 2005; Li et al. 2005; Jiang et al. 2008; Song et al. 2008).

Pesticide transformation occurs through degradation in the atmosphere, soil, and water and by metabolization in plants and other organisms. Pesticides are degraded by physico-chemical and biological processes, such as photolysis, hydrolysis, oxidation–reduction, and biological degradation. The rates of degradation of some pesticides are relatively high, and their residues remain in the environment for a short time. Some molecules degrade completely in the environment reaching the mineralization. Although part of this process is caused by chemical or physico-chemical reactions, such as hydrolysis and photolysis, microbiological catabolism and metabolism are generally the main mineralization processes.

Degradation rates of pesticides and their metabolites are among the most essential parameters required in evaluating their potential environmental exposure. Approaches used in calculating the degradation parameters from experimental data can significantly affect the resulting degradation rates and thus the environmental risk assessments and evaluation of a pesticide (Rice et al. 2007).

In soil, biological degradation (or biodegradation)—oxidation, reduction, hydrolysis, and their conjugations, mediated by microorganisms—is the most efficient in degrading pesticide residues. Biodegradation is more active in soil root zone, mainly due to the presence of aerobic bacteria, which are more efficient in degrading pesticides, higher content of organic matter and better soil–water–air relationships for this biota. Depending on soil moisture content, hydrolysis may be important, especially when combined with other processes, such as biodegradation.

In acid soils, there is predominance of fungi, which are less efficient in degrading organic chemicals. In close to neutral or slightly alkaline conditions, there is a predominance of bacteria and actinomycetes. Other factor has a great weight, which is the adaptability of soil populations to the substrate.

In water and in the atmosphere, degradation happens preferentially by chemical processes: hydrolysis, photolysis, and oxidation–reduction. Hydrolysis is the main means of degradation of pesticides in water bodies, and the H_3O^+ availability of the medium interferes in this process, due to the interaction with physical and chemical characteristics inherent to the molecule. Another interfering factor is temperature, which has also great influence on hydrolysis rates.

In photolysis, light causes the breakdown of chemical bonds, at first by means of photochemical reactions. Indirect photolysis can also occur, where light acts as a catalyst for other physico-chemical processes, especially in water. Photolysis is considered as the process of transformation with the greatest spectrum of action, because it reaches any pesticide that is on surface of plants, soil and, water.

Oxidation–reduction (or redox) process mainly acts on chemical changes that the pesticide undergoes in photodegradation or biodegradation reactions, catalyzed, respectively, by light or microorganisms. However, in some very special situations, these reactions can occur alone and are related to environments without light and in absence of microorganisms, in deep layers of soil or in groundwater.

When it is not complete, degradation products or metabolites may also be of importance to environment and human health because some of them have the same or higher ecological and human toxicity than the original molecules.

11.4 Pesticide Environmental Behavior and Fate Modeling

Modeling environmental behavior of pesticides began in the 1970s. Since then, their role has increased in importance, as it is an economically and scientifically based way of studying and predicting the behavior and fate of these chemicals in the environment.

According to Boesten (2000), modeling attempts to generalize the knowledge of pesticide behavior in the field by identifying its most important properties, which can be measured in the laboratory. Thus, modeling can serve as a link between laboratory and field studies. Modeling has the advantage of being an economic and efficient way of assessing pesticide fate and considering the large number of relevant environmental scenarios (Scorza and Boesten 2005).

Starfield (1997) summarized the most important misconceptions of the role of models. Uncertainty about model inputs and outputs, lack of validation, and insufficient model description play important roles (Beissinger and Westphal 1998; Grimm et al. 2006). According to Schmolke et al. (2010), there are no standards for model development, documentation, application, and evaluation that would allow an easy judgment about the adequateness of models for specific risk assessment questions.

As noted by Scorza and Boesten (2005), environmental fate modeling of pesticides has become an important tool to assess the potential of contamination of water resulting from the agricultural use of pesticides. Pesticide models have been used increasingly to support authorities in decisions concerning the approval of pesticide registration at European Union level (FOCUS 1995, 1997) and in the USA (Russell et al. 1994). If models are used for registration decisions, their validation status should follow the intended purpose.

A model is a representation of a real system and presents some degree of simplification and abstraction, as well as limitations of use. As seen, several processes are involved in the environmental behavior of pesticides (e.g., drift, volatilization, leaching, runoff, sorption, degradation). Mathematical representation of each of these processes is variable and depends on the degree of detail required for the study.

Studies have shown that responses from complex ecological systems can often be represented by simpler mathematical models with few data, since incorporating the dominant process variables and identifying general patterns of pesticide behavior in the environment may not be serious problems (DeCoursey 1992). The volume of resources and time required to obtain and compile data and parameters needed for application of more complex, process-based models are challenges to be overcome (Foster and Lane 1987). For the purposes of screening or general management guidance, the use of simpler models that are less data intensive is justified (Di and Aylmore 1997).

A mathematical model can be used to describe or explain the pesticide behavior in the environment. Thus, there are mathematical models that are descriptive and others that are explanatory. Descriptive models are developed on an empirical basis and correspond to mathematical equations that represent experimental data in an acceptable way, without the concern of explaining the processes involved. Therefore, extrapolation of conclusions to conditions other than those in which the experiment was performed should be avoided. Thus, these models present a limitation in their predictive capacity. One of the main reasons for the use of descriptive models is that they require a small number of input data.

Explanatory models, supported by scientific knowledge, generally reflect biological or physico-chemical mechanisms observed or hypothesized and aim to explain the environmental behavior of pesticides, integrating the explanatory level and the level to be explained. For example, to explain the environmental behavior of a pesticide under field conditions—level to be explained—using laboratory data (e.g., half-life time, sorption coefficient) in combination with climate and soil data—explanatory level.

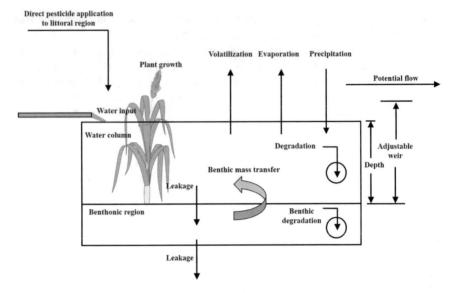

Fig. 11.4 Representation of the conceptual model of the pesticide in flooded applications model—PFAM. *Source* White et al. (2016)

The explanatory models are expressed initially by means of a conceptual model, which is elaborated based on knowledge of the processes of transport and transformation associated with environmental behavior. Thus, knowledge of pesticide environmental behavior processes is fundamental in the formulation of conceptual models in explanatory modeling.

According to Rice et al. (2007), conceptual models serve as a frame of reference for the discussion of environmental variables that are involved in determining the fate and transport of pesticides in the environment. In Fig. 11.4, as an example, it is shown the conceptual model of the Pesticide in Flooded Application Model—PFAM (White et al. 2016), which is used by the USEPA to estimate pesticide concentrations in surface water in flooded fields, such as rice paddies.

Boesten (2000) presented the steps for developing and testing an explanatory model of pesticide environmental behavior. As a first step, it is necessary to define the problem for which the model will be developed. After this definition, a conceptual model is elaborated along with the mathematical equations to describe each considered process, which will be translated into a computer program that solves the equations using a stable and reliable procedure. Next, input data are obtained by means of laboratory experiments. The next important step corresponds to field experiments to obtain observed data on environmental behavior of the pesticide, and that will be compared with the results of simulations. Finally, it is decided whether the explanatory model described and explained the environmental behavior of the pesticide in the field satisfactorily or not, based on laboratory, climate, soil, and crop data set. If the simulations have not been satisfactory, it is necessary to review all the testing steps, starting again with the conceptual model. It is important to emphasize that experimental methods for obtaining data in laboratory and in the field are important sources of error and, therefore, can influence the test of a mathematical model.

In addition to grouping models in descriptive or explanatory, they can be classified in deterministic versus stochastic, functional (or empirical) vs mechanistic, for management vs research purpose.

A deterministic model assumes that an input data set results in a single set of simulated values. On the other hand, a stochastic model assumes that input data can be represented by random variables (that can be represented by a distribution of probability). An example of a common random variable when studying environmental behavior of pesticides is soil hydraulic conductivity, which can lead to the spatial variation of leaching. Consequently, the simulated results correspond to ranges of values.

A mechanistic model uses, for description and explanation of the considered processes, mathematical equations based on chemical, physical, and/or biological processes, as they are currently understood. For example, for pesticide leaching, this type of model uses equation that combines the mass flow, diffusion, and dispersion mechanisms. Functional models give processes a simplified treatment to describe the processes.

A mechanistic model usually reflects observed or hypothesized biological or physical mechanisms and has parameters with real-world interpretation. In contrast, functional models selected for specific numerical properties are fitted to data; model parameters may or may not have real-world interpretation. When data quality is otherwise equivalent, extrapolation from mechanistic models (e.g., biologically-based dose–response models) often carries higher confidence than extrapolation using empirical models (USEPA 2011b).

Research models are used to enhance the understanding of a system and can be used to test hypotheses or to identify gaps in research that need to be studied. Management models are mainly used to aid in decision-making on the adoption of agricultural practices.

Although some authors have classified and differentiated some types of models of pesticide environmental behavior, Jarvis et al. (1995) mentioned that a very clear distinction between these types may be difficult in some cases. For example, one may have a model in which soil water flow is mechanically treated, while degradation of the pesticide in the soil is taken in a functional way. It is important to understand that giving a mechanistic treatment to a process may not be possible in some cases, due to lack of knowledge or because it is very difficult to parameterize the process.

As mentioned by Di Guardo et al. (2018), improvements in characterizing and quantifying physico-chemical properties and emission estimates and refinements in model formulation and parameterization have led to better comparability between model results and monitoring data, which in turn imparts greater confidence to the use of models within regulatory chemical risk assessment.

In the context of regulatory risk assessment, it is important to strike a balance between increasing model complexity and diversity driven by scientific progress, on the one hand, and the need for models that are practical and relatively easy to use, on the other hand. Ongoing scientific development and the presence of open scientific questions must not reduce the availability of practical tools for regulatory risk assessment out of 'paralysis by analysis.' Therefore, there is a continuous need for the translation of new scientific insights into the sphere of (often simplified) modeling tools suitable for regulatory risk assessment. This translation process is a task in itself and requires time and resources (Di Guardo et al. 2018).

As noted by Schmolke et al. (2010), it is important to bear in mind that the purposes of scientific publications and regulatory dossiers are quite different; in pure science, models are assessed with regard to their originality and the general insights they provide. In ecological applications, including pesticide registration, decision-making requires that all protocols and tools used in the risk assessments are transparent, standardized (to some degree), and sufficiently reliable to inform real-world decision-making.

11.5 Tools for Aquatic Fate-Exposure Characterization

As highlighted by Di Guardo et al. (2018), models are currently heavily relied upon to assess chemical hazard, exposure and risk, and to formulate chemical management strategies. In ERA of pesticides, characterization of exposure using mathematical models is usual and they have been used to estimate pesticide concentrations in habitats (environmental compartments), amounts in contact with organisms, and doses in dietary elements.

As pointed out by Dubus et al. (2002), the use of an inappropriate model will lead to a poor simulation and, although detailed information on individual models is usually widely available, little guidance is accessible to support model selection on the basis of objective criteria. Guidance such as that generated by Pennell et al. (1990) would be useful. Del Re and Trevisan (1995) identified a number of criteria for selecting models, however these were generic. Vanclooster et al. (2000b) presented an assessment of twelve models simulating water movement and pesticide transport in soil, with the purpose of describing the problems encountered in using these models, and to introduce a Good Modeling Practice.

Aquatic models are used to estimate pesticide concentrations in water, and the estimated concentrations are used to assess exposure to aquatic organisms. Aquatic fate-exposure models currently adopted by the USEPA in ERA (USEPA 2017) are:

- Pesticide in Water Calculator—PWC (USEPA 2016).
- Tier I Rice Model (USEPA 2007).
- Pesticide in Flooded Applications Model—PFAM (Young 2013).
- Kow-based Aquatic Bioaccumulation Model—KABAM (USEPA 2009).

PWC is an updated version of Surface Water Concentration Calculator—SWCC. It can simulate both surface water and groundwater and has an improved volatilization routine and more batch run capabilities. PWC is, in fact, a specialist system and comprises a graphical user interface, a field model (PRZM), and a water body model (Variable Volume Water Model—VVWM).

PRZM is a continuous simulation model that considers interactions of pesticides in surface runoff (in water and on eroded sediment), advection in percolating water, molecular diffusion, dispersion, uptake by plants, sorption to soil, and biological and chemical degradation (Carsel et al. 1985). PRZM is a standard model to be used in environmental risk and exposure assessments by the USEPA, and it is included in the list of recommended regulatory models for pesticide registration.

VVWM (Young 2016) simulates the processes that occur in the water body by using the runoff and spray drift loading generated by PRZM to estimate the fate, persistence, and concentration of a pesticide in a water body on a day-to-day basis. As such, the model accounts for volatilization, sorption, hydrolysis, biodegradation, and photolysis of the pesticide. In VVWM, the user can vary its volume on a daily scale and to include sediment burial, although these features are only used for higher tiered assessments.

Tier I Rice Model is used to estimate surface water exposure from the use of pesticides in rice paddies. As a screening-level model, it provides short- and long-term concentrations that can be used for both aquatic organisms and drinking water exposure assessments.

PFAM is used to estimate surface water exposure from the use of pesticides in flooded fields. In comparison with the Tier 1 Rice Model, PFAM allows for more advanced evaluation of pesticide use on flooded agricultural areas, accounting for water and pest management practices and for degradation in soil and aquatic environments, as well as for post-processing of discharged paddy waters to a user-defined receiving water.

KABAM is used by USEPA to estimate potential bioaccumulation of hydrophobic organic pesticides in freshwater aquatic food webs and subsequent risks to mammals and birds via consumption of contaminated aquatic prey.

Some models or specialist systems are no longer used by USEPA; nevertheless, they can be used in other countries, such as: Screening Concentration in Ground Water—SCI-GROW (Barrett 1997); EXAMS-PRZM Exposure Simulation Shell—EXPRESS (Crowe and Mutch 1994; Burns 2006); FQPA Index Reservoir Screening Tool—FIRST (USEPA 2008); Generic Estimated Environmental Concentration—GENEEC (Parker et al. 1995); and Surface Water Concentration Calculator—SWCC (Fry et al. 2014).

As a screening tool, SCI-GROW provides conservative estimates of pesticide concentrations in groundwater and considers minimal chemical parameters. Results provided from it are a single fixed-point estimate of exposure that is assumed to represent the entire USA and does not consider regional differences in soil type and climate conditions or areas with greater vulnerability (Rice et al. 2007). If SCI-GROW-based assessment results indicate that pesticide concentrations in drinking water exceed levels of concern, the ability to refine the assessment is limited. As a more refined approach was desired by both regulators and industry to assess potential groundwater contamination, SCI-GROW has been replaced by PRZM-GW.

The PRZM-GW (PRZM for groundwater) is used as a screening-level and refined tool for risk assessment purposes. In general, PRZM-GW provides conservative estimates of pesticide concentrations in groundwater; however, it also allows for region-specific scenario development in cases where refined drinking water assessments are needed. The PRZM-GW is a 1-dimensional finite difference solution to the fate and vertical transport of pesticides through the crop root zone. In the PRZM-GW conceptual model, pesticides are applied to a field of uniform crop with negligible runoff and erosion and maximum infiltration of precipitation and irrigation water. The field sits atop a standardized soil profile and shallow unconfined aquifer with a 100-cm drinking water well screen. Average groundwater concentrations in the well screen zone simulated on a daily time step provide the estimated concentrations used in exposure and risk assessments (Padilla et al. 2017).

EXPRESS is a shell for PRZM and EXAMS, and it was designed to facilitate a rapid and consistent assessment of aquatic pesticide exposure on a variety of crops. PRZM (PRZM per se and VADOFT) combines differing root zone and vadose zone characteristics into a single simulation to predict pesticide transport and transformation in the crop root and unsaturated zones, and edge-of-field losses in runoff waters and eroded sediment. Daily edge-of-field loadings of pesticides dissolved in runoff waters and sorbed to entrained sediment, as predicted by PRZM, are discharged into a standard water body simulated by the EXAMS model. The EXAMS model accounts for subsequent hydrologic transport, volatilization, sorption, hydrolysis, biodegradation, and photolysis of the pesticide.

EXAMS (Burns et al. 1982) is an interactive software application for formulating aquatic ecosystem models and rapidly evaluating the fate, transport, and exposure concentrations of synthetic organic chemicals including pesticides, industrial materials, and leachates from disposal sites.

FIRST tool is used to assess exposure to pesticides in drinking water. Using a few basic chemical parameters (e.g., half-life in soil) and pesticide label application information, FIRST estimates peak values (acute) and long-term (chronic) average concentrations of pesticides in water. It is based on the linked PRZM and EXAMS models and is a single-event process.

GENEEC is a screening model to predict environmental concentrations of pesticides in surface water for aquatic exposure assessments. It is simpler than the PRZM and EXAMS models in its treatment of hydrology. In GENEEC it is assumed that a single large rainfall/runoff event occurs that removes a large quantity of pesticide from the field to the water all at one time. Longer-term, multiple-day average concentration values are calculated based on the peak day value and subsequent values considering degradation processes. A 'standard agricultural field-farm pond' scenario is used, which assumes that rainfall onto a treated, 10-ha agricultural field causes pesticide-laden runoff into a 1-ha, 2-m-deep water body.

SWCC estimates pesticide concentrations in water bodies that result from pesticide applications to land. It is designed to simulate the environmental concentration of a pesticide in the water column and sediment. USEPA has used SWCC for tier II estimations of pesticide concentrations in surface waters for drinking water and aquatic exposure assessments. It uses PRZM and VVWM.

Spray drift can contribute to the amount of pesticide in surface water bodies. AGDISP™—Agricultural Dispersal (Bilanin et al. 1989) and AgDRIFT® (Teske et al. 2000) are examples of models designed to predict spay drift from application sites. AGDISP™ is a 'first-principles' science-based model that predicts spray drift from application sites. The model was developed by the USDA Forest Service. AGDISP™ was designed to optimize agricultural spraying operations and has detailed algorithms for characterizing the release, dispersion, and deposition over and downwind of the application area. This model can be used in estimating downwind deposition of spray drift from aerial and ground boom applications. In addition, it can be used in estimating downwind deposition of spray drift from forestry applications.

AgDRIFT® is a modified version of AGDISP™ model and has the capability to assess a variety of spray drift conditions from agricultural applications and off-site deposition of liquid formulation of pesticides. This model can be used in estimating downwind deposition of spray drift from aerial, ground boom, and orchard/vineyard airblast applications.

In the EU, the currently approved fate-exposure models for surface water and groundwater are:

- MACRO (Jarvis 1991, 1994; Larsbo and Jarvis 2003, 2005).
- Pesticide Root Zone Model—PRZM (Carsel et al. 1984, 1985, 1997, 1998; Mullins et al. 1993).
- Pesticide Emission Assessment at Regional and Local Scales—PEARL (Leistra et al. 2000, 2001; Tiktak et al. 2000; Tiktak et al. 2002).
- Pesticide Leaching Model—PELMO (Klein 1995, 2011).
- Surface Water Tool for Exposure Predictions Steps 1 and 2—STEPS_ONE_TWO (FOCUS 2001a).
- Surface Water Scenarios Help—SWASH (Roller et al. 2003).
- Toxic Substances in Surface Waters—TOXSWA (Adriaanse 1996; Beltman and Adriaanse 1999).

MACRO is a physically based 1-dimensional numerical model of water flow and reactive solute transport in field soils. It calculates coupled unsaturated-saturated water flow in cropped soil, including the location and extent of perched water tables, and can also deal with saturated flow to field drainage systems. The model accounts for macropore flow, with the soil porosity divided into two flow systems or domains (macropores and micropores) each characterized by a flow rate and solute concentration. Richards' equation and the convection-dispersion equation are used to model soil water flow and solute transport in the soil micropores, while a simplified capacitance-type approach is used to calculate fluxes in the macropores. Exchange between the flow domains is calculated using approximate, physically based, expressions based on an effective aggregate half-width. Additional model assumptions include first-order kinetics for degradation in each of four 'pools' of pesticide in the soil (micro- and macropores, solid/liquid phases), together with an instantaneous sorption equilibrium and a Freundlich sorption isotherm in each pore domain.

PEARL is a 1-dimensional numerical model of pesticide behavior in the soil-plant system. Water flow in soil is described by Richards' equation including a range of possible lower boundary conditions (e.g., groundwater levels that fluctuate in response to the rainfall input). Soil evaporation and plant transpiration is calculated via multiplying a reference evapotranspiration rate with soil and crop factors. For the FOCUS scenarios, crop growth is simulated with a simple growth model that assumes a fixed length of the growing season. Heat flow in soil is described with Fourier's law. The thermal properties are a function of porosity and water content and are therefore a function of time and soil depth. PEARL is based on: (i) the convection-dispersion equation including diffusion in the gas phase with a temperature dependent Henry coefficient, (ii) a two-site Freundlich sorption model (one equilibrium site and one kinetic site), (iii) a transformation rate that depends on water content, temperature, and depth in soil, (iv) a passive plant uptake rate. The model includes formation and behavior of transformation products and describes also lateral pesticide discharge to drains. PEARL does not simulate preferential flow. Volatilization from the soil surface is calculated assuming a laminar air layer at the soil surface. It uses an explicit finite difference scheme that excludes numerical dispersion.

PELMO is a 1-dimensional simulation model simulating the vertical movement of pesticides in soil by chromatographic leaching, based on PRZM (Carsel et al. 1984). After the release of the first version of PELMO, runoff routines were upgraded, and routines for estimating the volatilization of pesticides were added, as well as a complementary tool in order to enable the transformation of the applied active ingredient to possible degradation products.

In Table 11.1 characteristics of MACRO, PEARL, PELMO, and PRZM models are briefly presented.

Garratt et al. (2002) pointed out that an important difference in models is the descriptions of water flow. PRZM and PELMO use a capacity approach to water flow, whereby water in excess of field capacity in any layer moves down to the next layer within that time step (usually 1 day). MACRO, in turn, uses Richards' equation to describe water flow, whereby water flow is determined by differences in water potential and soil hydraulic conductivity. MACRO is capable of simulating by-pass or macropore flow of water, as may occur in structured soils, by splitting the soil pore volume into a fast water flow domain and a slow water flow domain. If the slow-flowing domain is not sufficient for water flow through the matrix, then excess water is transported in the fast domain. Diffusive and dispersive fluxes are handled differently by these models: PRZM adopts numerical procedures based on soil layer thickness, while MACRO calculates these fluxes based on user-specified diffusivity and dispersivity parameters. PELMO uses a combination of both methods. PRZM, PELMO, and MACRO have modules to describe plant uptake of water.

Instantaneous, linear sorption, and first-order degradation can be simulated by all these three models (PRZM, PELMO, and MACRO). PELMO and MACRO simulate time-dependent change in sorption, and PELMO and MACRO simulate non-linear (Freundlich) sorption. PELMO can also simulate higher-order degradation pathways. PRZM (version 3), PELMO, and MACRO can consider the effect of soil moisture content and soil temperature on pesticide degradation (Garrat et al. 2002).

Table 11.1 Characteristics of the models MACRO, PEARL, PELMO, and PRZM

	MACRO	PEARL	PELMO	PRZM
Model type	Single-porosity and dual-permeability model	Single-porosity model	Single-porosity model	Single-porosity model
Water flow	Micropores: Richards' equation Macropores: Convection (gravity flow)	Richards' equation	Capacity based	Capacity based
Pesticide processes	Leaching Sorption Degradation Plant uptake	Leaching Sorption Degradation Volatilization Plant uptake	Leaching Surface Runoff Sorption Degradation Volatilization Plant uptake	Leaching Surface Runoff Sorption Degradation Volatilization Plant uptake
Leaching	Matrix: Convection-dispersion equation Preferential flow: Convection (gravity flow)	Convection-dispersion equation	Convection-dispersion equation	Convection-dispersion equation
Surface runoff	–	–	Soil Conservation Service curve number technique	Soil Conservation Service curve number technique
Sorption	Matrix: Linear or Freundlich, instantaneous equilibrium and kinetic Preferential: Linear or Freundlich	Linear or Freundlich, instantaneous and non-equilibrium	Freundlich, option for increase of sorption with time and automated pH-dependence	Freundlich, option for increase of sorption with time
Degradation	First-order kinetics, separate rate coefficients for solid and liquid phases, function of soil temperature and moisture	First-order kinetics	First-order kinetics, constant rate with soil temperature and moisture	First-order kinetics, with options for bi-phasic degradation, and effects of soil temperature and moisture
Volatilization	–	Soil and canopy	Simple model using Fick's and Henry's law	Combination of results from previous research

(continued)

Table 11.1 (continued)

	MACRO	PEARL	PELMO	PRZM
Plant uptake	Empirical sink terms	A function of transpiration	Simple model based on soil concentrations	Simple model based on soil concentrations

As summarized by Giannouli and Antonopoulos (2015), MACRO and PEARL account for a complete water balance including water flow, canopy interception and root water uptake, seepage to drains and groundwater. PEARL can also account for pesticide volatilization from soil or plant canopy, while MACRO does not take into account this process. Crop growth in both models is described by a simple model where both leaf area index and root depth are a function of the development stage of the crop. In MACRO, leaf area index and root depth follow a logistic curve, when in PEARL both are linear. Potential evapotranspiration is calculated using the equation of Penman–Monteith (Allen et al. 1998) or can be pre-calculated and provided by the user and can be partitioned into actual evaporation and transpiration through reduction functions. Finally, both models assume that runoff happens when infiltration capacity is exceeded.

STEPS_ONE_TWO is a stand-alone tool for the derivation of PEC values in water and sediment based on the chosen scenario. The tool requires a minimum of input values, and it is designed to evaluate both active substances and metabolites. At Step 1, inputs of spray drift, runoff, erosion, and/or drainage are evaluated as a single loading to the water body and 'worst-case' surface water and sediment concentrations are calculated. At Step 2, inputs are evaluated as a series of individual loadings comprising drift events followed by a loading representing a runoff, erosion, and/or drainage event four days after the final application. This assumption is similar to that developed by the USEPA in their GENEEC model (Parker et al. 1995). Degradation is assumed to follow first-order kinetics in soil, surface water, and sediment, and the user also has the option of using different degradation rates in surface water and sediment.

SWASH is an overall user-friendly shell, encompassing a number of individual tools and models, involved in Step 3 calculations for the FOCUS surface water scenarios. Its main functions are: (i) maintenance of a central pesticides properties database for use in MACRO, PRZM, and TOXSWA; (ii) preparation of other input for these models, notably application patterns, application methods, and dosages; (iii) creation of projects, containing all Step 3 FOCUS runs required for use of a pesticide on a specified crop; (iv) calculation of spray drift deposition onto ditch, stream, and pond-like water bodies (Drift Calculator); and (v) provision of an overview of crop and water body combinations in each scenario, of the extent of each scenario and of the installed versions of each model, including its shell and database. The SWASH tool may also be useful for refined risk assessment calculations of Step 4 by editing the input files prepared by SWASH outside the shells of the MACRO, PRZM, and TOXSWA models.

TOXSWA is a quasi-2-dimensional numerical model of pesticide behavior in a small surface water system, including sediment. It describes the behavior of pesticides in a water body at the edge-of-field scale (i.e., a ditch, pond, or stream adjacent to a single field). TOXSWA considers four processes: (i) transport, (ii) transformation, (iii) sorption, and (iv) volatilization. In the water layer, pesticides are transported by advection and dispersion, while in the sediment, diffusion is included as well. The transformation rate covers the combined effects of hydrolysis, photolysis, and biodegradation, and it is a function of temperature. It does not simulate formation of metabolites. Sorption to suspended solids and to sediment is described by the Freundlich equation. Sorption to macrophytes is described by a linear sorption isotherm. Pesticides are transported across the water-sediment interface by diffusion and by advection. FOCUS-TOXSWA handles transient hydrology and pesticide fluxes resulting from surface runoff, erosion and drainage as well as instantaneous entries via spray drift deposition.

In the EU, the Drift Calculator (FOCUS 2001b, 2015) is used to estimate the spray drift contribution to the amount of pesticide in surface water bodies. It is embedded in the SWASH shell and is run automatically in the background to provide drift values for use in TOXSWA simulations of aquatic fate. For Step 3 FOCUS calculations, types of water bodies associated with each scenario and default distances between crop and water have been set by the FOCUS SW group. The user enters the crop type, the application rate, and the number of applications per season in SWASH, and the SWASH shell automatically runs the Drift Calculator and generates a report for each TOXSWA run. It is also useful for Step 4 calculations for the FOCUS surface water scenarios.

In lower tiers of the assessment in the European pesticide authorization (FOCUS 2000, 2001a), numerical pesticide fate models such as MACRO, PRZM, PEARL, and PELMO are used in combination with a limited number of standard scenarios to assess the PECs, as reported by Tiktak et al. (2013). A scenario consists of a combination of soil, climate, and crop parameters to be used in modeling. Standardized scenarios are needed because they increase the uniformity of the regulatory evaluation process by minimizing the influence of the methodology applied and of the person that performs the PEC calculation and because they make exposure assessments and their interpretation easier for regulators and industry (Boesten 2000; Vanclooster et al. 2004).

Many of the models capable of simulating the behavior of pesticides present temporal and, mainly, spatial limitations. An important feature of modeling pesticide behavior is that input data used in models have spatial distribution, which greatly affects processes involved and their interactions. Therefore, spatial characterization of pesticide behavior and fate in the environment requires tools that can effectively handle georeferenced data. Simulation models can be more effective with visualization and spatial analysis through a geographic information system—GIS. Simulation interpretation and understanding of environmental contamination processes are enhanced by spatial visualization of model results (Engel et al. 1997), while advanced spatial analysis facilitates and improves simulation (Stoorvogel 1995; Campbell et al. 2000). The need for enhancing the understanding of the temporal and spatial dynamics of ecosystem exposure within ERA procedures has recently been pointed out in the literature (e.g., Di Guardo and Hermens 2013; Ghirardello et al. 2014). GeoPEARL (Tiktak et al. 2003) and EuroPEARL (Tiktak et al. 2006) were developed to evaluate the risk related to leaching on a geographical basis.

Most of the models or shells developed for ERA of pesticides in developed countries are data demanding and adjusted to specific regional or local scenarios. In developing countries, there is a need for simpler models for the first tier or step of ERA. The following models have been used in Brazil for ERA purpose (IBAMA 2012) concerning aquatic organisms and drinking water: SCI-GROW, GENEEC, and ARAquá.

The ARAquá model (Spadotto et al. 2010; Spadotto and Mingoti 2014), which has been developed in Brazil, estimates pesticide concentrations in surface and groundwater bodies from agricultural uses and compares them with parameters of toxicity to aquatic organisms and water quality for human consumption. Equations in ARAquá software are conservative in making estimates, and therefore, it is to be used only in screening tier of ERA of pesticides. For this feature, the estimates are made so that only environmentally safe combinations of pesticide–soil–weather may be exempted from more refined calculations in successive tiers of assessment. The registration by user of conditions of weather and terrain and properties of soil and pesticide allows calculations for specific situations, in addition to those pre-registered that follow with the software. Thus, ARAquá proves adaptable to any environmental conditions and limited availability of data.

11.6 A Case Study of ERA of Pesticides with Limited Availability of Input Data

As a case study of ERA of pesticides when few input data are available, the ARAquá tool was used to estimate surface water and groundwater concentrations in Brazilian scenarios.

The estimate of pesticide concentration in surface water in the ARAquá tool is made considering a 'standard pond' scenario, as defined by Parker et al. (1995), which assumes that rainfall onto a treated, 10-ha agricultural field causes pesticide-laden runoff into a 1-ha, 2-m-deep water body.

The estimated pesticide loss in surface runoff M (μg) is based on the simplified formula for indirect loadings caused by runoff—SFIL (OECD 1999), as follows:

$$M = D \times c \times f \times \exp\left(-3 \times \frac{0.693}{t^{1}/_{2}}\right) \times \frac{100}{1 + Kd} \qquad (11.1)$$

where D is dose of the pesticide (g ha^{-1} converted to μg ha^{-1}); c is runoff coefficient, which is the ratio of volume of drained water and precipitation volume; f is a cor-

Table 11.2 Values of runoff coefficient c for agricultural fields, depending on terrain slope and soil texture

Terrain slope (%)	Soil texture		
	Sandy	Medium	Clayey
0–5	0.30	0.50	0.60
5–10	0.40	0.60	0.70
10–30	0.52	0.72	0.82

Source USDA (1991)

rection factor, product of three components: f_1, f_2, and f_3; half-life of the pesticide in soil is represented as $t\frac{1}{2}$ (*days*) and its sorption coefficient as Kd (mL g^{-1}).

Table 11.2 shows some values of runoff coefficient for agricultural fields. These values were originally recommended by the Soil Conservation Service—SCS (currently known as Natural Resources Conservation Service—NRCS) of the USDA (USDA 1991). In ARAquá, it was initially registered a scenario with c of 0.60; however, other values can be used.

The component f_1 of the f factor takes into account the terrain slope in the field around the standard pond. If the slope is greater than or equal to 20%, $f_1 = 1$; on the other hand, if the slope is less than 20%, $f_1 = 0.02153d + 0.001423d^2$, where d is slope (%). In ARAquá, it was registered a scenario with $d = 10\%$, which can be changed in other data settings.

The component f_2 considers the interception of pesticides by crop canopy, and it is calculated by: $f_2 = 1 - i/100$, where i is interception by plants (%). It was considered that there is no crop plants, where $i = 0$, and thus $f_2 = 1$. However, the user can enter manually other value of i.

The component f_3, in turn, is related to the presence buffer zone between the pesticide application field and the standard pond. It is calculated by: $f_3 = 0.83^w$, where w is width of vegetated buffer zone (*m*). No buffer zone was initially assumed, which implies that $w = 0$, and thus $f_3 = 1$. Users can register other scenarios with different values of w. The overall correction factor is dimensionless and calculated as: $f = f_1 \times f_2 \times f_3$.

In SFIL, it was assumed that three days after pesticide application a runoff event happens; therefore, it estimates the proportion of pesticides applied is lost in water solution in the runoff event. The losses are then calculated for 10 ha (M_{10}), and the pesticide concentration in the pond $EECsw$ (µg L^{-1}) is estimated by:

$$EECsw \cong \frac{M10}{v} \tag{11.2}$$

where v is water volume in the standard pond (2×10^7 L).

The mathematical calculation of pesticide in soil leaching is based on the set of equations presented by Rao et al. (1976) and Rao et al. (1985). The generalized form is used for soils with several layers (or horizons). The expression presented by Rao et al. (1985) is:

$$AF = \exp(-tr \times k) \tag{11.3}$$

where AF is attenuation factor (*mass/mass*), tr represents travel time (day), and k is degradation constant of the pesticide in soil (1 day^{-1}).

Pesticide half-life $t\frac{1}{2}$ (day) is related to k according the equation:

$$k = \frac{0.693}{t^{1}/_{2}} \tag{11.4}$$

Travel time (day) is calculated by the following equation:

$$tr = \left(\frac{L \times FC}{q}\right) \times RF \tag{11.5}$$

where L represents distance to the groundwater body (m), FC is soil field capacity (*volume/volume*), and q stands for net recharge rate (mm year^{-1} converted to m^3 m^2 day^{-1}).

Retardation factor RF (*dimensionless*) of the pesticide movement is obtained by the equation:

$$RF = 1 + \frac{(BD \times OC \times K_{oc})}{FC} \tag{11.6}$$

where BD is soil density (g cm^{-3}), OC is soil organic carbon content (*mass/mass*), and K_{oc} is organic carbon pesticide partitioning coefficient (mL g^{-1} or cm^3 g^{-1}).

Estimate pesticide mass (μg) reaching the groundwater body is given by: $m = D \times AF$, where D is pesticide doses (g ha^{-1} converted to μg ha^{-1}).

Pesticide concentration in groundwater $EECgw$ (μg L^{-1}) is estimate by means of the expression:

$$EECgw \cong \frac{m}{p \times d \times a} \tag{11.7}$$

where p represents aquifer porosity (*volume/volume*), d is mix depth in groundwater body (assumed to 2 m), and a represents the area of 1 ha (10,000 m^2).

The estimated environmental concentration—EEC—of pesticides in surface water and groundwater can be compared with toxicity parameters. The determination of the EEC corresponds to the characterization of environmental exposure.

For surface water, ARAquá calculates the peak EEC in the pond water for a given pesticide, under certain conditions; thus, results cannot be used to characterize the chronic risk to aquatic plants and animals. In order to estimate the acute risk, the effect concentration for 50% of test population—EC$_{50}$—is adopted. The risk quotient—RQ—(Urban and Cook 1986), as a function of hazard and exposure, is calculated dividing the EEC value by the acute toxicity parameter of each pesticide. In order to assist in the interpretation of RQ, it is used the level of concern—LOC,

Table 11.3 Parameters of toxicity and level of concern—LOC for ERA

Risk presumption	Toxicity parameter	Level of concern (LOC)
Aquatic animals		
Acute high risk	LC_{50} or EC_{50}	0.5
Acute restricted use	LC_{50} or EC_{50}	0.1
Acute endangered species	LC_{50} or EC_{50}	0.05
Chronic risk	NOAEC	1.0
Aquatic plants		
Acute high risk	EC_{50}	1.0
Acute endangered species	EC_{05} or NOEC	1.0

Source USEPA (2016)

Note LC_{50} = lethal concentration for 50% of test population: that causes 50% mortality of the population of organisms subjected to the test conditions in exposure time (24–96 h); EC_{50} = effect concentration for 50% of test population: causing acute effect (e.g., immobility) of 50% of the population of organisms subjected to the test conditions in exposure time (24–48 h); EC_{05} = effect concentration for 5% of test population; NOAEC = no observed adverse effects concentration: the highest concentration that does not cause a statistically significant deleterious effect on test conditions at the time of exposure (7 days). NOEC = no observed effect concentration: the highest concentration that causes no statistically discernible effect

which is the limit value of RQ. Table 11.3 shows parameters of toxicity and the LOC values used by the USEPA (USEPA 2016).

The compliance with drinking water standards of the USEPA and the Lifetime Health Advisory from USEPA (2012) is considered, and the chronic RQ for drinking water is calculated using the maximum values for the pesticide drinking water standard. LOC of 1.0 for drinking water is adopted, and standards are considered for both groundwater and surface water.

In this case study, agricultural areas with annual crops (mainly soybean, corn and cotton) in the Cerrado biome in the central-west region of Brazil were turned to account as scenarios of use of herbicides in order to estimate their concentrations in surface and groundwater bodies.

Information on scenarios of surface water and groundwater in agricultural areas of the Cerrado biome, central-west region of Brazil, is provided in Tables 11.4 and 11.5. Properties of the soils are presented in Table 11.6. Soils with the largest total crop areas in that region are, according to the Brazilian System of Soil Classification, Latossolos (Oxisols in the USDA Soil Taxonomy and Ferralsols in the FAO Soil Classification) and Neossolos (Entisols in the USDA Soil Taxonomy and Arenosols in the FAO Soil Classification).

Table 11.4 Information on agricultural areas for surface water scenarios in the Cerrado biome, central-west region of Brazil

	Scenario for surface water	
	A	B
Soil type	Latossolos (Oxisols, Ferralsols)	Neossolos (Entisols, Arenosols)
Terrain slope	12%	3%
Runoff coefficient	0.8	0.3

Table 11.5 Information on agricultural areas for groundwater scenarios in the Cerrado biome, central-west region of Brazil

	Scenario for groundwater			
	I	II	III	IV
Soil type	Latossolos	Latossolos	Neossolos	Neossolos
Precipitation (mm/year)	1250	1750	1250	1750
Evapotranspiration (mm/year)	900	900	900	900
Soil-saturated zone porosity (v/v)	0.630	0.630	0.433	0.433
Water table depth (cm)	100	100	100	100

Table 11.6 Soil properties in the scenarios of the Cerrado biome, central-west region of Brazil

Layer depth (cm)	Field capacity (v/v)	Soil density (g cm^{-3})	Organic carbon content (m/m)
Latossolos (Oxisols)			
0–25	0.46	0.93	0.021
25–145	0.47	1.03	0.008
Neossolos (Quartzpsaments)			
0–50	0.08	1.42	0.007
50–250	0.14	1.50	0.002

In Table 11.7, information on the considered herbicides is shown, and the properties of their behavior in soil are in Table 11.8.

Concentration of the herbicides estimated by using the ARAquá in the two scenarios for surface water is presented in Table 11.9, and the respective values of RQ are in Table 11.10.

Estimated concentration and RQ values of the herbicides in the four scenarios for groundwater are presented in Tables 11.11 and 11.12.

Results shows that the highest concentration values in surface water were estimated in the scenario with Latossolos in the Cerrado biome, central-west region of Brazil. RQ values show that the risk of the use of diuron, ametrine, or pyroxasulfone is very high for aquatic plants in those conditions and should be assessed in refined higher tier that is less conservative. The ecological risk of those herbicides for aquatic plants also needs attention in scenarios with Neossolos in that region.

Table 11.7 Dose and ecotoxicological data of the considered herbicides

Herbicide	Dose (g ha^{-1})	Aquatic animals EC50 (μg L^{-1})	Aquatic plants EC50 (μg L^{-1})	Drinking water standard (μg L^{-1})
2,4-D Amine	1080	260	210	30
Ametrine	3000	1700	3.6	60
Atrazine	2500	720	22	2
Bentazon	960	48,000	5400	200
Diuron	3200	160	2.4	90
Pyroxasulfone	240	2200	0.38	(NA)
Trifluralin	1068	8.4	15.3	10

NA Not available

Table 11.8 Properties of considered herbicides in soil

Herbicide	Organic carbon sorption coefficient (mL g^{-1})	Soil half-life (days)
2,4-D Amine	20	10
Ametrine	300	60
Atrazine	100	60
Bentazon	34	20
Diuron	480	90
Pyroxasulfone	223	22
Trifluralin	8000	60

Table 11.9 Estimated concentration values for different herbicides in two scenarios for surface water

Herbicide	Estimated concentration (μg L^{-1}) in surface water	
	Scenario A	Scenario B
2,4-D Amine	114.48	8.93
Ametrine	73.56	10.85
Atrazine	144.35	16.49
Bentazon	93.54	8.11
Diuron	52.30	8.33
Pyroxasulfone	7.12	0.99
Trifluralin	1.13	0.21

Table 11.10 Risk quotient (RQ) values for different herbicides in two scenarios for animals and plants in surface water

Herbicide	RQ			
	Scenario A		Scenario B	
	Aquatic animals	Aquatic plants	Aquatic animals	Aquatic plants
2,4-D Amine	0.44	0.55	0.03	0.04
Ametrine	0.04	20.40	0.01	3.01
Atrazine	0.20	6.56	0.02	0.75
Bentazon	<0.01	0.02	<0.01	<0.01
Diuron	0.33	21.80	0.05	3.47
Pyroxasulfone	<0.01	18.70	<0.01	2.60
Trifluralin	0.14	0.07	0.03	0.01

Table 11.11 Estimated concentration values for different herbicides in four scenarios for groundwater

Herbicide	Estimated concentration ($\mu g\ L^{-1}$)			
	Scenario I	Scenario II	Scenario III	Scenario IV
2,4-D Amine	<0.001	<0.001	<0.001	0.100
Ametrine	<0.001	<0.001	<0.001	0.013
Atrazine	<0.001	0.081	0.032	6.760
Bentazon	<0.001	<0.001	0.001	0.817
Diuron	<0.001	<0.001	<0.001	0.009
Pyroxasulfone	<0.001	<0.001	<0.001	<0.001
Trifluralin	<0.001	<0.001	<0.001	<0.001

Table 11.12 Risk quotient (RQ) values for different herbicides in four scenarios for drinking standard in groundwater

Herbicide	RQ			
	Scenario I	Scenario II	Scenario III	Scenario IV
2,4-D Amine	<0.001	<0.001	<0.001	0.003
Ametrine	<0.001	<0.001	<0.001	<0.001
Atrazine	<0.001	0.040	0.016	3.380
Bentazon	<0.001	<0.001	<0.001	0.004
Diuron	<0.001	<0.001	<0.001	< 0,001
Pyroxasulfone	(–)	(–)	(–)	(–)
Trifluralin	<0.001	<0.001	<0.001	<0.001

(–) Not considered (drinking water standard not available)

The estimated concentrations in groundwater were low, except for atrazine in the scenario with Neossolos and net water recharge of 850 mm per year, in the Cerrado biome, central-west region, Brazil. Atrazine presented in those conditions the highest RQ value and its estimated concentration is above the drinking water standard. Higher refined tiers need to be carried out in order to verify if the risk is high even in a less conservative assessment.

This is a case study using a selected tool and available dataset. Thus, results presented here do not substitute the ERA conducted by official agencies.

11.7 Final Considerations

As noted by Rice et al. (2007), many advances have been made in assessing the environmental fate of organic compounds. Modeling will probably continue to play a major role in pesticide exposure with the implementation of new models able to handle always more relevant processes. These models could also be used to design the most effective mitigation measures or combination of measures to be implemented. Ecological effect modeling will also need to be further developed and implemented for ERA.

Development and use of mathematical models, formation of databases, and creation of computerized tools can make simulations of environmental concentrations of pesticides more agile and less subject to operational errors, while ensuring the necessary scientific basis.

According to Di Guardo and Hermens (2013), some potential research areas to improve the current predictive approaches for estimating exposure are related to: the need to enlarge the applicability of models to chemicals other than the classical nonpolar (such as ionized, polar, and nanomaterials), the capability of predicting space and time variable concentrations, to describe more realistic ecological scenarios, to further improve terrestrial and aquatic bioaccumulation models and the need of integrating exposure and effects modeling approaches.

As pointed out by Boivin and Poulsen (2017), current practices of regulatory ERA are based on experimentations and modeling tools originating from the academic board. ERA of pesticides is in perpetual improvement, and it does constantly benefit from academic inputs. Even if academic and regulatory constraints, objectives, and schedules are known not to be the same, it has been obvious in recent years that collaboration between them has been fruitful for both sides and would benefit to human and environment protection.

Modeling enables both the identification of lack of information on retention, transformation, and transport processes, as well as the deepening of knowledge about the environmental behavior of pesticides under different conditions. Simulation models can also be used as didactic support in this area of knowledge.

References

ACP/ECP, American Crop Protection, European Crop Protection (1999) Framework for the ecological risk assessment of plant protection products [S.l.]: ACP/ECP. p 52. (ACP/ECP. Technical Monograph, 21)

Adriaanse PI (1996) Fate of pesticides in field ditches: the TOXSWA simulation model. DLO Winand Staring Centre for Integrated Land, Soil and Water Research, Wageningen, the Netherlands, 241 p (SC-DLO Report 90)

Allen RG, Pereira LS, Raes D, Smith M (1998) Crop Evapotranspiration guidelines for Computing Crop Water Requirements. FAO Irrigation and Drainage Paper 56, Rome, Italy

Barrett M (1997) Initial tier screening of pesticides for groundwater concentration using the SCI-GROW model. US Environmental Protection Agency, Washington, DC. Available at https://archive.epa.gov/oppefed1/web/html/scigrow_description.html

Beissinger SR, Westphal MI (1998) On the use of demographic models of population viability in endangered species management. J Wildlife Manage 62:821–841

Belluck DA, Benjamin SL, Dawson T (1991) Groundwater contamination by atrazine and its metabolites: risk assessment, policy and legal implications. In: Somasundaram L, Coats JR (eds) Pesticide transformation products: fate and significance in the environment. American Chemical Society, Washington, DC, pp 254–273

Beltman WHJ, Adriaanse PI (1999) User's manual TOXSWA 1.2. Simulation of pesticide fate in small surface waters. DLO Winand Staring Centre for Integrated Land, Soil and Water Research, Wageningen, the Netherlands, 112 p (SC-DLO Technical Document 54)

Bereswill R, Golla B, Streloke M, Schulz R (2012) Entry and toxicity of organic pesticides and copper in vineyard streams: erosion rills jeopardize the efficiency of riparian buffer strips. Agr Ecosyst Environ 146:81–92

Bilanin AJ, Teske ME, Barry JW, Ekblad RB (1989) AgDISP: the aircraft spray dispersion model, code development and experiment validation. T ASAE 32(1):327–334

Boesten JJTI (2000) From laboratory to field: uses and limitations of pesticide behaviour models for the soil/plant system. Weed Res 40(1):123–138

Boesten JJTI, Kopp H, Adriaanse PI, Brock TCM, Forbes VE (2007) Conceptual model for improving the link between exposure and effects in the aquatic risk assessment of pesticides. Ecotox Environ Safe 66:291–308

Boivin A, Poulsen V (2017) Environmental risk assessment of pesticides: state of the art and prospective improvement from science. Environ Sci Pollut R 24:6889–6894

Brock TCM, Arts GHP, Maltby L, Van den Brink PJ (2006) Aquatic risks of pesticides, ecological protection goals and common aims in EU legislation. Integr Environ Asses 2:20–46

Burns LA (2006) The EXAMS-PRZM exposure simulation shell, user manual for EXPRESS. US Environmental Protection Agency, Athens, GA (EPA/600/R-06/095)

Burns LA, Cline DM, Lassiter RP (1982) Exposure analysis modeling system (EXAMS): user manual and system documentation. US Environmental Protection Agency (EPA-600/3-82-023)

Calabrese EJ, Baldwin LA (1993) Performing ecological risk assessments. Lewis Publishers, Boca Raton, FL, p 257

Campbell KR, Bartell SM, Shaw JL (2000) Characterising aquatic ecological risks from pesticides using a diquat dibromide case study II: approaches using quotients and distributions. Environ Toxicol Chem 19(3):760–774

Carriquiriborde P, Mirabella P, Waichman A, Solomon K, Van den Brink PJ, Maund S (2014) Aquatic risk assessment of pesticides in Latin America. Integr Environ Asses 10(4):539–542

Carsel RF, Imhoff JC, Hummel PR, Cheplick JM, Donigian AS (1998) PRZM-3: a model for predicting pesticide and nitrogen fate in the crop root and unsaturated soil zones: user's manual for release 3.12. National Exposure Research Laboratory. Office of Research and Development, US Environmental Protection Agency, Athens, GA

Carsel RF, Imhoff JC, Hummel PR, Cheplick JM, Donigian JS (1997) PRZM-3, a model for predicting pesticide and nitrogen fate in crop root and unsaturated soil zones: user's manual for release 3.0. Athens, GA, USEPA

Carsel RF, Mulkey LA, Lorber MN, Baskin LB (1985) The pesticide root zone model (PRZM): a procedure for evaluating pesticide leaching threats to groundwater. Ecol Model 30(1–2):49–69

Carsel RF, Smith CN, Mulkey LA, Dean JD, Jowise, P (1984) User's manual for the pesticide root zone model (PRZM): release 1. Athens, GA: USEPA, 219 p (EPA-600/3-84-109)

Clark JR, Lewis MA, Pait AS (1993) Pesticide inputs and risks in coastal wetlands. Environ Toxicol Chem 12:2225–2233

Cornejo J, Hermisín MC, Celis R, Cox L (2005) Methods to determine sorption of pesticides and other organic compounds. In: Bened JA, Carpena RM (eds) Soil—water—solute process characterization. CRC Press, Boca Raton, pp 435–463

Crowe AS, Mutch JP (1994) An expert systems approach for assessing the potential for pesticide contamination of ground water. Ground Water 32(3):487–498

CWQG, Canadian Water Quality Guidelines (1999) Task force on water quality guidelines of the Canadian council of resource and environment ministers, Ottawa, ON

De Laender F, van den Brink PJ, Janssen CR, Di Guard A (2014) The ChimERA project: coupling mechanistic exposure and effect models into an integrated platform for ecological risk assessment. Environ Sci Pollut R 21:6263–6267

DeCoursey DG (1992) Developing models with more detail: do more algorithms give more truth? Weed Technol 6(2):709–715

Del Re AAM, Trevisan M (1995) Selection criteria of xenobiotic leaching models in soil. Eur J Agron 4:465–472

Devos Y, Gaugitsch H, Gray AJ, Maltby L, Martin J, Pettis JS, Romeis J, Rortais A, Schoonjans R, Smith J, Streissl F, Suter GW II (2016) Advancing environmental risk assessment of regulated products under EFSA's remit. EFSA J 14(S1):s0508

Di Guardo A, Gouin T, MacLeod M, Scheringer M (2018) Environmental fate and exposure models: advances and challenges in 21st century chemical risk assessment. Environ Sci-Proc Imp 20(1):58–71

Di Guardo A, Hermens JLM (2013) Challenges for exposure prediction in ecological risk assessment. Integr Environ Assess Manag 9(3):4–14. https://doi.org/10.1002/ieam.1442

Di HJ, Aylmore LAG (1997) Modeling the probabilities of groundwater contamination of pesticides. Soil Sci Soc Am J 61:17–23

Dubus IG, Beulke S, Brown CD (2002) Calibration of pesticide leaching models: critical review and guidance for reporting. Pest Manag Sci 58:745–758 (online: 2002)

ECOFRAM, Ecological Committee on FIFRA Risk Assessment Methods (1999) ECOFRAM Aquatic and Terrestrial Final Draft Reports, US environmental protection agency USEPA. Available at www.epa.gov/oppefed1/ecorisk/index.htm

Engel T, Hoogenboom G, Jones JW, Wilkens PW (1997) Aegis/win: a computer program for the application of crop simulation models across geographical areas. Agron J 89(6):919–928

FAO, Food and Agriculture Organization of the United Nations (1989) Revised guidelines on environmental criteria for the registration of pesticides. Rome, 51 p

Finizio A, Villa S (2002) Environmental risk assessment for pesticides—a tool for decision making. Environ Impact Asses 22:235–248

Flury M, Flüher H, Jury WA, Leuenberger J (1994) Susceptibility of soils to preferential flow of water: a field study. Water Resour Res 30(7):1945–1954

FOCUS, Forum for Co-ordination of Pesticide Fate Models and their Use (2001a) FOCUS Surface Water Scenarios in the EU Evaluation Process under 91/414/EEC. Report of the FOCUS Working Group on Surface Water Scenarios, 221 p (EC Document Reference SANCO/4802/2001-rev.1)

FOCUS, Forum for Co-ordination of Pesticide Fate Models and their Use (2001b) FOCUS Surface Water Scenarios in the EU Evaluation Process under 91/414/EEC. Report of the FOCUS Working Group on Surface Water Scenarios, 245 p (EC Document Reference SANCO/4802/2001-rev.2)

FOCUS, Forum for Co-ordination of Pesticide Fate Models and their Use (2015) Generic Guidance for FOCUS Surface Water Scenarios. FOCUS Working Group on Surface Water Scenarios, version: 1.4. 367 p

FOCUS, Forum for Co-ordination of Pesticide Fate Models and their Use (2000) FOCUS groundwater scenarios in the EU review of active substances. Ground water Scenarios Workgroup. FOCUS Report, European Commission, 202 p (EC Document Reference SANCO/321/2000 rev.2)

FOCUS, Forum for Co-ordination of Pesticide Fate Models and their Use (2005) Landscape and mitigation factors in aquatic risk assessment. Vol. 1. Extended summary and recommendations. Report of the FOCUS working group on landscape and mitigation factors in ecological risk assessment, 133 p (EC Document reference SANCO/10422/2005)

FOCUS, Forum for Co-ordination of Pesticide Fate Models and their Use (1995) Leaching models and EU registration, 123 p (EC document reference 4952/VI/95)

FOCUS, Forum for Co-ordination of Pesticide Fate Models and their Use (1997) Surface water models and EU registration of plant protection products, 231 p (EC document reference 6476/VI/96)

Forbes VE, Calow P, Sibly RM (2008) The extrapolation problem and how population modeling can help. Environ Toxicol Chem 27:1987–1994

Forbes VE, Hommen U, Thorbek P, Heimbach F, Van den Brink P, Wogram J, Thulke H-H, Grimm V (2009) Ecological models in support of regulatory risk assessments of pesticides: developing a strategy for the future. Integr Environ Asses 5:167–172

Foster GR, Lane LJ (1987) Beyond the USLE: advancements in soil erosion prediction. In: Boersma LL (ed) Future developments in soil science research. Soil Science of America Society, Madison, pp 315–326

Fry M, Milians K, Young D, Zhong H (2014) Surface Water Concentration Calculator User Manual. Environmental Fate and Effects Division, Office of Pesticides, United States Environmental Protection Agency, 21 p (USEPA/OPP 734F14001)

Garratt JA, Capri E, Trevisan M, Errera G, Richard M, Wilkins RM (2002) Parameterisation, evaluation and comparison of pesticide leaching models to data from a Bologna field site, Italy. Pest Manag Sci 58:3–20 (online: 2002)

Ghirardello D, Morselli M, Otto S, Zanin G, Di Guardo A (2014) Investigating the need for complex versus simple scenarios to improve predictions of aquatic ecosystem exposure with the SoilPlus model. Environ Pollut 184:502–510

Giannouli DD, Antonopoulos VZ (2015) Evaluation of two pesticide leaching models in an irrigated field cropped with corn. J Environ Manage 150:508–515

Gilliom RJ (2007) Pesticides in U.S. streams and groundwater. Environ Sci Technol 41:3408–3414

Girling AE, Tattersfield L, Mitchell GC, Crossland NO, Pascoe D, Blockwell SJ, Maund SJ, Taylor EJ, Wenzel A, Janssen CR, Jüttner I (2000) Derivation of predicted no-effect concentrations for lindane, 3,4-dichloroaniline, atrazine and copper. Ecotox Environ Safe 46:148–162

Gonçalves CM, Da Silva JCGE, Alpendurada MF (2007) Evaluation of the pesticide contamination of groundwater sampled over two years from a vulnerable zone in Portugal. J Agr Food Chem 55(15):6227–6235

Grimm V, Berger U, Bastiansen F, Eliassen S, Ginot V, Giske J, Goss-Custard J, Grand T, Heinz S, Huse G, Huth A, Jepsen JU, Jørgensen C, Mooij WM, Müller B, Pèer G, Piou C, Railsback SF, Robbins AM, Robbins MM, Rossmanith E, Rüger N, Strand E, Souissi S, Stillman RA, Vabø R, Visser U, DeAngelis DL (2006) A standard protocol for describing individual-based and agent-based models. Ecol Model 198:115–126

Hageman KJ, Simonich SL, Campbell DH, Wilson GR, Landers DH (2006) Atmospheric deposition of current-use and historic-use pesticides in snow at national parks in the Western United States. Environ Sci Technol 40:3174–3180

Hallberg GR (1989) Pesticide pollution of groundwater in the humid Unites States. Agr Ecosyst Environ 26:299–367

Harman-Fetcho JA, Hapeman CJ, McConnell LL, Potter TL, Rice CP, Sadeghi AM, Smith RD, Bialek K, Sefton KA, Schaffer BA (2005) Pesticide occurrence in selected South Florida canals and Biscayne Bay during high agricultural activity. J Agr Food Chem 53:6040–6048

Hart A (2001) Probabilistic risk assessment for pesticides in Europe: implementation and research needs. European workshop on probabilistic risk assessment for the environmental impacts of plant protection products. The Netherlands

Hoffman RS, Capel PD, Larson SJ (2000) Comparison of pesticides in eight U.S. urban streams. Environ Toxicol Chem 19:2249–2258

Hornsby AG, Wauchope RD, Herner AE (1996) Pesticide properties in the environment. Springer-Verlag, New York, NY

Hull RN, Kleywegt S, Schroeder J (2015) Risk-based screening of selected contaminants in the Great Lakes Basin. J Great Lakes Res 41:238–245

IBAMA, Instituto Brasileiro do Meio Ambiente e dos Recursos Naturais Renováveis (2012) Avaliação de risco ambiental de agrotóxicos no Ibama. DIQUA/CGASQ, Brasília, IBAMA [Portuguese]. Available at http://ibama.gov.br/phocadownload/agrotoxicos/avaliacao/2017/2017-07-25-avaliacao_risco_ambiental_agrotoxicos_ibama_2012-ARA.pdf

Jarvis NJ (1991) MACRO—A model of water movement and solute transport in macroporous soils. Uppsala: Swedish University of Agricultural Sciences, 58 p (Reports and Dissertations, 9)

Jarvis NJ (1994) The MACRO model (Version 3.1). Technical description and sample simulations. Department of Soil Science, Swedish University of Agricultural Science, Uppsala, Sweden, 51 p (Reports and Dissertations 19)

Jarvis NJ, Bergström LF, Brown CD (1995) Pesticide leaching models and their use for management purposes. In: Roberts TR, Kearney PC (eds) Environmental behaviour of agrochemicals. Wiley, New York, pp 185–220

Jiang L, Huang J, Liang L, Zheng PY, Yang H (2008) Mobility of prometryne in soil as affected by dissolved organic matter. J Agr Food Chem 56(24):11933–11940

Klein M (1995) PELMO: pesticide leaching model. Fraunhofer Institute, Schmallenberg, p 103

Klein M (2011) User Manual PELMO (Pesticide Leaching Model) Version 4.0; Fraunhofer Institute: Schmallenberg, Germany

Koskinen WC, Rice PJ, Anhalt JA, Sakaliene O, Moorman TB, Arthur EL (2002) Sorption-desorption of "aged" sulfonylaminocarbonyltriazolinone herbicides in soil. J Agr Food Chem 50:5368–5372

Kreuger J (1998) Pesticides in stream water within an agricultural catchment in southern Sweden, 1990–1996. Sci Total Environ 216:227–251

Laabs V, Amelung W, Pinto A, Altstaedt A, Zech W (2000) Leaching and degradation of com and soybean pesticides in an Oxisol of the Brazilian Cerrados. Chemosphere 41:1441–1449

Laabs V, Amelung W, Pinto A, Zech W (2002a) Fate of pesticides in tropical soils of Brazil under field conditions. J Environ Qual 31(1):256–268

Laabs V, Amelung W, Pinto A, Zech W, Wantzen M, da Silva CJ, Wolfgang Zech W (2002b) Pesticides in surface water, sediment, and rainfall of the northeastern Pantanal Basin. Brazil. J Environ Qual 31(5):1636–1648

Larsbo M, Jarvis NJ (2005) Simulating solute transport in a structured field soil: uncertainty in parameter identification and predictions. J Environ Qual 34(2):621–634

Larsbo M, Jarvis NJ (2003) MACRO 5.0. A model of water flow and solute transport in macroporous soil. Technical Description. Department of Soil Sciences, Swedish University of Agricultural Sciences, Uppsala

Leistra M, van der Linden AMA, Boesten JJTI, Tiktak A, van den Berg F (2001) PEARL model for pesticide behaviour and emissions in soil-plant systems: description of the processes, Alterra Rep 13. Wageningen University and Research Centre, Wageningen, The Netherlands, p 115

Leistra M, van der Linden, AMA, Boesten JJTI, Tiktak A, van den Berg F (2000) PEARL Model for Pesticide Behaviour and Emissions in Soil-plant Systems. Description of Processes. Alterra Report 013. Alterra, Wageningen, The Netherlands

Lennartz B (1999) Variation of herbicide transport parameters within a single field and its relation to water flux and soil properties. Geoderma 91(3–4):327–345

Leu C, Singer H, Stamm C, Muller SR, Schwarzenbach RP (2004) Variability of herbicide losses from 13 fields to surface water within a small catchment after a controlled herbicide application. Environ Sci Technol 38:3835–3841

Levanon D, Codling EE, Meisinger JJ, Starr JL (1993) Mobility of agrochemicals through soil from two tillage systems. J Environ Qual 22(1):155–161

Li K, Xing B, Torello WA (2005) Effect of organic fertilizers derived dissolved organic matter on pesticide sorption and leaching. Environ Pollut 134(2):187–194

Lorenz S, Rasmussen JJ, Süß A, Kalettka T, Golla B, Horney P, Stähler M, Hommel B, Schäfer RB (2017) Specifics and challenges of assessing exposure and effects of pesticides in small water bodies. Hydrobiologia 793:213–224

Malaj E, von der Ohe PC, Grote M, Kühne R, Mondy CP, Usseglio-Polatera P, Brack W, Schäfer RB (2014) Organic chemicals jeopardize the health of freshwater ecosystems on the continental scale. P Natl Acad Sci USA 111:9549–9554

Manahan SE (ed) (1992) Toxicological chemistry, 2nd edn. Lewis Publishers, Ann Arbor, MI

Margni M, Rossier D, Crettaz P, Jolliet O (2002) Life cycle impact assessment of pesticides on human health and ecosystems. Agr Ecosyst Environ 93:379–392

Matallo MB, Spadotto CA, Luchini LC, Gomes MAF (2005) Sorption, degradation, and leaching of tebuthiuron and diuron in soil columns. J Environ Sci Heal B 40(1):39–43

Matsumura F (ed) (1985) Toxicology of insecticides, 1st edn. Plenum Press, New York, p 598

Meyer MT, Kalkoff SJ, Scribner EA (2006) Comparison between the transport of isoxaflutole and its degradates to triazine and acetanilide herbicides in ten Iowa rivers. Proceedings of the 231st National Meeting of the American Chemical Society, Atlanta

Miyamoto M, Tanaka H, Katagi T (2008) Ecotoxicological risk assessment of pesticides in aquatic ecosystems. R&D Report, Sumitomo Kagaku, p 18p

Mullins JA, Carsel RF, Scarbrough JE, Ivery AM (1993) PRZM-2, a model for predicting pesticide fate in the crop root and unsaturated zones: user's manual for release 2.0. Athens: United States Environmental Protection Agency (EPA/600/R-93/046)

Munns Jr WR (2006) Assessing risks to wildlife populations from multiple stressors: overview of the problem and research needs. Ecol Soc 11:23

NRC, National Research Council (1983) Risk assessment in the federal government: managing the process. National Academy Press, Washington DC

NRC, National Research Council (1996) Understanding risk: informing decisions in a democratic society. National Academy Press, Washington DC

OECD, Organization for Economic Co-operation and Development (1982) OECD Hazard Assessment Project, STEP System Group: final report. Stockholm

OECD, Organization for Economic Co-operation and Development (1999) Indirect load. In: OECD (ed), Annex 2. Report of phase 1 of the aquatic risk indicators project, pp 28–32

Padilla L, Winchell M, Peranginangin N, Grant S (2017) Development of groundwater pesticide exposure modeling scenarios for vulnerable spring and winter wheat-growing areas. Integr Environ Asses 13(6):992–1006

Papastergiou A, Papadopoulou-Mourkidou E (2001) Occurrence and spatial and temporal distribution of pesticide residues in groundwater of major corn-growing areas of Greece. Environ Sci Technol 35:63–69

Parker RD, Jones RD, Nelson HP (1995) GENEEC: A screening model for pesticide environmental exposure assessment. In: Proceedings… international exposure symposium on water quality modeling, American Society of Agricultural Engineers, pp 485–490

Pastorok RA, Bartell SM, Ferson S (2002) Ecological modeling in risk assessment: chemical effects on populations, ecosystems, and landscapes. Lewis, Boca Raton, FL, USA

Pennell KD, Hornsby AG, Jessup RE, Rao PSC (1990) Evaluation of five simulation models for predicting aldicarb and bromide behavior under field conditions. Water Resour Res 26(11):2679–2693

Posthuma L, Suter II GW, Traas TP (eds) (2002) Species sensitivity distributions in ecotoxicology. Lewis publishers

Rabiet M, Margoum C, Gouy V, Carluer N, Coquery M (2010) Assessing pesticide concentrations and fluxes in the stream of a small vineyard catchment—effect of sampling frequency. Environ Pollut 158:737–748

Rao PSC, Davidson JM, Hammond LC (1976) Estimation of nonreactive and reactive solute front locations in soils. In: Hazard: wastes res Symp 1976. Proc., pp 235–241 (EPA-600/19-76-015)

Rao PSC, Hornsby AG, Jessup RE (1985) Indices for ranking the potential for pesticide contamination of groundwater. Soil Crop Sci Soc Fl 44:1–8

Reichenberger S, Amelung W, Laabs V, Pinto A, Totsche KU, Zech W (2002) Pesticide displacement along preferential flow pathways in a Brazilian Oxisol. Geoderma 110(1–2):63–86

Rice PJ, Rice PJ, Arthur EL, Barefoot AC (2007) Advances in pesticide environmental fate and exposure assessments. J Agric Food Chem 55:5367–5376

Roller JA, Van den Berg F, Adriaanse PI (2003) Surface water scenarios help (SWASH) Version 2.0. Technical Documentation version 1.3. Alterra-rapport 508. Wageningen, Alterra Green World Research, the Netherlands

Russell MH, Layton RJ, Tillotson PM (1994) The use of pesticide leaching models in a regulatory setting: an industrial perspective. J Environ Sci Heal A 29:1105–1116

Schmolke A, Thorbek P, Chapman P, Grimm V (2010) Ecological models and pesticide risk assessment: current modeling practice. Environ Toxicol Chem 29(4):1006–1012

Schulz R (2004) Field studies on exposure, effects, and risk mitigation of aquatic nonpoint-source insecticide pollution: a review. J Environ Qual 33(2):419–448

Scorza PJ Jr, Boesten JJTI (2005) Simulation of pesticide leaching in a cracking clay soil with the PEARL model. Pest Manag Sci 61:432–448

Scorza RP (2002) Pesticide leaching in macroporous clay soils: field experiment and modeling. 234 f. Thesis (Doctoral). Wageningen University and Research Centre, Wageningen

Scott GI, Baughman DS, Trim AH, Dee JC (1987) Lethal and sublethal effects of insecticides commonly found in nonpoint source agricultural runoff to estuarine fish and shellfish. In: Vernberg WB, Calabrese A, Thurberg FP, Vernberg FJ (eds) Pollution physiology of estuarine organisms. University of South Carolina Press, Columbia, SC, pp 251–273

Seiler AP, Brenneisen P, Green DH (1992) Benefits and risks of plant protection products—possibilities of protecting drinking water: case atrazine. Water Supp 10:31–42

SETAC, Society for Environmental Toxicology and Chemistry (1994) Pesticide risk and mitigation. Final Report of the Aquatic Risk Assessment and Mitigation Dialog Group, SETAC Foundation for Environmental Education, Pensacola, FL, USA, p 220

SETAC, Society of Environmental Toxicology and Chemistry (1987) Research priorities in environmental risk assessment. Report of a workshop held in Breckenridge, CO. Washington, DC: SETAC

Solomon K (2010) Ecotoxicological risk assessment of pesticides in the environment. Hayes' Handbook of Pesticide Toxicology. Chapter 56:1191–1217. https://doi.org/10.1016/B978-0-12-374367-1.00056-2

Solomon KR (1996) Overview of recent developments in ecotoxicological risk assessment. Risk Anal 16(5):627–633

Solomon KR, Sibley P (2002) New concepts in ecological risk assessment: where do we go from here? Mar Pollut Bull 44:279–285

Song NH, Chen L, Yang H (2008) Effect of dissolved organic matter on mobility and activation of chlorotoluron in soil and wheat. Geoderma 146(1/2):344–352

Spadotto CA, Hornsby AG (2003) Organic compounds in the environment: soil sorption of acidic pesticides: modeling pH effects. J Environ Qual 32(3):949–956

Spadotto CA, Hornsby AG, Gomes MAF (2005) Sorption and leaching potential of acidic herbicides in Brazilian soils. J Environ Sci Heal B 40(1):29–37

Spadotto CA, Mingoti R (2014) Base técnico-científica do ARAquá 2014: software para avaliação de risco ambiental de agrotóxico. Campinas: Embrapa Gestão Territorial, 6 p (Embrapa Gestão Territorial. Circular Técnica, 2) (Portuguese)

Spadotto CA, Moraes DAC, Ballarin AW, Laperuta Filho J, Colenci RA (2010) ARAquá: software para avaliação de risco ambiental de agrotóxico. Campinas: Embrapa Monitoramento por Satélite, 15 p (Boletim de Pesquisa e Desenvolvimento, 7) (Portuguese)

Starfield AM (1997) A pragmatic approach to modeling for wildlife management. J Wildlife Manage 61:261–270

Stehle S, Schulz R (2015) Agricultural insecticides threaten surface waters at the global scale. P Natl Acad Sci USA 112:5570–5575

Stephenson GR, Solomon KR (2007) Pesticides and the environment. Canadian Network of Toxicology Centres Press Guelph, Ontario, Canada

Stoorvogel JJ (1995) Linking GIS and models: structure and operationalization for a Costa Rican case study. Neth J Agr Sci 43:19–29

Strassemeyer J, Daehmlow D, Dominic AR, Lorenz S, Golla B (2017) SYNOPS-WEB, an online tool for environmental risk assessment to evaluate pesticide strategies on field level. Crop Prot 97:28–44

Streissl F (2010) Potential role of population modeling in the regulatory context of pesticide authorization. In: Thorbek P, Forbes VE, Heimbach F, Hommen U, Thulke H-H, Van den Brink PJ, Wogram J, Grimm V (eds) Ecological Models for Regulatory Risk Assessments of Pesticides: Developing a Strategy for the Future. SETAC, Pensacola, FL, USA, pp 97–104

Suter GW, Barnthouse LW, Bartell SM, Mill T, Mackay D, Patterson S (1993) Ecological risk assessment. Lewis Publishers, Boca Raton, FL, p 538

Taghavi L, Probst J, Merlina G, Marchand A, Durbe G, Probst A (2010) Flood event impact on pesticide transfer in a small agricultural catchment (Montoussé at Auradé, south west France). Int J Environ An Ch 90:390–405

Teklu BM, Adriaanse PI, Ter Horst MMS, Deneer JW, Van den Brink PJ (2015) Surface water risk assessment of pesticides in Ethiopia. Sci Total Environ 508:566–574

Teske ME, Bird SL, Esterly DM, Curbishley TB, Ray SL, Perry SG (2000) AgDRIFT: a model for estimating near-field spray drift from aerial applications. Environ Toxicol Chem 21(3):659–671

Tiktak A, Boesten JJTI, Egsmose M, Gardi C, Klein M, Vanderborght J (2013) European scenarios for exposure of soil organisms to pesticides. J Environ Sci Heal B 48(9):703–716

Tiktak A, Boesten JJTI, van der Linden AMA, Vanclooster M (2006) Mapping ground water vulnerability to pesticide leaching with a process-based metamodel of EuroPEARL. J Environ Qual 35:1213–1226

Tiktak A, van den Berg F, Boesten JJTI, van Kraalingen D, Leistra M and van der Linden AMA (2000) Manual of FOCUS PEARL version 1.1.1, RIVM Rep 711 401 008, RIVM, Bilthoven, The Netherlands, 144 p

Tiktak A, Van den Berg F, Boesten JJTI, Van Kraalingen D, Leistra M, Van der Linden AMA (2002) Manual of FOCUS PEARL version 1.1.1. Bilthoven: RIVM/Alterra

Tiktak A, van der Linden AMA, Boesten JJTI (2003) The GeoPEARL model: description, applications and manual. RIVM Report 716601007. RIVM, Bilthoven, The Netherlands

UNEP, United Nations Environment Program (2009) Existing sources and approaches to risk assessment and management of pesticides, particular needs of developing countries and countries with economies in transition, 95 p

USDA, United States Department of Agriculture, Soil Conservation Service (1991) Peak discharge (other methods), study guide. Engineering, Hydrology Training Series, Module 206D, 27 p

Urban DJ, Cook NJ (1986) Standard evaluation procedure for ecological risk assessment. Springfield, VA: National Technical Information Service. Hazard Evaluation Division, Office of Pesticide Programs, US Environmental Protection Agency, Washington, DC (NTIS PD 86-247-657)

USEPA, United State Environmental Protection Agency (1979) Toxic substances control act, discussion of premanufacture testing policies and technical issues: request for comment. Fed Reg 44:16240–16292

USEPA, United States Environmental Protection Agency (1992) Framework for ecological risk assessment. Risk Assessment Forum. Washington, DC (EPA/630/R-92/001).

USEPA, United States Environmental Protection Agency (1998) Guidelines for ecological risk assessment. Risk Assessment Forum. Washington, DC (EPA/630/R-95/002F)

USEPA, United States Environmental Protection Agency (2001) Risk assessment guidance for superfund: volume III—part A, process for conducting probabilistic risk assessment. Washington, DC (EPA 540-R-02-002)

USEPA, United States Environmental Protection Agency (2007) Tier I rice model—version 1.0—guidance for estimating pesticide concentrations in rice paddies. Washington, DC: USEPA. Available at https://www.epa.gov/pesticide-science-and-assessing-pesticide-risks/tier-i-rice-model-version-10-guidance-estimating

USEPA, United States Environmental Protection Agency (2008) FIRST: a screening model to estimate pesticide concentrations in drinking water version 1.1.1. Available at https://archive.epa.gov/epa/pesticide-science-and-assessing-pesticide-risks/first-version-111-description.html

USEPA, United States Environmental Protection Agency (2009) User's guide and technical documentation KABAM version 1.0 (Kow (based) Aquatic BioAccumulation Model). Washington, DC: USEPA. 123 p

USEPA, United States Environmental Protection Agency (2011a) Guidance for the development of conceptual models for a problem formulation developed for registration review, Washington. Available at https://www.epa.gov/pesticide-science-and-assessing-pesticide-risks/guidance-development-conceptual-models-problem

USEPA, United States Environmental Protection Agency (2011b) Integrated risk information system (IRIS) glossary. Office of Research and Development/National Center for Environmental Assessment/Integrated Risk Information System. Available at https://www.epa.gov/iris

USEPA, United States Environmental Protection Agency (2012) Technical overview of ecological risk assessment: risk characterization. Available at https://www.epa.gov/pesticide-science-and-assessing-pesticide-risks/technical-overview-ecological-risk-assessment-risk

USEPA, United States Environmental Protection Agency (2016) Pesticide in Water Calculator User Manual for Versions 1.50 and 1.52. Washington, DC: USEPA, 23 p

USEPA, United States Environmental Protection Agency (2017) Models for pesticide risk assessment. Available at https://www.epa.gov/pesticide-science-and-assessing-pesticide-risks/models-pesticide-risk-assessment

USGS, United States Geological Survey (2006) National water quality assessment (NAWQA) program. Available at http://pubs.usgs.gov/fs/2006/3101/

Van de Zande JC, Porskamp HAJ, Michielsen JMGP, Holterman HJ, Huijsmans JMF (2000) Classification of spray applications for driftability, to protect surface water. Asp Appl Biol 57:57–64

Van den Brink P, Baveco JM, Verboom J, Heimbach F (2007) An individual-based approach to model spatial population dynamics of invertebrates in aquatic ecosystems after pesticide contamination. Environ Toxicol Chem 26:2226–2236

Van den Brink PJ, Baird DJ, Baveco JMH, Focks A (2013) The use of traits-based approaches and eco(toxico)logical models to advance the ecological risk assessment framework for chemicals. Integr Environ Asses 9:47–57

Van Jaarsveld JHA, Van Pul WAJ (1999) Modeling of atmospheric transport and deposition of pesticides. Water Air Soil Poll 115(1–4):167–182

Vanclooster M, Boesten JJTI, Tiktak A, Jarvis NJ, Kroes JG, Munoz-Carpena R, Clothier BE, Green SR (2004) On the use of unsaturated flow and transport models in nutrient and pesticide management. In: Feddes RA, de Rooij GH, van Dam JC (eds) Unsaturated-zone modelling: progress, challenges and applications, vol 6, pp 331–361. Wageningen UR Frontis Series, Wagenignen University: Wagenignen, the Netherlands

Wauchope RD (1978) The pesticide content of surface water draining from agricultural fields—a review. J Environ Qual 7:459–472

Westman WE (1985) Ecology: impact assessment and environmental planning. Wiley, New York

White K, Biscoe M, Fry M, Hetrick J, Orrick G, Peck C, Ruhman M, Shelby A, Thurman N, Vilanueva P, Young DF (2016) Development of a conceptual model to estimate pesticide concentrations for human health drinking water and guidance on conducting ecological risk assessments for the use of pesticides on rice. USEPA, Washington DC, p 112

WHO, World Health Organization (2004) IPCS Harmonization Project–IPCS Risk Assessment Terminology, Geneva

Williams PRD, Hubbell B, Weber E, Fehrenbacher C, Hrdy D, Zartarian V (2010) An Overview of Exposure Assessment Models Used by the US Environmental Protection Agency. In Hanrahan, G. Modelling of Pollutants in Complex Environmental Systems, Volume II. ILM Publications, pp 61–131

Wogram J (2010) Regulatory challenges for the potential use of ecological models in risk assessments of plant protection products. In: Thorbek P, Forbes VE, Heimbach F, Hommen U, Thulke H-H, Van den Brink PJ, Wogram J, Grimm V (eds) Ecological models for regulatory risk assessments of pesticides: developing a strategy for the future. SETAC, Pensacola, FL, USA, pp 27–32

WWF, World Wildlife Fund (1992) Improving aquatic risk assessment under FIFRA: report of the aquatic effects dialogue group, pp 23–24

Young D (2016) The variable volume water model—revision A. Environmental Fate and Effects Division, Office of Pesticide Programs, U.S. Environmental Protection Agency, 36 p (USEPA/OPP 734S16002)

Young DF (2013) Pesticides in flooded applications model (PFAM): conceptualization, development, evaluation, and user guide. Washington, DC: USEPA, 61 p (EPA-734-R-13-001)

Chapter 12
Management of Agrochemical Residues in the Environment

Sílvio Vaz Jr. and Luciano Gebler

Abstract This chapter deals with the more relevant strategies for the management of agrochemical residues in soil and water. Furthermore, the most advanced treatment technologies will be explored.

Keywords Hazardous residues · Technologies of treatment · Environmental chemistry

12.1 Introduction

The management of agrochemical residues on the environment is paramount to assure that these compounds will not pollute air, soil, water, and food. Besides, it is a form to reduce the risk to the human and animal health from the use of these hazardous compounds.

According to the World Health Organization (2010), pesticides—a very representative agrochemical class—can be classified based on their toxicity and human hazard. Table 12.1 presents this classification.

From this ranking, the pesticide molecules are classified as I_a, I_b, II, III, and U—extremely hazardous, highly hazardous, moderately hazardous, slightly hazardous, and unlikely to present acute hazard, respectively. Table 12.2 presents examples of classified compounds and their physicochemical and toxicological characteristics.

These compounds demand technological approaches to treat their presence on the environment to avoid their deleterious effects.

S. Vaz Jr. (✉)
Brazilian Agricultural Research Corporation, National Research Center for Agroenergy (Embrapa Agroenergy), Embrapa Agroenergia, Parque Estação Biológica, s/n, Av. W3 Norte (final), Brasilia, DF 70770-901, Brazil
e-mail: silvio.vaz@embrapa.br

L. Gebler
Brazilian Agricultural Research Corporation, National Research Center for Grape and Wine (Embrapa Grape and Wine), Embrapa Uva e Vinho, R. Livramento, 515 - Conceição, Bento Gonçalves, RS 95701-008, Brazil

© Springer Nature Switzerland AG 2019
S. Vaz Jr. (ed.), *Sustainable Agrochemistry*,
https://doi.org/10.1007/978-3-030-17891-8_12

Table 12.1 Classification of pesticides according to their hazard

WHO class	LD_{50} for the rat (mg kg^{-1} body weight)	
	Oral	Dermal
I_a, extremely hazardous	<5	<50
I_b, highly hazardous	5–50	50–200
II, moderately hazardous	50–2,000	200–2,000
III, slightly hazardous	Over 2,000	Over 2,000
U, unlikely to present acute hazard	5,000 or higher	5,000 or higher

Source World Health Organization (2010)

This classification is the most well-known and required labeling of pesticide marketing bottles worldwide for its ease by relying solely on the LD50, but when the analysis is about the risks to the environment, there are classifications that take into account factors such as negative impact on the environment, including toxicity to animals or plants, nonbiodegradability, tendency to accumulate, and possibility of harmful chemical reactions (UNEP 2018). As a result, the labels also make explicit the degree of environmental risk that the pesticide presents in the form of phrases expressing the degree of danger to the environment in high, medium, or low.

12.2 Chemical Constitution of Water and Soil, and Their Interaction with Agrochemical Residues

Agrochemicals, such as pesticides, can achieve soil and water directly and indirectly, which depends on their source and route of transport. Figure 12.1 depicts this dynamic in the environment. Furthermore, the chemical constitution of the environmental matrixes, mainly soil and water, is paramount for the agrochemical residues movement in these matrixes.

We can consider two sources of agrochemical residues—agriculture and poultry. For the pathway, we can have a lot of them, but pest control and fertilization and feces and urine can be highlighted based on the large amount of agricultural areas and cattle worldwide. Then, surface water, groundwater, and soil can be achieved by those residues.

12.2.1 Soil Composition and Properties

Soil is one of the most complex environmental matrixes due to the chemical constitution of its organic and inorganic components, and its physical states—soils are

Table 12.2 Representative compounds in the I_a, I_b, II, III, and U classes

Common name	CAS n^0	Chem. type	Phys. state	Mains use	LD_{50} (mg kg^{-1})
Class I_a (extremely hazardous)					
Aldicarb	116-06-3	C	S	I-S	0.93
Brodifacoum	56073-10-0	CO	S	R	0.3
Chlorethoxyfos	54593-83-8	OP	L	I	1.8
Disulfoton	298-04-4	OP	L	I	2.6
Hexachlorobenzene	118-74-1	OC	S	FST	10,000
Parathion	56-38-2	OP	L	I	13
Phosphamidon	13171-21-6	OP	L	I	7
Terbufos	13071-79-9	OP	L	I-S	2
Class I_b (highly hazardous)					
Acrolein	107-02-8	–	L	H	29
Carbofuran	1563-66-2	C	S	I	8
Methiocarb	2032-65-7	C	S	I	20
Pentachlorophenol	87-86-5	–	S	I, F, H	80
Strychnine	57-24-9	–	S	R	16
Thiometon	640-15-3	OP	Oil	I	120
Class II (moderately hazardous)					
Alachlor	15972-60-8	–	S	H	930
Ametryn	834-12-8	T	S	H	110
Carbaryl	63-25-2	C	S	I	300
Carbosulfan	55285-14-8	C	L	I	250
Deltamethrin	52918-63-5	PY	S	I	135
Dicamba	1918-00-9	–	S	H	1,707
Diquat	2764-72-9	BP	S	H	231
Endosulfan	115-29-7	OC	S	I	80
Permethrin	52645-53-1	PY	L	I	500
Class III (slightly hazardous)					
Atrazine	1912-24-9	T	S	H	2,000

(continued)

Table 12.2 (continued)

Common name	CAS n⁰	Chem. type	Phys. state	Mains use	LD_{50} (mg kg^{-1})
Bacillus thuringiensis (Bt)	68038-71-1	–	S	I	>4,000
Chlorpyrifos-methyl	5598-13-0	OP	S	I	>3,000
Diuron	330-54-1	–	S	H	3,400
Glyphosate	1071-83-6	–	S	H	4,230
Malathion	121-75-5	OP	L	I	2,100
Metoxuron	19937-59-8	–	S	H	>3,200
Thiabendazole	14879-8	–	S	F	3,330
Class U (unlike to present acute hazard)					
Anthraquinone	84-65-1	–	S	RP (birds)	>5,000
Carbendazim	10605-21-7	–	S	F	>10,000
Chlorpropham	101-21-3	C	S	PGR	>5,000
Imazaquin	81335-37-7	–	S	H	>5,000
Simazine	122-34-9	T	S	H	>5,000
Tiocarbazil	36756-79-3	TC	L	H	10,000
Validamycin	37248-47-8	–	S	F	>10,000

Source Adapted from World Health Organization (2010). **Chemical type**: *BP* Bipyridylium derivative; *C* Carbamate; *CO* Coumarin derivative; *OC* Organochlorine compound; *OP* Organophosphorus compound; *PY* Pyrethroid; *T* Triazine derivative; *TC* Thiocarbamate. **Physical state**: *L* Liquid; *S* Solid. **Main use**: *F* Fungicide; *FST* Fungicide, for seed treatment; *H* Herbicide; *I* Insecticide; *PGR* Plant growth regulator; *R* Rodenticide; *RP* Repellant (species); *-S* Applied to soil: not used with herbicides

formed by chemical substances in the solid, liquid, and gaseous states. Then, soils have a natural tendency to interact with different pollutants.

Table 12.3 presents the chemical composition of soil samples of the red latosol type, common in the southwest region of Brazil. Anyway, it is very important to know what kind of soil is been worked, because there are a large variety of different soils at different regions, since those very light (high levels of sand and few clay) to those highly clayey soils, that affect the pesticide residue behavior into the ground.

The results presented are typical for red latosol, which was formed under strong weather conditions in hot and humid regions containing low concentration of silicate minerals and high concentration of FeO, Fe_2O_3, and Al_2O_3. According to Weber et al. (2005), this type of soil presents variable electric charges on the surface. The

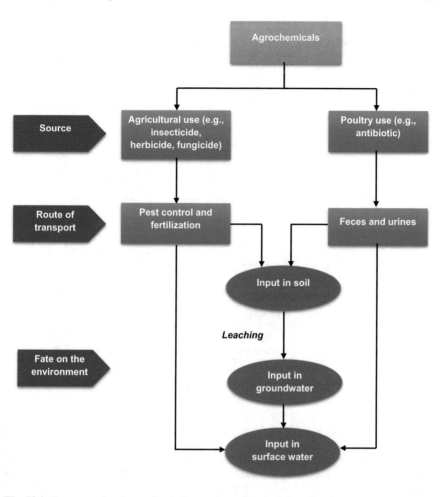

Fig. 12.1 Sources and pathways for the input of agrochemicals in soil and water

distribution of organic matter (OM) and the mineral fraction in layers in the soil is as follows:

- Horizon O (surface): OM in decomposition (0.3 m of depth).
- Horizon A: OM accumulated mixed with the mineral fraction (0.6 m of depth).
- Horizon B: clay accumulation, Fe^{3+}, Al^{3+}, and low OM content (approximately 1 m of depth).
- Horizon C: materials from rock mother.

Thus, it is expected that the higher the concentration of OM, especially of humic acids present in it, the greater the retention capacity of metallic cations in soils, especially in Horizon O, which leads to a reduction in the transport of metallic pollutants in the soil, as the humic substances act as strong complexing agents due to

Table 12.3 Results of physicochemical analysis of samples of red latosol, collected at different depths

ID	pH (at CaCl$_2$ solution)	OM (g dm^{-3})	CTC (mmolc dm^{-3})
ST01-0.3m	6	23	87
ST01-0.6m	5.9	24	79
ST01-1.4m	6.1	12	87
ST02-0.3m	4.4	7	56
ST02-0.6m	4.6	2	42
ST03-0.3m	5.2	19	49
ST03-0.6m	5	4	46
ST04-0.3m	5.5	10	54
ST04-0.6m	5.6	9	57
ST05-0.3m	6.1	19	62
ST05-0.6m	5.7	5	50
ST06-0.3m	5.6	15	66
ST06-0.6m	4.3	16	102
ST07-0.3m	5.4	30	66
ST07-0.6m	5.8	25	83
ST08-0.3m	6	27	106
ST08-0.6m	6	30	102

OM Organic matter; *CEC* Cation exchange capacity. Analyses carried out according to the methods from the Brazilian Agricultural Research Corporation (1997)
Source Vaz (2010)

the presence of binder sites formed by carboxylic and phenolic groups (Clapp et al. 2001). Therefore, a higher concentration, for example, of bivalent metal cations in the samples of Horizons O and A of the soil is expected, considering the effect of the presence of silicate compounds in the metal retention, where a greater value of cation exchange capacity (CEC) of the soil denotes a higher availability of binding sites for the metal after the exit of cations or protons associated with these silicates, due to the negative surface charge of the latter.

Figure 12.2 shows two soils with high clay content: one in the central region and the other in the south of Brazil with clay contents that exceed 60% in its composition. The color and clarity of the soil are due to the different types of metallic oxides, weathering intensity, and percentage of organic matter in its composition.

According to the University of the West of England (2013), the main pollutants observed in soil are:

- Asbestos;
- Dioxin and dioxin-like chemicals;
- Metals: cadmium, lead, mercury;
- Non-metal: arsenic;

Fig. 12.2 Different soils from two distinct regions of Brazil containing high levels of clay. *Source* Sílvio Vaz and Conte et al. (2016)

- **Persistent organic pollutants (POP):** polychlorinated biphenyls (PCBs), poly-brominated biphenyls, polychlorinated dibenzofurans (PCDFs), polycyclic aromatic hydrocarbons (PAHs), organophosphorus and **carbamate insecticides, herbicides**, organic fuels (gasoline, diesel), and pharmaceuticals and their metabolites.

12.2.2 Water Composition and Properties

When we talk about water as an environmental matrix, we must consider it in the plural, since we are dealing with two distinct types, but which strongly correlate: *surface water* and *groundwater*. It is expected a variation in the water composition according to climate and environmental conditions.

The surface water is that found in rivers, lakes, seas, and oceans, while groundwater is that found in the aquifers. Drinking water and wastewater are subclassifications related to surface water and groundwater according to their use. Table 12.4 shows the main characteristics of each of them.

It may be noted that for most of the ions listed above there is an increase in their concentration when considering groundwater relative to surface water. As observed by Snoeyink and Jenkins (1996), groundwater, which has a higher concentration of CO_2 gas, is in greater contact with rocks and soil, which leads to a longer dissolution time. The carbonic acid (H_2CO_3) produced by the solubilization of CO_2 when in contact with these materials leads to the solubilization of the minerals, releasing their constituent ions.

Table 12.4 Typical surface water and groundwater composition in the USA, according to Snoeyink and Jenkins (1996)

Chemical species	Concentration in surface water ($mg\ L^{-1}$)	Concentration in groundwater ($mg\ L^{-1}$)
SiO_2	1.2	10
Fe^{3+}	0.02	0.09
Ca^{2+}	36	92
Mg^{2+}	8.1	34
Na^+	6.5	8.2
K^+	1.2	1.4
HCO_3^-	119	339
SO_4^{2-}	22	84
Cl^-	13	9.6
NO_3^-	0.1	13
Total dissolved solids	165	434
Total hardness as $CaCO_3$	123	369

A large amount of suspended material can be found mainly in surface waters. Clay, sand, and organic matter are examples of particles in suspension.

There are also a large amount of microorganisms present in water, highlighting bacteria such as coliforms and cyanobacteria, which often compromise the quality of water, especially surface water.

Important analytical parameters for the monitoring of water quality are:

- Electrical conductivity (EC): It provides information on the distribution of ionic species in the medium, with the conductivity being directly proportional to the concentration of these species.
- Dissolved oxygen (DO): O_2 gas has a low solubility in water, with a reduction in its concentration indicating its consumption by the chemical oxygen demand (COD) for the formation of oxidized species, as well as consumption by biochemical demand (BOD) due to the activity of the metabolism of present microorganisms—groundwater has DO values much smaller than surface water.
- pH: Its value indicates the concentration of H^+ in the medium—a pH value around 6 is the most common in drinking waters; however, there are variations due to the presence of organic or inorganic species.
- Redox potential (E_h): It indicates the oxidizing or reducing characteristic of the medium, having a direct correlation with the pH values.
- Presence of organic compounds: determination of petroleum derivatives, **agrochemicals,** and organochlorine compounds produced by treatment processes, which are the main xenobiotics observed in waters, among others.

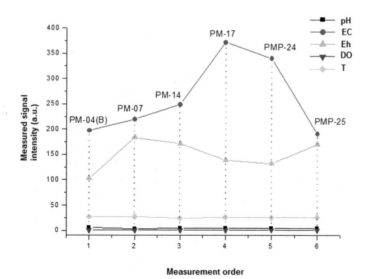

Fig. 12.3 Behavior of physicochemical parameters measured in situ with the use of multiparameter probe in groundwater located in the southeast region of Brazil. PM = monitoring well; PMP = deep monitoring well. EC in $\mu S\ m^{-1}$, E_h in mV, DO in mg L^{-1}, and T in °C. *Source* Vaz (2013)

- Presence of toxic metals: Cadmium, mercury, chromium, etc., are also, in most cases, xenobiotic species.

Figure 12.3 shows an example of in situ measurements for the physicochemical parameters mentioned above. We can observe that EC and E_h presented the main variation in their values, probably due to the variability in the concentration of ionic species presented in each monitoring well.

Figure 12.4 shows a water body (surface water) that can also be monitored using the analytical parameters as shown in Fig. 12.3.

Other water-related matrices are:

- Sediments: naturally occurring material such as rocks, sand, and silt, in contact with water bodies, such as rivers.
- Sewage or municipal wastewater.

Fig. 12.4 Small creek (surface water). *Source* Sílvio Vaz Jr

Table 12.5 Categorization of sources of chemicals in drinking water, according to the World Health Organization (2017)

Source	Examples
Naturally occurring chemicals (including naturally occurring algal toxins)	Rocks and soils (e.g., calcium, magnesium, but also arsenic and fluoride, cyanobacteria in surface water)
Chemicals from agricultural activities (including pesticides)	**Application of manure, fertilizer, and pesticides; intensive animal practices**
Chemicals from human settlements (including those used for public health purposes, e.g., vector control)	Sewage and waste disposal, urban runoff, fuel leakage
Chemicals from industrial activities	Manufacturing, processing, and mining
Chemicals from water treatment and distribution	Water treatment chemicals; corrosion of and leaching from storage tanks and pipes, by-products of chemical treatment

- Sludge: a residual semi-solid material from industrial, water, or wastewater treatment processes.
- Wastewater: result of a domestic, industrial, commercial, or **agricultural activities**, with negative impacts on the human health and environment.

The World Health Organization (2017) established the occurrence of chemicals in drinking water (Table 12.5).

12.3 Degradation of Pesticide Molecules

From their chemical structures and their environmental fate, the reactions suffered by pesticide molecules are *hydrolysis, photolysis, oxidation, and reduction*. Hydrolysis is defined as the solvolysis by the water, or a reaction with a solvent (water) involving the rupture of one or more bonds in the reacting solute (e.g., pesticide molecule). Photolysis is the cleavage of one or more covalent bonds in a molecular entity (as pesticide molecule) resulting from absorption of light, or a photochemical process in which such cleavage is essential. Oxidation involves the loss of one or more electrons to an oxidative species, whereas reduction involves the gain of one or more electrons from a reductive species—these are common processes for metallic species because of their electroactive property part (International Union of Pure and Applied Chemistry 2018).

Hydrolysis (Scheme 12.1) occurs when the pesticide molecule is in aqueous and vapor phase—with or without the presence of an acid—and photolysis (Scheme 12.2) when the molecule is exposed to the electromagnetic radiation ($h\nu$), as ultraviolet (UV) and visible radiation. These are the main reactions associated with the degradation of pollutants in environmental matrixes (Don et al. 2017; Balmer et al. 2000). Redox processes for water treatment, such as Fenton reaction (Scheme 12.3), are

Scheme 12.1 Hydrolysis
reaction of the
trimethylamine, a basic
nitrogen compound

$$(CH_3)_3N + H_2O \longrightarrow (CH3)_3NH^+ + OH^-$$

Scheme 12.2 Photolysis
reaction of the nitrous acid in
the indoor atmosphere

$$HNO_2 + h\nu \longrightarrow {}^.OH + NO$$

Scheme 12.3 Fenton
reaction to generate the
hydroxyl radical and further
reactions of degradation of
organic molecules

$$Fe^{2+} + H_2O_2 \longrightarrow Fe^{3+} + {}^.OH + OH^-$$

$$RX + {}^.OH \longrightarrow {}^.RX^+ + OH^-$$

$$RH + {}^.OH \longrightarrow {}^.R + H_2O$$

$$RHX + {}^.OH \longrightarrow RHX(OH)$$

applied combining photolysis with oxidation reaction by a hydroxyl radical (${}^.OH$) formation (Mirzaei et al. 2017); this radical will attack and degrade the organic molecule, such as a pesticide.

Degradation of a certain molecule under environmental condition—e.g., a pollutant in soil or air—is a frequent process that involves, mainly, *biodegradation* and *photodegradation*. In the first case, the International Union of Pure and Applied Chemistry (2018) defines it as the breakdown of a substance catalyzed by enzymes in vitro or in vivo. This may be characterized for purposes of hazard assessment as follows:

i. Primary: alteration of the chemical structure of a substance resulting in loss of a specific property of that substance.

ii. Environmentally acceptable: biodegradation to such an extent as to remove undesirable properties of the compound. This often corresponds to primary biodegradation, but it depends on the circumstances under which the products are discharged into the environment.

iii. Ultimate: complete breakdown of a compound to either fully oxidized or reduced simple molecules (such as carbon dioxide/methane, nitrate/ammonium, and water). It should be noted that the products of biodegradation can be more harmful than the substance degraded.

Photodegradation is defined by the International Union of Pure and Applied Chemistry (2018) as the photochemical transformation of a molecule into lower molecular weight fragments, usually in an oxidation process. This term is widely used in the destruction (oxidation) of pollutants by UV-based processes.

Biodegradation is most common for soil and water and related matrices (e.g., sewage, wastewater, etc.) while photodegradation for air or atmosphere.

Table 12.6 Functional groups that undergo microbial reduction

Reactant	Process	Product
$R-\overset{\overset{O}{\|}}{C}-H$	Aldehyde reduction	$R-\overset{\overset{H}{\|}}{\underset{\underset{H}{\|}}{C}}-OH$
$R-\overset{\overset{O}{\|}}{C}-R'$	Ketone reduction	$R-\overset{\overset{OH}{\|}}{\underset{\underset{H}{\|}}{C}}-R'$
$R-\overset{\overset{O}{\|}}{S}-R'$	Sulfoxide reduction	R-S-R'
R-SS-R'	Disulfide reduction	R-SH, R'-SH
$\underset{H}{\overset{R}{\diagdown}}C=C\underset{\diagup R'}{\overset{\diagdown H}{}}$	Alkene reduction	$R-\overset{\overset{H}{\|}}{\underset{\underset{H}{\|}}{C}}-\overset{\overset{H}{\|}}{\underset{\underset{H}{\|}}{C}}-R'$
R-NO$_2$	Nitro group reduction	R-NO, R-NH$_2$ $R-N\underset{OH}{\overset{H}{\diagup}}$

Adapted from Manahan (2000)

The photodegradation of organic pollutants in surface water at $\lambda > 290$ nm depends on the season, pH value, humic acids, and nitrate ion—humic substances and/or nitrate/nitrite ions can serve as photosensitizers. Apart from photodegradation, the fate of such chemicals in the aquatic environment may be influenced by hydrolysis reactions in an unclear mode (Koumaki et al. 2015).

Regarding the biodegradation products, Table 12.6 describes examples of them.

These reactant groups are very common in pesticide molecules. Then, biodegradation is a class of processes of treatment to be considered, for instance, in wastewater treatment.

Furthermore, as seen in soil and water composition and properties, pollutant degradation is a field of study inside the environmental chemistry, a branch of chemistry dedicated to understand the dynamic and transformation of chemical species on the environment and their influence on the life and environmental quality.

12.4 Technologies for Water and Soil Treatment

Firstly, we can consider water due to its vital significance to the life.

Water, when compared against soil, is expected to have a less complexity in its chemical composition. A good family of technologies for water treatment is the advanced oxidation processes (AOPs). Figure 12.5 illustrates the classes of AOPs, the large family of treatment processes for organic pollutants, as pesticides. In Fig. 12.6, we can see an AOP treatment plant.

As an example of the AOP use for pesticide removal in water, Yang et al. (2018) applied the vacuum ultraviolet/ultraviolet (VUV)-derived process to remove aldicarb, alachlor, chloroneb, methiocarb, and atrazine. Pesticides were degraded by >90%,

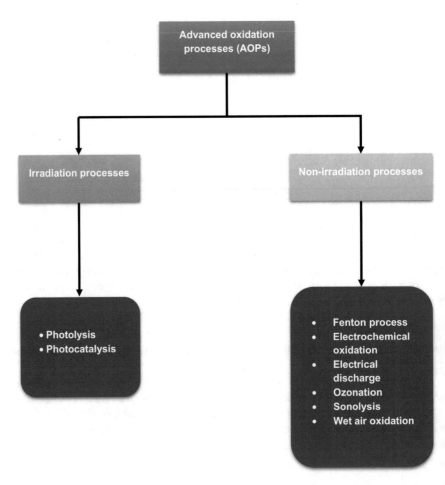

Fig. 12.5 Classes of advanced oxidation processes (AOPs). Adapted from Hisaindee et al. (2013)

Fig. 12.6 An AOP plant for groundwater treatment for photolysis of organic compounds by UV radiation. Courtesy of TrojanUV

and the toxicity of the pesticide solutions was reduced by 33.7–74.8% after the treatment.

Technologies for soil treatment commonly comprise two categories, in situ and ex situ. In the first category, the polluted soil will be treated without removal to a certain place dedicated to it; in the second category, the polluted soil will be removal to the treatment plant. Table 12.7 presents both technologies and the advantages and disadvantages of them.

We need to keep in mind that the pesticide molecule can suffer those degradation processes described in Sect. 12.3 under environmental conditions. Then, it should be taken into account for the choice of the best technology.

Laboratory and field treatability studies must be performed under site-specific conditions before the technology application for the remediation. Furthermore, the choice of the best remediation technology and its use depends on the correct analytical monitoring and control.

A most detailed description of analytical methods for environmental matrixes can be obtained in the SW-846 compendium for hazardous waste test methods of the US-Environmental Protection Agency (2018b). One example is the Method 8085 which comprises the compound-independent elemental quantitation of pesticides by gas chromatography with atomic emission detection (GC-AED). This method analyzes a large number of compounds presented in Table 12.2 in Sect. 12.1.

Table 12.7 Examples of in situ and ex situ technologies for soil treatment polluted with pesticides

Technology	Process type	Advantage	Disadvantage
In situ			
Lasagna™ process based on electro-osmosis	Chemical	• Combines the advantages of different types of techniques (biological and physicochemical) • Compared to other in situ technologies, it has shown greater accessibility in low-permeability soils such as clays, silts, and fine sands	• Limited knowledge on the treatment chemistry and application procedures • Possible diffusion from untreated zone during treatment may impede effectiveness of treatment
Soil flushing		• Is applicable to a wide range of pollutants • Can effect a rapid and adequate cleanup of newly deposited pollutants, such as those from an accidental spill	• The additives for flushing could remain in low amounts in the soil and need to be monitored • Only useful when the solution can be contained and recaptured • Low-permeability or heterogeneous soils are difficult to treat • Above-ground separation and treatment costs for recovered fluids can drive the economics of the process
Soil vapor extraction thermally enhanced		• The cost of the use of this technology is typically less than for incineration • Applicable to a wide range of pollutants • Readily available equipment for onsite or offsite treatment • All pollutants are under a vacuum during operation and the possibility of pollutant release is greatly reduced	• Is highly dependent upon the specific soil and chemical properties of the polluted media • Soil with highly variable permeability may result in uneven delivery of airflow to the polluted regions • Performance in extracting certain pollutants varies depending upon the maximum temperature achieved in the process selected • A suitable off-gas treatment system for contaminated vapors removed from the subsurface needs to be installed • Hot air injection has limitations due to low heat capacity of air • Soil that is tight or has high moisture content has a reduced permeability to air and requires more energy to increase vacuum and temperature. A potential explosion hazard exists from concentrated fumes released from the vacuum unit

(continued)

Table 12.7 (continued)

Technology	Process type	Advantage	Disadvantage
Bioventing	Biological	• Uses readily available and easily installed equipment • Creates minimal disturbance to site operations • Requires short treatment times • Cost competitive • Easily combines with other technologies • May not require costly off-gas treatment	• High pollutant concentrations may initially be toxic to microorganisms • High soil moisture or low-permeability soils reduce bioventing performance • Extremely low soil moisture content may limit biodegradation and the effectiveness of bioventing • Cannot always achieve very low cleanup standards • Permits generally required for nutrient supply • Only treats unsaturated zone soils • Monitoring of vapor at the soil surface may be required
Enhanced bioremediation		• Frequently allows the site to be used during the cleanup • Cost competitive	• Very high pollutant concentrations may be toxic to microorganisms • Under anaerobic conditions, contaminants may be degraded to products that are more hazardous than the original pollutants • Safety precautions must be used when handling hydrogen peroxide • Low-permeability soils are difficult to treat • Both biotic and abiotic sinks for oxygen can increase costs, as well as, operation and maintenance duration • Use of amended oxygen can produce an increase in biological growth near the injection well reducing the diffusion of oxygen in the remaining polluted site and the input of nutrients. • Concentrations of hydrogen peroxide—used to enhance the rate of aerobic biodegradation, greater than 100–200 ppm in groundwater—inhibit the activity of microorganisms

(continued)

Table 12.7 (continued)

Technology	Process type	Advantage	Disadvantage
Landfarming		• Is extremely simple and inexpensive • Requires no extensive process controls • Relatively unskilled personnel can perform the technique • Certain pollutants can be completely removed from the soil	• Requires an extensive amount of space and time • Certain pollutants cannot be reduced to sufficiently low levels • Runoff must be collected and may require treatment • Can incorporate polluted soil into soil that is uncontaminated, creating a larger volume of polluted material • Conditions affecting biological degradation of pollutants (e.g., temperature and rainfall) are largely uncontrolled and may increase the time to complete remediation • The depth of treatment is limited to the depth of achievable tilling
Natural attenuation		• For simple cases, this option is very inexpensive • Minimal disturbance to the site operations • Less generation or transfer of remediation wastes • Less intrusive as few surface structures are required	• Extensive site characterization and monitoring are typical to ensure there is no risk to the outside environment prior to completion of the remediation • The naturally occurring depollution process may not achieve the required cleanup levels • This method may take a long time • Pollutants may migrate (erosion, leaching, and volatilization) before they are degraded or transformed • Cannot be effective where constituent concentrations are high
Phytoremediation		• Has low projected costs for polluting candidate soils • Is a very low-tech method since implementing it requires little more than basic agriculture techniques	• Its operating characteristics and costs for large-scale implementation have not been fully assessed • The roots of plants can effectively clean soil to a limited depth • Plant residues may need to be disposed of as hazardous waste or be further treated • Degradation by-products may be mobilized in groundwater or bio-accumulated in animals • If pollutant concentrations are too high, plants may die • It may be seasonal depending on location

(continued)

Table 12.7 (continued)

Technology	Process type	Advantage	Disadvantage
Ex situ			
Thermal treatment [in association among one or more or without association: desorption, hot gas, pyrolysis, off-gas treatment (e.g., alkali bed reactors)]	Physical	• Generally require shorter time periods • There is more certainty about the uniformity of treatment because of the ability to screen, homogenize, and continuously mix the contaminated media	• Require excavation of soils, which increases costs and engineering for equipment, permitting, and materials handling worker safety issues

Source Adapted from Eugris (2018) and US-Environmental Protection Agency (2018a)

12.5   Strategies for Management of Pesticide Residues

Taking into account the content of Table 12.2, we can note that pesticide residues are, in essence, hazardous waste. As a waste, it can be discarded, abandoned, neglected, released, or designated as a waste material, or one that may interact with other substances to be hazardous. On the other hand, there are environmental laws that regulate and control these types of residues and their treatment. Then, a concise and reliable strategy for the management of pesticide residues is desirable.

The main problem in dealing with pesticide residues is the great diversity of chemical species, organic or not, divided into different families, often antagonistic to each other, unlike other categories of pollutants whose treatment strategy is directed specifically to one or more few molecules (Mackay et al. 1997; Baird 2002).

For example, in the same day and place, it is possible to use two pesticides, classified as herbicides (for weed control), one of them being cationic and another anionic. Even if a ban on mixing in the spray tank is considered, there is still a risk that other substances are not initially expected to be produced on the floor or soil at the place of preparation because of accidental spills. This would already call for different waste treatment strategies either alone or together.

The same can be said about pesticides of the groups of insecticides, fungicides, growth regulators, among others, taking into account that there may be mixtures between different groups (e.g., fungicides + insecticides) during working with such substances.

It should further be considered that molecules of the active ingredients of the pesticides undergo degradation steps until their complete elimination (Mackay et al. 1997), and that many of them result in substances which may be more dangerous or chemically resilient than the original molecule of the pesticide.

A generic action to manage these residues should comprise:

- Classification of hazardous substances and residues/wastes, according to ASTM International (American Society for Testing and Methods 2018);
- Determination of the sources of wastes/residues, considering industries, retailers, and farmers (the ending consumer);
- Classify and understand the degradation routes of the different categories of pesticides;
- Understanding of the reactivity of the substance(s).

This information will give support to define the best strategy and the best technology to be applied.

This knowledge is not easily obtainable and available, and is maintained as an industrial secret in some countries where most of the laws about pesticide determine that the only recognized way to eliminate its residues is blast furnaces (temperatures ranging from 800 to 1,200 °C). An example of an installation used for this purpose is the blast furnaces of the cement industry (Fig. 12.7).

However, despite being the safest method, the logistics involved in achieving this goal are inefficient and extremely expensive, especially in countries with large

Fig. 12.7 Rotary blast furnace from a Mexican cement industry used to give final disposal to empty vials and other collected pesticide residues. In the photo, the red arrow points the rotary blast furnace into action. *Source* Luciano Gebler

territorial extensions and low availability of such ovens, such as the USA, Brazil, China, Russia, Australia, India, African countries (e.g., Morocco), among others, being necessary for the search of new technical solutions applicable to the place of use of the pesticides or near it.

12.6 Tools for Residue Management and Treatment in Agricultural Systems

There is no impediment to the application of traditional pesticide residue management techniques in agricultural systems (physicochemical methods, hydrolysis, ozonization, and oxidative processes, such as activated sludge, Fenton, and photo-Fenton).

However, agricultural systems have extreme difficulty in controlling waste and effluents, because they are open systems with little or no access to treatment equipment or energy in most of the major food-producing regions and, mainly, the lack of personnel trained to handle the system. In addition, in most agricultural countries, the required waste treatment costs can generate an economic deficit in the final result of the productive activity, putting its existence at risk.

Fig. 12.8 Drying ditch in use on a farm in Italy. *Source* Luciano Gebler

The solution found was the massive adoption of the principles of natural attenuation, solarization, phytoremediation, and biodegradation as a tool for the management and treatment of pesticide residues in agricultural areas in the world.

Solarization and evaporation systems such as drying ditch (or drying bed) are widely used in the world, especially in regions with a high incidence of solar energy, allowing the management of large volumes of effluents in a relatively easy way (Fig. 12.8).

However, many pesticides, or their degradation products, are resilient to the action of photolysis and hydrolysis that occur in these physical reactors, leaving, at the end of the drying period, large amount of the active ingredients in the residual sludge and making it more toxic than the components individual from which they originate.

The next step in the management of these waste sludge would be the collection and shipping for incineration at authorized sites. This will depend on the logistic of the use of the products, the size of the property and the volume of pesticides used in the farm, the rainfall regime of the region, and the management of this sludge in the property.

A common practice in the management of pesticide residues in agricultural systems is the application of the principles of phytoremediation and landfarming, disposing the waste in large agricultural areas in order to provide environmental deconcentration, maximizing the exposure of the waste to the environment, and taking

Fig. 12.9 Biobed ready for operation on a farm in Sweden. *Source* Luciano Gebler

advantage of the effect of rhizosphere of planted crops. However, in most cases, the process does not involve the prior quantification of the concentration of residues to be dispersed in the area, generating risks to the environment.

A solution to this problem, practiced in some European countries in the last 30 years, is the application of pesticide residues in fixed-bed reactors, called biobeds (Fig. 12.9), or their adaptations, in order to utilize the processes of accelerated microbial degradation and, subsequently, dispose of landfarming material in the farm area (Torstensson and Castillo 1997; Castillo et al. 2008).

This has favored that the waste quickly disappears from the environment without causing major damage to the rural property area, even if it is not destroyed by "industrial" methods, eliminating the logistics required for transport to points of collection and destruction, and their associated costs.

12.7   Case Study

The work with bioreactors in the form of biobeds has been officially adopted by Sweden, England, Poland, the Netherlands, France, Switzerland, Guatemala, among others, and its potential has been tested in works in Germany, Italy, Chile, Costa Rica, China, Greece, Belgium, Brazil, and several others (Fogg et al. 2003; Spliid et al. 2006; Vischetti et al. 2008; de Roffignac et al. 2008; Wenneker et al. 2008; Karanasios et al. 2010; Sniegowski et al. 2011; Diez et al. 2013; Gao et al. 2015; Gebler et al. 2015a; Holmsgaard et al. 2017).

Fig. 12.10 Biobed installed in the south of Brazil, with a buffer tank to receive the pesticide residues before spreading on the substrate and transparent plastic cover to avoid excess rainwater in the reactor but allowing the passage of sunlight. *Source* Luciano Gebler

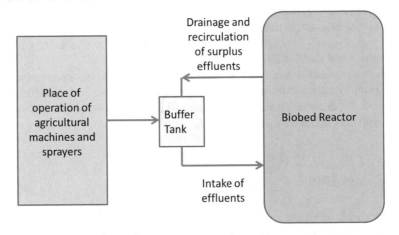

Fig. 12.11 Flowchart demonstrating the movement of the source effluent (site of management of agricultural machines and sprayers) to the biobed reactor for treatment, including the recirculation apparatus and the buffer tank, in case of excess effluent in the bottom of the reactor. *Source* Luciano Gebler

Fig. 12.12 Example of biobed component substrate (composite mixture of agricultural soil, wheat straw, and peat) during the field tests in a Brazilian fruit company. *Source* Luciano Gebler

Biobed is a simple structure, operating for over 30 years in Sweden (Torstensson and Castillo 1997) and in England (Fogg et al. 2003). Originally, a ground ditch, impermeable or not, is filled with a mixture of agricultural soil, straw, and peat, on which a vegetation cover is planted, where pesticide residues and machine washing effluents are discarded (Fig. 12.10) (Castillo et al. 2008; de Roffignac et al. 2008; Gebler et al. 2015b).

The collection and treatment system foresees that the effluent enters the system, and in case of excess liquid, they are drained and reapplied in the reactor. This occurs in places where there are high volumes of rainfall throughout the year or in cases where there is a requirement for work with lined-up reactors, where the system of

Fig. 12.13 Phase of ecotoxicological tests with earthworms prior to biobed recommendation and installation in Brazil. *Source* Luciano Gebler

inputs and outputs must be constantly monitored and there may be a pre-storage (buffer tank) (Fig. 12.11).

The organic compound of soil, straw, and peat that fills the biobed is called "substrate" or "biomix" (Fig. 12.12), being used for quite long period without the need of replacement, and planned to be discarded in the form of landfarming when its life span useful ends, closing the entire treatment cycle in the property without the need for higher costs or transport or disposal risks (Torstensson 2000; Fogg et al. 2004).

The color of the substrate and its texture is dependent on its components. Coloration and luminosity, higher or lower content of clay and humus, type of straw to be used (corn, sugarcane bagasse, wheat, etc.), and the type of peat can make substrates from very different places near each other visually, but always with same or similar efficiency.

It is also necessary to ensure that the residues do not impact the environment, and the control can be made via ecotoxicological tests (Fig. 12.13), using soil indicators (earthworms, collembola, indicator plants) or through chemical analyzes, using appropriate methodologies for extraction and quantification from the biobed substrate (Gebler et al. 2015a; Vareli et al. 2018).

References

American Society for Testing and Methods (2018) Waste management standards. https://www.astm. org/Standards/waste-management-standards.html. Accessed Aug 2018

Baird C (2002) Química ambiental. Bookman, Porto Alegre

Balmer ME, Goss K-U, Schwarzenbach RP (2000) Photolytic transformation of organic pollutants on soil surfaces—an experimental approach. Environ Sci Technol 34:1240–1245

Brazilian Agricultural Research Corporation. Embrapa (1997) Manual de métodos de análise de solo [Guidelines for soil analysis], 2nd edn. National Research Center for Soils, Rio de Janeiro

Castillo MDP, Torstensson L, Stenström J (2008) Biobeds for environmental protection from pesticide use—a review. J Agric Food Chem 56:6206–6219

Clapp CE, Hayes MHB, Senesi N, Bloom PR, Jardine PM (eds) (2001) Humic substances and chemical contaminants. Soil Society of America, Madison

Conte ED, Dal Magro T, Gebler L (2016) Boas práticas de manejo de solo, plantas daninhas e agricultura de precisão. EDUCS, Caxias do Sul

de Roffignac L, Cattan P, Mailloux J, Herzog D, Bellec FL (2008) Efficiency of a bagasse substrate in a biological bed system for the degradation of glyphosate, malathion and lambda-cyhalothrin under tropical climate conditions. Pest Manag Sci 64:1303–1313

Diez MC, Tortella GR, Briceño G, Castillo MP, Díaz J, Palma G, Altamirano C, Calderón C, Rubilar O (2013) The influence of novel lignocellulosic residues in a biobed biopurification system on the degradation of pesticides applied in repeated high doses. Electron J Biotechnol. http://www.ejbiotechnology.info/index.php/ejbiotechnology/article/view/v16n6-17/ 1793. Accessed Nov 2018

Don H, Qiang Z, Lian J, Qu J (2017) Degradation of nitro-based pharmaceuticals by UV photolysis: kinetics and simultaneous reduction on halonitromethanes formation potential. Water Res 119:83–90

Eugris (2018) Further description. http://www.eugris.info/FurtherDescription.asp?e=26&Ca=2& Cy=0&T=In. Accessed Aug 2018

Fogg P, Boxall ABA, Walker A (2003) Degradation of pesticides in biobeds: the effect concentration and pesticide mixtures. J Agric Food Chem 51:5344–5349

Fogg P, Boxall ABA, Walker A, Jukes A (2004) Leaching pesticides from biobeds: effect of biobed depth and water loading. J Agric Food Chem 52:6217–6227

Gao W, Liang J, Pizzul L, Feng XM, Zhang K, Castillo MP (2015) Evaluation of spent mushroom substrate as substitute of peat in Chinese biobeds. Int Biodeterior Biodegratation 98:107–112

Gebler L, Pizzutti IR, Cardoso CD, Klauberg Filho O, Miquelluti DJ, Santos RSS (2015a) Biorreactors to organize the disposal of phytosanitary effluents of Brazilian apple production. Chem Eng Trans 43:343–348

Gebler L, Pizzutti IR, Magro TD, Santos RSS, Cardoso CD, Klauberg Filho O (2015b) Sistema Biobed Brasil: Tecnologia para Disposição Final de Efluentes Contaminados com Agrotóxicos Originados na Produção de Frutas de Clima Temperado. Embrapa Uva e Vinho, Bento Gonçalves

Hisaindee S, Meetani MA, Rauf MA (2013) Application of LC-MS to the analysis of advanced oxidation process (AOP) degradation of dye products and reaction mechanisms. TrAC Trends Anal Chem 49:31–44

Holmsgaard PN, Dealtry S, Dunon V, Heuer H, Hansen LH, Springael D, Smalla K, Riber L, Sørensen SJ (2017) Response of the bacterial community in an on-farm biopurification system, to which diverse pesticides are introduced over an agricultural season. Environ Pollut 229:854–862

International Union of Pure and Applied Chemistry (2018) IUPAC compendium of chemical terminology—the gold book. http://goldbook.iupac.org/index.html. Accessed Aug 2018

Karanasios E, Tsiropoulos NG, Karpouzas DG, Ehaliotis C (2010) Degradation and adsorption of pesticides in compost-based biomixtures as potential substrates for biobeds in Southern Europe. J Agric Food Chem 58:9147–9156

Koumaki E, Mamais D, Noutsopoulos C, Nika M-C, Bletsou AA, Thomaidis NS, Eftaxias A, Stratogianni G (2015) Degradation of emerging contaminants from water under natural sunlight:

the effect of season, pH, humic acids and nitrate and identification of photodegradation by-products. Chemosphere 138:675–681

Mackay D, Shiu W, Ma K (1997) Illustrated handbook of physical-chemical properties and environmental fate for organic chemicals. CRC Press, Boca Ranton

Manahan SE (2000) Environmental chemistry. CRC Press, Boca Ranton

Mirzaei A, Chen Z, Hghighat F, Yerushalmi L (2017) Removal of pharmaceuticals from water by homo/heterogonous Fenton-type processes—a review. Chemosphere 174:665–688

Sniegowski K, Bers K, Van Goetem K, Ryckeboer J, Jaeken P, Spanoghe P, Springael D (2011) Improvement of pesticide mineralization in on-farm biopurification system by bioaugmentation with pesticide-primed soil. FEMS Microbiol Ecol 76:64–73

Snoeyink VL, Jenkins D (1996) Química del água [Water chemistry]. Limusa, México City

Spliid NH, Helweg A, Heinrichson K (2006) Leaching and degradation of 21 pesticides in a full-scale model biobed. Chemosphere 65:2223–2232

Torstensson L (2000) Experiences of biobeds in practical use in Sweden. Pestic Outlook 11:206–211

Torstensson L, Castillo MP (1997) Use of biobeds in Sweden to minimize environmental spillages from agricultural spray equipment. Pestic Outlook 8:24–27

United Nations Environment Programme—UNEP (2018) Guidance on chemicals legislation: overview. https://www.unenvironment.org/zh-hans/node/14392. Accessed Nov 2018

University of the West of England (2013) Science for environment policy in-depth report: soil contamination: impacts on human health. Report produced for the European Commission DG Environment. http://ec.europa.eu/environment/integration/research/newsalert/pdf/IR5_en.pdf. Accessed Oct 2018

US-Environmental Protection Agency (2018a) Thermal treatment: ex situ. https://clu-in.org/techfocus/default.focus/sec/Thermal_Treatment%3A_Ex_Situ/cat/Overview/. Accessed Aug 2018

US-Environmental Protection Agency (2018b) https://www.epa.gov/hw-sw846/sw-846-compendium. Accessed Aug 2018

Vareli CS, Pizzutti IR, Gebler L, Cardoso CD, Gai DH, Fontana M (2018) Analytical method validation to evaluate dithiocarbamates degradation in biobeds in South of Brazil. Talanta 184:202–209

Vaz S Jr (2013) Química analítica ambiental [Environmental analytical chemistry]. Embrapa, Brasília

Vaz Júnior S (2010) Estudo da sorção do antibiótico oxitetraciclina a solos e ácidos húmicos e avaliação dos mecanismos de interação envolvidos. [Study of the antibiotic oxytetracycline sorption to soils and humic acids and evaluation of the interaction mechanisms involved]. PhD thesis, University of São Paulo, São Carlos, Brazil. 2010. http://doi.org/10.11606/T.75.2010.tde-30062010-155624

Vischetti C, Monaci E, Cardinali A, Perucci P (2008) The effect of initial concentration, co-application and repeated applications on pesticide degradation in a biobed mixture. Chemosphere 72:1739–1743

Weber OLS, Chitolina JC, Camargo AO, Alleoni LRF (2005) Cargas elétricas estruturais e variáveis de solos tropicais altamente intemperizados [Structural and variable electrical charges of highly weathered tropical soils]. Rev Bras Cienc Solo 29:867–873

Wenneker M, Beltman WH, De Werd HAE, Van De Zande JC (2008) Identification and quantification of point sources of surface water contamination in fruit culture in the Netherlands. Aspects Appl Biol 84:369–375

World Health Organization (2010) The WHO recommended classification of pesticides by hazard and guidelines to classification: 2009. World Health Organization, Geneva

World Health Organization (2017) Chemical mixtures in source water and drinking-water. World Health Organization, Geneva

Yang L, Li M, Li W, Jiang Y, Qiang Z (2018) Bench- and pilot-scale studies on the removal of pesticides from water by VUV/UV process. Chem Eng J 342:155–162